Open Space

Information Policy Series

Edited by Sandra Braman

The Information Policy Series publishes research on and analysis of significant problems in the field of information policy, including decisions and practices that enable or constrain information, communication, and culture irrespective of the legal silos in which they have traditionally been located as well as state-law–society interactions. Defining information policy as all laws, regulations, and decision-making principles that affect any form of information creation, processing, flows, and use, the series includes attention to the formal decisions, decision-making processes, and entities of government; the formal and informal decisions, decision-making processes, and entities of private and public sector agents capable of constitutive effects on the nature of society; and the cultural habits and predispositions of governmentality that support and sustain government and governance. The parametric functions of information policy at the boundaries of social, informational, and technological systems are of global importance because they provide the context for all communications, interactions, and social processes.

Open Space

The Global Effort for Open Access to Environmental Satellite Data

Mariel John Borowitz

The MIT Press
Cambridge, Massachusetts
London, England

This book was set in ITC Stone Sans Std and ITC Stone Serif Std by Toppan Best-set Premedia Limited. Printed and bound in the United States of America.

Library of Congress Cataloging-in-Publication Data

Names: Borowitz, Mariel.
Title: Open space : the global effort for open access to environmental satellite data
  / Mariel Borowitz.
Description: Cambridge, MA : The MIT Press, [2018] | Series: Information policy |
  Includes bibliographical references and index.
Identifiers: LCCN 2017019414 | ISBN 9780262037181 (hardcover : alk. paper)
Subjects: LCSH: Astronautics in earth sciences. | Environmental monitoring—
  Remote sensing. | Climatic changes—Remote sensing. | Communication policy. |
  Information resources management.
Classification: LCC QE33.2.R4 B677 2018 | DDC 550.28/4—dc23 LC record
available at https://lccn.loc.gov/2017019414

10 9 8 7 6 5 4 3 2 1

To family, especially my husband, Jeff, who also provided endless support and a bottomless cup of coffee.

Contents

Series Editor's Introduction

Sandra Braman

The story is complex. We tend to think of "openness" as an on/off condition—but often, as Mariel Borowitz's comprehensive analysis of access to satellite data from around the world finds, the most successful way to maximize usage of very large datasets is to blend diverse types of access. The industrial economy has given way to an information economy—but as Borowitz's history of efforts to commercialize satellite data shows, that doesn't mean that every attempt to make a market in information will succeed. There is wide agreement on the importance of making publicly available data that will help communities recuperate from problems such as those raised by the environmental crisis—but as we learn from *Open Space*, short-term or spot sharing won't do the trick. In a lot of systems, data become available only when users specifically request it based on particular needs—but a great many of the most valuable uses of data develop only after the information is freely available.

With all of today's talk about and experimentation with open access to data, to publications, and to decision-making procedures, it will be a surprise to most of us to learn that one of the earliest references to "open data" as a policy issue showed up in debate over what to do with data gathered by satellites. Indeed, Mariel Borowitz's in-depth analysis of meteorological and space agencies in the United States, Europe, and Japan found that they all began operations assuming they would provide open access. The fact that a mix of approaches is currently in use, both for those satellite-sensing systems presented in full case studies and in the rest of the 35 countries that conducted remote sensing by the beginning of 2016 that are all analyzed in *Open Space*, is evidence of just how difficult it is to achieve effective and meaningful access to data for all who desire it, in a manner in which they can use it.

A variety of factors figure in. There is an interplay between expectations, reputations, innovation, economics, and the uses to which data are put. Implications of developments are not always obvious and a change in any one factor can affect the relative weighting and relationships among the rest. The trend toward storing ever-larger satellite datasets in clouds provides an example. The US government chose to move its satellite data to private sector clouds because doing so was more cost-effective and efficient than trying to developing its own cloud-based storage. It might seem that such a move would inevitably shift costs to users who then must access the data through a private sector provider, but Borowitz points out that that isn't so—whatever funds a government entity had been devoting to storage and access on its own servers could be used in an arrangement with companies to ensure that user access and use of data is free.

Borowitz's research was inspired by the problem of how to get access to satellite data that is of such value to environmentalists. By its conclusion, though, this clear and insightful analysis, informed by theories and research from multiple disciplines, provides a model for understanding the complexities of open access to data of any type, and from any source. *Open Space* concludes with real gold—a range of realistic, feasible policy recommendations for data access that maximizes access and openness for countries, situations, and data of different types. These, too, can be used as models for access to data from sources other than satellites.

Unusually, and usefully, in her conclusions Borowitz notes things that users as well as data providers and the governments standing behind them whether through intimate or arm's-length relationships can do. We can make sure to always attribute sources of data, a practice that runs counter to the big data trend of ignoring or losing provenance altogether in an important and necessary way. We can analyze and publicize the value of data that are openly accessible. We can be consistent in our reputational treatment of those who do and do not make their data available for use by environmentalists and others concerned about social issues of such importance to us all. And, whether or not particular leaders and communities have turned their backs on facts as the basis of their decision-making, we can demand the data, and use them.

Preface: How to Use This Book

This book lies at the intersection of a number of fields and interest areas, including climate change, space activity, and information policy. The material covered may be of interest to scholars in these fields, policy-makers, government officials, and general interest readers. However, these different readers may find the greatest value in different parts of this book. Some may find that the level of detail presented in one part or chapter exceeds their needs or interest. Therefore, here I include a brief summary of the structure of the book and provide some suggestions to each of these groups on how to best proceed in using this book.

Chapter 1 introduces the two central mysteries that the book seeks to solve, describes the methods used to examine and answer these questions, and provides a summary of the book's key findings and their implications.

Part I focuses on the creation of a model of data sharing development and provides a review of key information policy literature. Chapter 2 gives an overview of the model and lays out key definitions, while chapters 3 and 4 delve into the literature on relevant organizations and institutions, and contextual elements of data sharing, respectively. Part I ends with a table summarizing the key elements of the model.

Part II includes the case studies used to explore the mysteries laid out in the introduction. It begins with case studies of two international organizations (chapters 5 and 6) that aim to give the reader insight not only into the history of data sharing debates within these entities, but also into broad international developments that affected the individual agencies. The next eight case studies (chapters 7–14) form the core of this part, reviewing developments and changes in data sharing policies in the space and meteorological agencies in the United States, Europe, and Japan. Finally, chapter 15 provides summaries of data sharing policy developments in satellite-operating agencies in the BRICS nations: Brazil, Russia, India, China, and South Africa.

Part III presents the book's conclusions and implications, with chapter 16 dedicated to applying the model of data sharing development and marshaling evidence from the case studies to explain the mysteries presented in chapter 1, and chapter 17 examining the implications of these findings for future satellite data sharing and for the open data movement as a whole.

Scholars focused on information policy and open data development are likely to find significant value in part I, which provides a thorough literature review and describes in detail the development of a new model for examining the development of data sharing policies. These readers will benefit from reading a selection of the case studies presented in part II, perhaps focusing on case studies related to a particular nation, or similar types of agencies across multiple countries. Each case study chapter includes a short introduction at the beginning and a summary at the end that can be reviewed to determine which chapters the reader might want to read in more detail. These readers can then jump to part III and the conclusions and implications.

Scholars interested in space policy or space history may choose to skip the in-depth literature review and data sharing policy model development in part I, instead focusing only on the introductory material at the outset of the part, which provides a useful overview of the model, and the table presented at the end of chapter 4, which summarizes the key components of the model. These readers may enjoy reading the full set of case studies in part II, which provide comprehensive reviews of space agency organization and decision-making related to Earth observations and data sharing in a variety of time periods and countries. The conclusions and implications in part III will also be of interest.

Policy-makers, government officials, and others focused on practical issues of data sharing policy development will benefit from reading the introductory material in part I, and from reviewing the table at the end of this part, in chapter 4, which is designed to be useful as a practical tool in thinking through policy development. Those interested in better understanding the logic underlying the guidance in the table—for example, the economic theory underlying guidance on maximizing economic efficiency of data policies—can read the relevant portions of part I in more detail.

These readers should look at case studies in part II according to their own interest, referring to the introductions and summaries in each chapter for quick overviews, or they may choose to skip this historical part altogether, jumping directly to part III, particularly the forward-looking policy implications discussed in chapter 17.

General interest readers or those interested in broader issues of policy development, science and society, climate change, or the environment may find value in reading the full book. However, it is useful to note that nearly every chapter begins with introductory material and ends with a summary of key points, so readers always have the option to review this material to determine whether they wish to read the full text of that chapter.

Operating at the intersection of multiple fields, this book includes information that will be novel or useful to a broad range of individuals. It is my hope that each of these readers can utilize the book in the way that most directly meets their needs.

Acknowledgments

Thank you to the many current and retired government officials, and the space, climate, and open data experts who provided their time and effort to make this book possible. Thank you also to the friends, family, and colleagues who read early drafts and provided feedback—you have helped to make the book what it is.

1 Two Mysteries

Enduring Satellite Data Sharing Gaps

In 2014, the Intergovernmental Panel on Climate Change stated that without additional mitigation efforts, "warming by the end of the 21st century will lead to high to very high risk of severe, widespread, and irreversible impacts globally," including loss of life, extinction of plant and animal species, ecosystem degradation, and economic losses.[1] Understanding and addressing climate change requires precise, continuous monitoring of a multitude of variables related to the land, oceans, and atmosphere sustained over long periods of time.

With their unique vantage point and ability to provide continuous global coverage, satellites play an important role in collecting data about the climate. Satellites offer comprehensive global coverage that can't be matched by in situ observation systems, particularly in remote or sparsely populated areas, including the oceans, the Arctic, and the Antarctic, all of which are especially important for understanding climate change. Interference from the atmosphere prohibits accurate measurements on the ground, so satellites offer the only option for measuring Earth's energy balance—the incoming energy from the sun and outgoing thermal and reflected energy from the Earth—a foundational climate measurement.[2]

Since the beginning of the space age, 35 nations have been involved in the operation of an Earth observation satellite, and more than 450 unclassified Earth observation satellites have been successfully launched, 173 of which were government-owned satellites actively operating as of the beginning of 2016. Yet with so many variables to be collected, so precisely, and over such long periods, there are still gaps in our ability to adequately monitor climate change.[3]

More surprising is that these gaps are exacerbated by a lack of international data sharing. While some nations make data from their unclassified, government-owned Earth observation satellites freely available, others restrict access to this type of data. This means that even the data that is collected is, in some cases, not contributing to global understanding of this important environmental issue.

This problem has not gone unnoticed. It has been referenced in reports and statements of international organizations for decades. The first IPCC Assessment Report in 1990 recommended improving systematic observation of climate variables with both satellite and surface-based instruments on a global basis and expressed the need to facilitate international exchange of climate data.[4] The United Nations Framework Convention on Climate Change (UNFCCC), negotiated at the Rio Earth Summit in 1992, called on parties to "support international and intergovernmental efforts to strengthen systematic observation ... and to promote access to, and exchange of, data and analysis."[5]

A number of existing international organizations developed policies expressly calling on their members to support these efforts. The Committee on Earth Observation Satellites (CEOS) developed data principles in the early 1990s encouraging countries to provide data to global change researchers.[6] The World Meteorological Organization (WMO) adopted a policy in 1995 urging its members to "strengthen their commitment to the free and unrestricted exchange of meteorological and related data and products," noting the importance of these data for climate and other issues. The WMO Global Climate Observing System (GCOS) went further, arguing that global environmental concerns were an overriding justification for the unrestricted international exchange of data and calling for full and open sharing at the lowest possible cost to users.[7]

Ten years after the Rio Earth Summit, which had resulted in the UNFCCC, the United Nations hosted a follow-up conference to focus on implementing the initiatives developed there, including improving international data sharing.[8] In support of this effort, in 2003, the G8 called for coordination of global observation strategies and improved data sharing.[9] This resulted in the creation of the Group on Earth Observations (GEO), a new international organization in which international data sharing was front and center as a key goal.[10] Focusing on needs for climate and other environmental issues,

GEO has worked to increase environmental data sharing for more than a decade.

Despite these high-level calls to action and the efforts of multiple international organizations, as of the beginning of 2016, data from less than half of unclassified government satellites was made freely available. The majority of the data was subject to fees and/or restrictions on access or redistribution. This is a problem because even relatively low fees or minimal restrictions can significantly limit data use, particularly for climate researchers, who often need access to large volumes of data, collected by an array of instruments, spanning long time periods. When one or more of the sources requires a detailed application or proposal for initial or continued access to the data, or when one or more sources imposes limitations on the use or redistribution of the data, it can significantly slow scientific progress and limit impact. A 2011 article in *Science* stated that "about half of the international modeling groups are restricted from sharing digital climate model data beyond the research community because of governmental interest in the sale of intellectual property for commercial applications."[11]

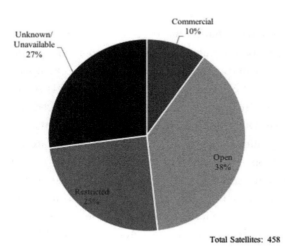

Total Satellites: 458

Figure 1.1
Satellite data sharing policies for unclassified government Earth observation satellites launched between 1957 and 2016.

Sharing of environmental data could also have benefits well beyond the issue of climate change. Satellite data plays an important role in disaster monitoring and relief, and limitations due to data sharing restrictions have repeatedly been cited as impediments to these efforts.[12] Data sharing can also contribute to improving food security, sustainable agriculture, and public health surveillance, monitoring and managing biodiversity and ecosystem issues, and fostering better management of natural resources.[13] Given the highly publicized importance of data sharing for monitoring climate change, and its potential to contribute to a wide array of environmental issues, it is a puzzle that in many cases the space and meteorological agencies around the world that collect satellite data essential to addressing these issues do not share that data freely.

This is the first mystery to be examined in this book: Why do some government agencies choose to restrict access to data from unclassified government satellites while others make this data freely available to all, what are the ramifications of this restriction, and what can be done to improve the system?

Open Data Leaders

While a lack of global satellite data sharing has been identified as an international challenge, the open data community has pointed to space and meteorological agencies, particularly in the United States, as exemplars in the open data movement. The second mystery focuses on this seemingly contradictory situation. In 2009, President Obama announced the Open Government Directive (OGD) calling on executive departments and agencies to improve the quality of government information, create an institutionalized culture of open government, and publish government information online. Open data sets were to be registered on Data.gov.[14] Two years after Data.gov was launched, five agencies accounted for more than 99 percent of all datasets and applications hosted on the site. All three civil agencies that operate satellites—the National Aeronautics and Space Administration (NASA), the National Oceanic and Atmospheric Administration (NOAA), and the United States Geological Survey (USGS)—were members of this small group.[15]

In fact, compliance with the Open Government Directive had been relatively easy for these agencies, particularly with respect to their satellite data, because it was already openly available. NASA data was made fully available to all users in 1992, NOAA had been steadily increasing the amount of data provided free of charge since the 1990s, and USGS opened the Landsat archive for free, unrestricted access in 2008.[16] In fact, one of the earliest uses of the term "open data" in a policy context can be traced to the space community: agreements between NASA and international partners to develop ground stations for direct access to Landsat data in the early 1970s required that partners adopt an "open-data policy comparable to that of NASA and other US agencies participating in the program." The agreements required that catalogs of all data processed, as well as the data itself, be made publicly available as soon as practicable to the domestic and international community.[17]

The open data movement was not limited to the United States, and neither was open sharing of satellite data. The US Open Government Directive was followed by a wave of open data movements around the globe, with more than 44 countries developing open data platforms by 2013.[18] Within the space community, Brazil and China made access to data from the China-Brazil Environmental Resource Satellite (CBERS) series freely available online for Brazilian users beginning in 2004, and opened this data portal to global users in 2010.[19] The European Organization for the Exploitation of Meteorological Satellites (EUMETSAT) had made all of its archived data freely available as of 1998, and gradually increased the amount of operational data that was shared freely, as well.[20] The European Space Agency (ESA) made access to all of its Earth observing satellite data free as of 2012.[21] One year later, the Japan Aerospace Exploration Agency (JAXA) made access to the majority of its satellite data freely available.[22] Many other nations, including Argentina, Canada, India, Japan, Singapore, South Africa, France, Germany, Italy, and Sweden, also provide data from at least some of their satellites openly to all users, without charge or restrictions.

Perhaps even more interesting is the fact that these leaders in open data did not always make data freely available, and many programs followed a similar pattern: from free data provision, to more restrictive policies, and then back to free and open sharing. The degree of data restriction varied. The US Landsat program was fully privatized, shifting from government to private control in 1985. Under its commercial data policy, Landsat images cost thousands of dollars a scene for both research and commercial users alike. NOAA data was subject to fees designed to recover the costs of some systems and support its data centers. NASA never sold its data, but it did restrict access, allowing only noncommercial uses. Data from European meteorological satellites was shared freely for years before EUMETSAT decided to encrypt its transmissions and implement a fee for data access. ESA contracted with a private consortium to market and sell data from its satellites to commercial users and required research users to submit full proposals for data access. As mentioned above, all of these agencies eventually transitioned to providing all or most of their data on a free and open basis. A similar pattern is seen in Japan, although some significant barriers to data access remain.

What drove these agencies to move from open data sharing, to restrictions, and then back to open data policies? Why did the most recent transitions to open data often predate national-level open data initiatives? What can explain the similarities observed across diverse agencies and regions? The way in which publicly collected data is treated—whether it is sold, shared, or restricted—affects the economy, national security, and interactions between the public and government. How can understanding these trends help to understand open data efforts in other countries, agencies, and sectors?

The effort to understand and solve these mysteries forms the basis for this book. In the rest of this introduction, I describe the model of data sharing policy development that is created to analyze these issues and explain the selection of case studies and data collection methods used to generate the data needed to apply the model and solve the mysteries. The introduction then includes a brief summary of these conclusions and their broader implications for the future of satellite data sharing and open data in general.

A Model of Data Sharing Policy Development

Both of these mysteries center on questions regarding data sharing policy development and change. Therefore, the first step in this book is to develop a model of data sharing policy development. This model, fully articulated in part I of the book, centers on the role of the government agency as the primary actor. In practice, this is the level at which data sharing policies are typically written. Knowledge of the agency's mission and culture is thus key to understanding why policies take the shape that they do, but these agencies are not operating in a vacuum: they are influenced by both people and ideas.

These outside actors that influence data sharing policy can include other government officials, particularly those operating at a national level. Executive officials can announce initiatives, release national-level data sharing policies, produce reports, or propose budgets to attempt to influence agency data sharing policies. Legislators wishing to affect data sharing policy development can hold hearings, require reporting, or pass laws on the subject. Although these actions can and often do have an effect on data sharing policies, their impact is mitigated by agency decisions and actions with respect to implementation.

Nongovernmental actors can also have an impact, particularly groups that have a direct interest in the data policy—data users or potential competitors in data provision. These organizations have an incentive to lobby both the agency and national-level government actors to influence policy development. Intergovernmental groups also play an important role, sharing information across countries, emphasizing norms of behavior, and creating reputational benefits and costs for member nations.

The positions and actions of all of these groups have the potential to impact the data sharing policies. Their positions, as well as those of the agency, also depend on ideas: understandings of the security, economic, and normative attributes of the data. Individuals developing a data policy may consider whether release of data poses national security or privacy risks or benefits. Data has economic attributes that generally allow it to be treated either as a public good or as a commodity. Determining which of these policies is likely to maximize net social benefits can also play a role in policy development. Finally, agencies generally must contend with normative issues related to sharing of data, including views of the proper role of government or ethical imperatives related to data sharing. Agencies must consider which of the competing priorities will drive policy development.

Lastly, it is important to consider the role of technology in developing data policy. Changes in technology over time can significantly impact what type of data can be collected or produced and how it can be used. Technological advancements can change the security, economic, and normative attributes of the data. For example, the spread of technology can make a particular type of data more ubiquitous and thus less of a security concern, or it can allow greater precision that poses risks not present in earlier versions of the data. New capabilities for data processing and analysis may increase its value, or more efficient transmission methods can decrease the costs of sharing. Attention to the technology that underlies data collection, production, analysis, and distribution is important in understanding data sharing policy development.

By examining the agency in depth, considering how the agency was influenced by external actors, analyzing the attributes of the data itself, and taking into account advancements in technology, it is possible to understand why particular data sharing policies were put in place, and why changes in policy occur. To illustrate this, in part II of the book, the model is applied to a series of case studies in the Earth observation satellite sector. Satellite data sharing policies differ significantly over time, across agencies, and across countries and regions, offering a rich opportunity for examining how these policies are developed and changed and identifying the key factors in these developments. Data from the case studies, analyzed using this model, is used to address the two mysteries posed at the outset of this chapter.

Methodology

The core set of case studies includes seven agencies responsible for operating Earth observation satellites in the United States, Europe, and Japan. These include the US National Aeronautics and Space Administration (NASA), the US National Oceanographic and Atmospheric Association (NOAA), the United States Geological Survey (USGS), the European Space Agency (ESA), the European Organization for the Exploitation of Meteorological Satellites (EUMETSAT), the Japan Aerospace Exploration Agency (JAXA), and the Japanese Meteorological Agency (JMA). A brief overview of activities in the US defense and intelligence communities and commercial remote sensing is also provided for completeness, due to the interactions between these activities and their civil counterparts.

These agency-level case studies include a history of the development of data sharing policies, based on document analysis and interviews with current and past agency officials. Using the model framework, the case studies capture the official and unofficial mission and norms within the agency and the impact of these elements on data sharing policy development. They also incorporate the impact of national-level actors, intergovernmental and nongovernmental external actors, and understandings and debates regarding the contextual factors of the data itself. The case studies focus not only on which policies were adopted, but also on the circumstances, arguments, and motivations that drove policy choices.

The United States, Japan, and Europe were chosen because together they are responsible for nearly half of the unclassified government Earth observation satellites launched between 1957 and 2016, and roughly half of the satellites operating as of the beginning of 2016. The United States maintains the largest program (30 percent of satellites operating in 2016), and Europe had a medium-sized program (10 percent), similar in size to the Earth observation programs in China and India. Japan's program is the smallest of the three cases chosen, with a 5 percent share of global Earth observation satellites in 2016. Its program is similar in size to Italy and Russia today. Focusing on these three locations provides insight into policies that govern a large portion of relevant climate data while also providing some variation in program size and region.

It is worth noting that although its space and meteorological organizations act in many ways like other government agencies, the European agencies are also international organizations. Just like the other agencies being analyzed, ESA and EUMETSAT were created with specific missions and they have developed organizational cultures and norms over time. However, influence from national-level policy-makers can come from many different nations, and power dynamics created by the organizations' unique structures can affect both the extent of this influence and its form. In addition, Earth observation activities on the European level were paralleled by national-level activities, and in the case of France, Germany, and Italy, these domestic programs were relatively large. These issues are addressed in the European case studies and in the analysis. A brief summary of each of the national-level Earth observation programs in Europe is included in appendix A.

In addition to the in-depth case studies described above, the book includes a chapter with summaries of data sharing development in Brazil, Russia, India, China, and South Africa, sometimes referred to as the BRICS countries, also based on analysis of publicly available documentation and interviews. Although access to policy documentation and agency officials was more limited in some instances, these summaries provide important insight into data sharing policy developments in a broader range of environments. Appendix A includes short summaries of data sharing policies and practices for every nation that operated a civil Earth observation satellite up through the beginning of 2016. In addition to the qualitative summaries, the data policies of these nations were coded to allow quantitative descriptions of the state of global satellite data sharing. The methods used to develop the dataset and code the data sharing policies are described in detail in appendix B.

The book also includes case studies of two of the most active international organizations with regard to Earth observation satellite data sharing. The World Meteorological Organization (WMO) and the Group on Earth Observations (GEO) both include promotion of international satellite data sharing in their primary goals, and both have developed data sharing principles or guidelines. The impact of these organizations on agency decision-making is addressed within the agency-level case studies. The international-level case studies instead focus on the internal dynamics of these organizations to provide further insight into global data sharing debates and norm development on the global level. This helps to demonstrate the multidirectional influences in policy development, in which international organizations affect national policy development, and are in turn influenced themselves by these national organizations. These case studies are presented prior to the national case studies to provide readers with an understanding of global trends before delving into how these global trends affected specific agencies.

Solving the Mysteries

Analysis of these case studies demonstrates that understandings and uncertainties regarding the economic attributes of satellite data have been particularly important in the development of data sharing policies over time. These policies have largely been driven by (1) differences in agency and national-level policy-maker beliefs regarding the attributes of the data, (2) influence and information provided by nongovernment actors, and (3) changes in technological capabilities.

Early in the space age, satellite technology was new and the value of the data was unknown. During this time, data sharing was necessary for agencies to evaluate the usefulness of the data. One of the earliest satellite applications developed was in the area of meteorology, which had a long-standing culture of free and open international data exchange. The United States, Europe, and Japan all began their Earth observation programs with the development of meteorological satellites. During this time, data sharing directly supported the missions of the space and meteorological agencies as they attempted to understand and improve the technology and its applications. National-level policy-makers during this time were primarily focused on the national prestige value of promoting and sharing data from these advanced technologies.

By the 1980s, the value of Earth observation satellites had become clear. Space agencies had branched out to develop land, ocean, and other remote sensing technologies, and meteorological agencies had taken over control of operational weather satellite series. From the perspective of the space and meteorological agencies, the value of the data was validation that their development efforts had been successful. The data could now make significant contributions to their scientific and meteorological missions. Many national-level policy-makers interpreted the situation differently. Concerned with tight budgets and interested in improving efficiency, these actors believed the value of the data could be used to offset the high costs of developing satellite systems. They also hoped to spur the development of a vibrant commercial remote sensing sector, gaining advantages for their nation by moving quickly in this area.

Without evidence that these types of cost recovery and commercialization efforts could not be achieved, they were undertaken with the extent of commercialization efforts largely determined by the importance of satellite data sharing to the agency's mission and the perceived economic value of the data collected by the agency. Policies ranged from the full privatization of Landsat to restrictions on commercial use of NASA data. Most agencies adopted tiered policies that included market fees for commercial users and lower fees for researchers, in an attempt to balance these competing pressures.

International organizations and interest groups, particularly in the meteorological and scientific community, protested against these restrictions. They argued that limitations on data sharing could harm the ability to provide accurate weather forecasts that are needed to save lives and property. They noted that restrictions were significantly inhibiting the ability of scientists to use satellite data and share their results. On top of these challenges, the efforts to sell data were not generating significant revenue. Commercial remote sensing efforts failed to take off, instead remaining dependent on government support for their survival. This led to a decrease in pressure from national-level policy-makers to engage in these activities, and the transition back to open data sharing began. The ability to store and transmit data using computer and Internet technology further hastened the adoption of free data sharing policies. Having advocated for greater data sharing from the beginning because of its importance to their missions, space and meteorological agencies were ahead of the national-level trends with regard to open data sharing.

Despite this larger trend, ongoing differences in opinion regarding the economic attributes of data continue to be one of two primary reasons that Earth observation satellite data is not shared freely. The United States, Europe, Japan, and a number of other nations with advanced space capabilities continue to provide significant government support to promote commercial remote sensing, with the hopes that these efforts will save the government money and eventually result in the growth of a robust commercial remote sensing sector. Engaging in these cost recovery and commercialization efforts requires that access to the data be restricted to protect the ability to sell data.

The second reason that Earth observation is not shared is related directly to agency mission and seen primarily among smaller, newer Earth observation programs. Many of these programs focus on technology development and capacity building, rather than science or other environmental applications. Given this mission, data sharing, particularly international data sharing, is not a priority of these programs. National-level policy-makers and agency officials do not provide or allocate resources needed to develop and maintain data distribution systems.

Even some more well-established programs experience this mission priority issue. JMA, for example, invests significant efforts in making data from its weather satellite available for official use by national meteorological services in other nations. However, it has not made significant investments in providing that same data to researchers or others, so these users are only able to access data through intermediaries, or through an offline system, subject to a fee. Similarly, some data from early satellite systems in the United States and Russia is not accessible online because the investment needed to access, digitize, and curate that data is not seen as a high priority, particularly compared to the need to disseminate data from new, high data volume satellites.

The Future

The last chapter of this book examines how these trends in data sharing may continue in the future. For cases where agency missions and priorities are the reason for the lack of sharing, international organizations, interest groups, and potentially national-level policy-makers will have an important role to play. These organizations and actors can help to increase the reputational benefits of satellite data sharing, giving agencies an incentive to make the data available. They can bring users and agencies together, raising awareness of potential uses of the data and of the existence of the demand for data. Finally, these actors can provide technical and financial assistance in developing data dissemination systems that directly lower the barriers to doing so.

Economic realities favoring free and open sharing of Earth observation satellite data are likely to continue driving increased sharing in this area; however, there are opportunities for engagement with the commercial sector. Procurement of satellite systems or of satellite data from the private sector, under an open data license, is one method that could prove useful across many cases, particularly when the technology for data collection is stable and well understood. Public–private partnerships can also offer the potential for savings when there is a significant number of commercial data users and when the noncommercial uses of the data are narrowly focused. Innovative agreements that preserve the ability to share the data, perhaps on a limited basis or after a specified interval of time, can be mutually beneficial. However, in areas where the primary users of the satellite data are noncommercial actors and the potential uses of the data are broad, governments should resist efforts to turn over production to the commercial sector, limiting access to, and redistribution of, data. In these cases, any savings from such an arrangement would be outweighed by the loss of benefits caused by the decrease in data use.

The use of the model developed in this book to understand these complex issues in the area of satellite remote sensing demonstrates its value for evaluation more broadly. It can be used to examine existing policies or guide the development of new policies in other government programs. It can be used to identify likely challenges or the most productive ways forward. The elements of this model, combined with the broad technological trends in society, suggest that the open data movement will continue to gain momentum across agencies and countries. Understanding these issues can help to ensure this process is undertaken thoughtfully and efficiently.

Part I   A Model of Data Sharing Policy Development

Information sharing is an interdisciplinary issue, with relevant theory and literature in a broad range of fields. Some scholars focus on the people and organizations involved in data sharing policy development: the structures and dynamics of government organizations or the role of national initiatives.[1] Others focus primarily on the data itself, cataloging the incentives and barriers to sharing a particular type of data, or evaluating the success or failure of sharing within a particular field.[2] The model presented here pulls these threads together, incorporating concepts from the literature on organizational theory, bureaucratic autonomy, international organizations, information policy, and many other areas to address both the people involved and the specific data context. The resulting model provides a holistic view of the data sharing policy development process that can provide insight into the current state of government data sharing and help explain the variation observed among different countries, fields, and agencies.

The key element of this model is the inclusion of both people and ideas: the structure and dynamics of the organizations and entities that control and influence policy-making and the key attributes of the data about which policy is being made. The government agency is the primary actor in data sharing policy development, emphasizing the importance of the structure, processes, culture, and norms within the organization. The agency is influenced by the actions of a range of external actors, including national-level policy-makers, nongovernmental organizations, and intergovernmental organizations. The model identifies and explores the variety of levers by which these actors may influence data sharing policy development within the agency. However, the motivations for the actions of these external actors are considered largely exogenous to the model (e.g., the model notes

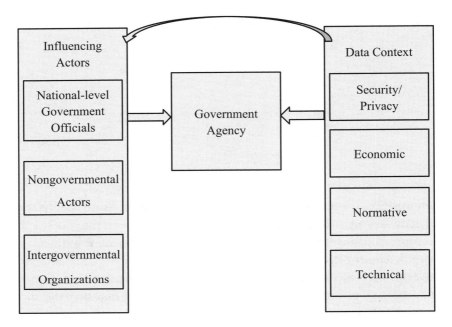

Figure I.1
Model of Data Sharing Policy Development

the ways in which the announcement of a national open data initiative by the president may influence an agency, but does not explore the reasons that an executive may choose to pursue this type of initiative).

The model also takes into account the power of ideas: the context of the data and its key attributes play an important role in the development of data sharing policies. The government collects and produces a wide range of data, from environmental observations to bus schedules to detailed accounting information. The security, economic, normative, and technical attributes of the data, and hence options and opinions regarding the appropriate data sharing policies for the data, vary significantly. This part looks at each of these elements in turn, reviewing the relevant literature and explaining the role of that element in the model.

2 Defining Data Sharing

To begin, it is important to define the concepts and terms that will be used in this book and that will bound the scope of the model. Two of the most common terms used are *data* and *information*. Formally, data can be defined as the "raw bits," the symbols that represent the properties of objects and events, and information is defined as data that is processed and organized to put it in context or make it more useful, creating value-added products, such as visualizations, specialized applications, or reports.[1] However, this distinction is problematic, because it differs significantly from common usage in policy development.

Most of what we think of as "data" has at least minimal processing—translating zeros and ones collected by a sensor into standard units, sanitizing a dataset to remove typos or outliers, or adding metadata to describe what has been collected. Minimally processed data remains an intermediate good: a good used as an input to the production of final goods, rather than a final product itself. It is this type of data to which many government data sharing policies refer. Because it does involve some processing, this type of data is sometimes referred to as information. Following common practice among those developing government data policies, this book will use the two terms interchangeably, both referring to minimally processed data.

This book deals specifically with data collected and/or produced by the government, sometimes referred to as publicly collected data or public sector information (PSI). This can include scientific data, such as that collected by space or environmental agencies, as well as data collected to carry out government functions, such as meteorological data collected to develop weather reports and warnings. Publicly collected data can also be about the government itself, providing politically important information about spending, future plans, or other government activities. It can include also benign information about government services or structures, such as public transit schedules, postal codes, or office hours.[2] Recognizing and addressing these distinctions is important, as discussed below.

This book will not deal with data sharing policies governing data collected by nongovernment entities supported by public funding, such as individual researchers supported by government grants. While there are many similarities in the development of data sharing policies for these grant-giving institutions, there are also unique considerations regarding the incentives and barriers for individual researchers that will not be addressed in detail here.[3] The book will also not address data sharing policy development within other entities, such as nongovernmental organizations, intergovernmental organizations, or companies, except as the policies of these organizations affect the development of government data sharing policies. Again, while there are many similarities, and those studying data sharing policy development may find this book informative in understanding these processes, these organizations face additional incentives and constraints in data sharing policy development that will not be explicitly discussed here.

With a working definition of data and information in hand, the next important term to define is data sharing policy. Once again, there are a series of terms used (often interchangeably) to refer to closely related ideas. Data sharing can refer to interpersonal, intraorganizational, and interorganizational data exchanges. In this usage, the term generally refers to a negotiated agreement between two or more entities for a mutual exchange of data or the sharing of data by one organization in exchange for products, services, or funds from another organization.[4] The term *data exchange* is generally used synonymously with this definition of data sharing. These types of negotiated arrangements are not the subject of this book. Instead, this book focuses on data sharing policies that are created by an agency on its own, without any official partner entities or formal external negotiation processes, sometimes referred to as data access policies. While the agency may consult with other organizations—domestic, international, government and nongovernment—it does not require that these organizations provide approval of the document.

A data sharing (or data access) policy defines what data will be shared, with whom (and/or for what purpose), at what price, and under what conditions the user can redistribute the data or applications developed using the data. These policies can vary significantly in terms of the overall level of openness and the particular types of restrictions. At one extreme, a data sharing policy covering top-secret information may limit access to only those with special credentials and an officially sanctioned "need to know." At the other extreme, an open data policy would support completely unrestricted access: anyone is free to use, reuse, and redistribute the data at no cost and for any purpose.[5] In this book, the term "data sharing policy" will generally refer to policies in which at least some access by external users is intended.

Data sharing policies specify the type of data to be shared and any key attributes of the data. The data may be near real time, minimally processed data, or the data may be transformed in some way: processed, aggregated, anonymized, truncated, or time delayed, for example. Policies may differentiate between types of users and/or uses: domestic, foreign, education, research, journalism, operational, policy, commercial, or others. Data sharing policies may provide access for free, or they may charge a fee for access to the data. Fees are sometimes set at the marginal cost of sharing the data to recoup the costs of distribution; the average cost of collecting and sharing the data, allowing for full cost recovery; or at a market rate based on the users' willingness to pay. The cost may be fixed or negotiated on a case-by-case basis, and it may differ depending on, for example, the type of data being shared and the individual or group interested in accessing the data.

Restrictions on access may also include requirements for user registration and/or the submission of an official request. Requests may be brief and approval routine, or they may involve more complex proposals to be reviewed using a lengthy, formal process before being selected. Some policies restrict redistribution of the data and/or products derived from the data. Like the policy itself, this may include restrictions on the types of users with whom the data (or products) can be shared, or the types of uses for which it can or cannot be shared (research, commercial, and other uses). Downstream users may be required to return to the original source of the data to request permission to access and use it. Policies often require proper credit be provided to the data provider.

Data sharing policies may be informal and defined via common practice, or they may be carefully delineated in an official document. The choice of license placed on the data, if any, also affects the clarity of the policy and the usability of the data.[6] An overview of the many types of potential restrictions is provided in Table 2.1.

Table 2.1
Elements of a Data Sharing Policy

What data is being shared?	• Type of data to be shared • Key attributes: Processing level, timeliness, aggregation, anonymization, and so on
Restrictions on Access	
With whom/for what purpose?	• Types of users: domestic, foreign, educators, researchers, journalists, nonprofit, commercial entities, and others • Types of uses: education, research, journalism, operational, policy, commercial, etc.
At what price?	• Free, marginal cost of sharing the data, average cost of collection and sharing, or market rate fixed or negotiated on a case-by-case basis • May be differentiated based on type of data and/or user or use
Using what process?	• User registration • Official request or detailed proposal • Routine approval or in-depth review
Restrictions on Redistribution	
Under what conditions can the data (or derivative products that use the data) be redistributed?	• Restrictions on the types of users or uses • Proper credit for data provider

3   People

Data sharing policies for most publicly collected data are developed by officials within the government agency that gathers or produces the data, and agency structure, processes, culture, and norms are key to understanding the creation of these policies. However, agencies cannot be considered in isolation: they are influenced by external actors with an interest in the data, including national-level policy-makers, nongovernmental actors, and intergovernmental groups. This chapter considers the role of each of these actors in turn.

Central Actor: Government Bureaucratic Agency

Bureaucracies are often given some discretion in the implementation of laws and regulations assigned to them by Congress or the president. Scholars have long realized that agency policy-making capability can also extend beyond this limited discretion agencies are granted by central planners, resulting in significant amounts of autonomy. This is particularly likely in cases, like information sharing, that deal with technical or highly specialized issues, where bureaucrats have greater expertise and understanding with regard to the subject matter compared to elected officials.[1]

Bureaucratic control over policy implementation makes agencies resistant to efforts by national-level policy-makers to encourage or enforce particular policies that do not align with agency incentives.[2] This autonomy in policy development and implementation, including resistance to outside efforts at control, has been highlighted often in the information-sharing literature, which often focuses on the role of the agency in developing policy. Nahon and Peled provide examples of reluctant agencies resisting or only superficially complying with open data initiatives in countries ranging from the United States to Estonia.[3]

Autonomy and Power

Maintaining autonomy, and the power that comes with it, is a primary concern for government agencies. For many, information is an important source of this power and autonomy.[4] Sharing this data, therefore, can affect power relationships.[5] Some agencies worry that adopting a data sharing policy that aligns with national or international standards will limit its freedom to act in the future,[6] or that increased transparency aimed to provide power to the citizenry will decrease its own autonomy and control.[7] New data policies may lead to new constituencies and the establishment of new expectations and requirements that will be difficult to change.[8] Some agencies are concerned that increased transparency will identify apparent or real redundancies among government organizations, potentially leading to a loss of some portion of the agency's tasks and control over information to another organization.[9]

Control over data and information has the potential to affect an agency's budget and reputation, two important factors in building and maintaining autonomy and power.[10] In some cases, agencies sell data to help recover costs and increase their budget. Agencies that do so are often reluctant to give up this income without seeing clear benefits to doing so.[11] Similarly, developing and maintaining the systems and expertise needed to share data can be expensive. Without a budget increase to cover these new costs and compensate for any losses in income from sales, agencies are unlikely to be supportive of free sharing.[12] Even with an increase in budget, bureaucrats may perceive the lack of revenue stream as a reduction in autonomy as the agency then becomes fully dependent on national-level policy-makers for their budget.[13]

Agencies may worry that a transition from cost recovery policies to free provision of data will result in a loss of political support, in addition to budgetary decreases from lost sales revenue. Some argue that agencies receive more recognition of the benefits of their data collection activities when they can point to revenues. Joffe reported that just the appearance of generating revenue, even if in reality it accounted for only a very small proportion of overall costs of data collection and sharing, was viewed positively by high-level decision-makers.[14] Without the ability to provide a dollar amount, agency officials may find it harder to explain and justify the importance or value of their data. This can result in a drop of support for their activities among policy-makers, and potentially further budget decreases.[15] This phenomenon may help to explain the finding of both Pinto and Onsrud, and Tulluch and Harvey, that larger, less resource constrained organizations seem more willing to share their data.[16] Larger agencies typically have greater political influence and visibility and are less dependent on revenue generation as proof of the value or importance of their activities.

However, there is a recognition by many that opening the data maximizes the number of users and creates more value overall, even if that value is not easily quantified. Joffe argues that agencies should put in place accounting mechanisms that capture the value created and money saved from use of the data, and use this information to advocate for larger budgets.[17] Tulluch and Harvey also emphasize the importance of communicating the value of data that is shared for free, although Martin points out that showing the return on investment on qualitative benefits such as increased transparency, more active citizenry, and higher participation in politics can be quite difficult.[18] Maximizing the number of users also maximizes the number of stakeholders in the agency's sustained provision of data, another possible source of political leverage.

In addition to the effects on its budget, agencies are also concerned with reputational effects. Perhaps most directly, some agencies worry that releasing their datasets will open up the agency to greater external evaluation and criticism.[19] While feedback that identifies and remedies inaccuracies is generally viewed as a good thing, Barry and Bannister (2014) found that some agency officials fear that real or perceived inaccuracies in the data, or misuse or misinterpretation by users, could disproportionately harm the reputation of the agency.[20] The potential for misuse by the press is also a common concern.[21]

Depending on prevailing norms and attitudes about open data provision in the government and general public, freely providing data could provide reputational benefits to an agency. In this way, freely sharing data is sometimes seen as a source of power.[22] Tulluch and Harvey state that agencies should always ensure that users know where the data came from, not only to get technical updates, but to ensure they know who to "thank and recognize."[23] However, Nahon and Peled noted that in cases where data is made available as part of a national effort, credit for providing this data is often directed toward the national-level policy-makers that announced the initiative, rather than the individual agencies that actually prepared and released the data, further decreasing incentives for agencies to provide data openly.[24]

Mission and Goals

One of the reasons that bureaucracies have autonomy in the creation of information-sharing policy is the specialized, technical nature of this activity. Agency officials with detailed knowledge of what data is collected and how it might be shared are considered best placed to develop policies on this issue. In other words, agency officials are most closely acquainted with the contextual attributes of the data and their implications for sharing, and best able to evaluate the policy options.

One way to examine this decision-making process is to view the agency as an instrumentally rational actor. It has goals, as defined by its official missions or directives, and it can identify alternative policy options and evaluate them according to these goals.[25] The implications of this framework are straightforward. If agencies do not see data sharing as an explicit goal of the agency, they are unlikely to support spending any significant time and resources on this effort.[26] For many agencies, however, data collection and sharing has some relation to their goals: Halonen reports that producing high quality data is increasingly seen as an important public service.[27] Agencies take into account the attributes of the data, and examine the effects of sharing this information in various ways.[28]

Of course, there are limits to the accuracy of this evaluation process. Bureaucracies are complex organizations that often have unclear or conflicting goals, and decision-makers often do not have full knowledge of all possible alternatives or their likely results. Organizations operate under conditions of bounded rationality, making decisions that are satisfactory, given the information and analysis capabilities available, rather than attempting to choose the best possible policy option.[29]

Culture and Norms

Agency culture and norms can also play an important role in agency decision-making. March and Olson (2007) suggest that many organizations use a "logic of appropriateness" rather than the "logic of consequences" assumed in a rational actor model. Under a logic of appropriateness, decision-makers are highly influenced by rules, routines, norms, and identities.[30] They attempt to match new situations to existing patterns of behavior. Unlike the instrumental view, which is primarily concerned with official agency goals, institutionalists call attention to the role of informal norms that are developed over the organization's history.

The role of culture is widely recognized in the information-sharing literature.[31] Constant et al. (1994) argue that organizational culture plays a significant role in encouraging data sharing by employees, and increasing the visibility of data sharing efforts encourages the enactment of sharing norms.[32] Drake et al. (2004) and Martin (2014) both find that agency culture is a key element in ensuring that information-sharing policies are sustainable.[33] Organizational norms are often closely related to the norms of the dominant profession in the organization. Dawes (1996) argues that the ethical standards of the dominant profession often dictate how information may be used.[34]

Agencies learn from experience over time, and may consider the experience of other, similar, organizations.[35] However, the path dependency that provides stability in values and norms and facilitates decision-making by a logic of appropriateness can also lead to historical bias and inflexibility.[36] This recalcitrance to change is noted in the information-sharing literature, as well.[37] A common observation in the literature is that government organizations often have a culture of secrecy and are naturally reluctant to release data, perhaps driven by the sensitivity to outside evaluation and criticism mentioned earlier.[38]

Barry and Bannister (2014) argue that changes in this culture must begin with senior management.[39] Other scholars agree that managers and organizational leaders have an important role to play in propagating norms of behavior related to data sharing. McDermott and O'Dell (2001) found that people within an organization are more likely to support data sharing if it is seen as important to senior management, and Harvey and Tulloch (2006) note that the attitudes or fundamental beliefs of those with authority over decision-making can be a very important factor in organizational action.[40]

Influencing Actors: National-Level Government Actors

Bureaucrats are central to the development of data sharing policies, but their policy-making options are not without limits. In fact, multiple scholars have found that bureaucratic autonomy, the bounds within which bureaucrats are free to make and implement policy according to their own preferences, can be significantly limited by actions taken by legislative, executive, and judicial officials.[41] The actions of legislative and executive officials may be driven by a desire to apply national-level priorities across agencies or to address national-level economic or other challenges. National-level policy-makers may place different values on the desires of certain constituencies or interest groups, such as businesses in a home district, than officials within a government agency. Therefore, when considering the development of data sharing policies, it is important to look at related trends, initiatives, and laws at a national level.[42]

Executive Controls

There are a number of avenues by which national-level policy-makers can affect bureaucratic action. This can be done through the use of political appointments—high-level agency officials who seek to implement executive initiatives. Other organizations with a focus on government-wide executive priorities, such as the Office of Science and Technology Policy or the Office of Management and Budget in the United States, may also implement new directives, coordinating initiatives, or budget recommendations favored by the executive.[43]

Open data initiatives in many countries were announced at the top executive or ministerial level, including President Barack Obama's announcement of the United States Open Government Directive in 2009.[44] The United Kingdom, China, Norway, Estonia, Denmark, and other nations have made similarly high-level announcements about data policies.[45] Although the overall success of the program is debated, the announcements undoubtedly had an effect in raising awareness of the issue of open data and mobilizing resources in at least some agencies.[46]

Legislative Ex Ante Controls

Legislators also have the power to influence the bureaucracy. Legislators may put in place laws that direct bureaucracies to take particular actions, or to report on policy options or choices before they are implemented. These are typically referred to as *ex ante controls*.[47] Just as the executive branch can release statements directing agencies to adopt particular data sharing policies, the legislative branch can attempt to initiate data sharing by passing laws requiring (or forbidding) particular types of changes. These actions are often necessary to provide visibility and push agencies to begin the process of reviewing and updating their policies.[48]

Whether the directives come from the executive or from the legislature, clarity is important. Halonen reports that one of the biggest problems for agency officials trying to implement open data initiatives in the United Kingdom was a lack of clear guidance from the central government. Agency officials weren't sure of which, or how much, data to release—and weren't even sure of the definition of open data.[49]

High-level directives and laws also have the potential to politicize the data sharing policy development process. The decision of what data is to be made available, when, and to whom can have political consequences. Releasing expenditure levels, crime statistics, or environmental data increases awareness of, and emphasizes, some problems over others. This politicization can overshadow broader objectives of data sharing initiatives.[50]

Legislative Ex Post Controls

Legislators can also impose e*x post controls*, passing new laws to reinforce or inhibit agency actions or increasing or decreasing budgets to influence agency behavior.[51] Laws prohibiting the release of data are generally effective, but can result in greater restrictions than intended. Barry and Bannister reported that uncertainty regarding the interpretation of the Data Protection Act was one of the greatest barriers to adopting open data policies in Ireland, with some agency officials concluding that the safest way to ensure compliance was to keep data restricted.[52] Laws requiring data be made available are harder to implement in a meaningful way if data provision doesn't align with the interest or culture of the target agency. Updating, or failing to update, laws has also affected data sharing policy efforts. Peled notes that the failure to make interagency data-trading practices illegal created an incentive for agencies to protect these valuable assets from inclusion in open data initiatives.[53]

Providing, or failing to provide, adequate budgetary support to offset lost data sales revenue and to support the technical and human resources needed to develop and maintain data sharing infrastructure, is one of the most effective tools national-level policy-makers have in influencing agency data sharing behavior. As mentioned earlier, data sharing can be a resource-intensive endeavor, and without adequate budgetary support, agencies often see little incentive or ability to undertake these efforts.[54]

Increasing Autonomy

The existence of so many sources of influence, some argue, is another potential source of bureaucratic autonomy. Agencies may get conflicting directions from executives and legislators, or even conflicting directions from different offices, committees, or individuals within each of these groups. In these cases, balancing these interests may allow the agency some room for independent action according to its own preferences.[55]

Influencing Actors: Nongovernmental Organizations

Actors outside the government also have the ability to influence policy. James Q. Wilson argued that the political environment occupied by an agency would play a large role in the extent to which it is influenced by outside interests. He identified four types of potential environments: "(1) a dominant interest group favoring [the agency's] goals; (2) a dominant interest group hostile to its goals; (3) two or more rival interest groups in conflict over its goals; or (4) no important interest group."[56]

Wilson referred to agencies in the first arrangement as "client agencies," and explained that they are often strongly influenced by this single, organized group. Much of their information and understanding of their mission may come from this group, and actions taken will likely favor this group. When the dominant interest group is hostile to the agency's goals, as in the second environment, the agency is at risk of capture, acquiescing to the demands of this group, and taking a more passive role in policy development. An agency with two or more rival interest groups will face both criticism and support for each decision, and is likely to support varying directives from national-level policy-makers as broader political situations change. Finally, an agency with no important interest groups is driven primarily by the character of its leader and his or her directions from national-level leadership.[57]

Interest Groups Supporting Open Data

Agencies, or subgroups within agencies, responsible for developing data sharing policies can fall into any one of these four environments. Plans to share data that has clear scientific or commercial applications may see support for free and open sharing from researchers, operational users, journalists, and value-added companies who want to use the data to carry out more research, improve understanding of issues, or develop new applications. Once these users incorporate data into their activities, continued provision becomes critical to their ability to conduct follow-on research or maintain a profitable business. Open data advocates often argue that the data could be useful to a very broad set of potential users, including the general public. Others argue that the stakeholders for open data are quite limited, including only those researchers and corporations with the time and expertise to find, analyze, and use the data.[58]

Data users are sometimes represented by professional societies or industry groups that can effectively advocate for or against particular data sharing policies. These can include national organizations, like the American Meteorological Society, or international organizations, such as the International Council for Science (ICSU) and its interdisciplinary Committee on Data for Science and Technology (CODATA).[59] As mentioned above, the norms developed and propagated within these organizations and the associated communities can be influential in the development of government data sharing policies. However, depending on the domain, groups of researchers, entrepreneurs, and small businesses may not be large, well organized, or politically active.

Over time, the general public may see secondary benefits from increased scientific knowledge, greater transparency, better public policy-making, and more products and services. At times, particular citizen groups interested in particular issue areas may advocate for the release of data.[60] With technology access and proficiency increasing among the citizens, direct engagement by the general public may increase.[61] Still, the benefits for the general public, though potentially great, are very widely dispersed, so the incentive for any individual citizen to advocate for this issue remains relatively low.[62]

Potential users may not even be aware of the existence or relevance of publicly collected data. One of the purported benefits of open data policies is the ability to increase the likelihood that data is used in new and innovative ways. Often this means reaching out to new types of users and promoting data discovery. By definition, this would mean that these users do not constitute an existing interest group advocating for free sharing of the data. If this outreach process does not occur, then low data use may be seen as an argument to decrease or discontinue data sharing efforts, despite the existence of potential value.[63]

Interest Groups Supporting Data Restrictions

These same user groups—researchers, operational users, and value-added companies—may also advocate against free and open sharing if they believe it will create unwanted competition. Researchers who have exclusive access to data as principal investigators, for example, may want to protect their ability to publish findings before other researchers gain access. For some, this is seen as an important incentive for investing years of effort in designing a data collection system or process. Similarly, high data costs or exclusive access licenses can be a significant barrier to entry to new value-added companies, privileging those that already exist.[64] These existing companies may thus oppose a transition to free and open data. Another potential interest group hostile to open data policies includes companies that sell data very similar to, or the same as, that collected by a government agency. If the government data is given away for free, these companies would be put out of business.

Influencing Actors: Intergovernmental Organizations

Data sharing policies are important precisely because data often has value for users far beyond the group that originally collected it. In many cases, the value of data extends beyond national borders. In a 1997 publication, the National Research Council observed "Data in science are universal—they have the same validity for scientists everywhere."[65] Joseph Stiglitz argued in 1999 that knowledge was a global public good, and that the international community has a responsibility for the creation and dissemination of this good, working through international institutions.[66] International organizations are recognizing this mutual interest in data sharing and raising this issue on a global level, with organizations such as the United Nations, World Bank, and Organization for Economic Cooperation and Development (OECD) actively working in this area. For this reason, it is important to consider the role of intergovernmental organizations in affecting the development of national data sharing policies.

Instrumental Benefits

As in the case of nongovernmental actors, discussed above, the ways in which international organizations operate and exert influence can be analyzed with an instrumental or normative frame. Neoliberal institutionalism emphasizes the ability of international organizations to enable cooperation for mutual benefit. These organizations lower transaction costs, providing a venue for states to share policy information and knowledge and to negotiate. States are also able to make credible commitments and monitor compliance of other nations with their own commitments. These interactions can provide stability and information that enables cooperation. The practical information that is shared can improve the quality of national policy-making. It can also change nations' interpretation of other nations' actions and their perception of their own self-interest.[67]

Norm Development and Propagation

The constructivist literature emphasizes the importance of international organizations in developing and propagating international norms: standards of appropriate behavior for actors with a given identity. In this view, international organizations base their actions on particular norms, and promote these norms in the international community through information sharing, open discourse, and other activities. They have the ability to influence states' understanding of what constitutes legitimate behavior. In addition to propagating norms, international organizations can put pressure on states not abiding by norms by drawing attention to bad behavior. The constructivist frame emphasizes both what states say and what they do as demonstrations of their understanding of, and compliance with, international norms.[68] Nations often discuss or justify their actions with reference to international norms. For example, if an international norm promotes open data sharing, you would expect to see nations that restrict data explaining or justifying their reasons for doing so.

Summary

The model of data sharing policy development presented here takes into account each potential impact of each of the actors described above.

Bureaucratic actors are central to data sharing policy development, and the agency mission, goals, culture, and norm should all be taken into account. The model also considers the ability of national-level policy-makers to play a role, with executives putting in place national priorities or initiatives or appointing high-level bureaucratic leadership with the goal of implementing particular policies. Legislative actors can pass laws allowing or prohibiting various types of data sharing, require reports or other information about these policies, and they control the allocation of budget needed to support data sharing systems. Nongovernmental actors—professional groups, companies, or citizens, for example—can also organize to lobby the agency according to their own interests. International organizations can provide venues for sharing information about policy design and impacts, and they can influence the perceived international norms in a given field. Together, these actors play a critical role in shaping government data sharing policies.

4   Ideas and Technology

Data sharing policy cannot be understood by understanding the actors and their motivations alone. The specific type of data under consideration, and its security, normative, and economic attributes have important effects on the incentives and actions of these actors. The technology for collecting, producing, analyzing, and distributing data also plays a role in the possibilities and preferences for data sharing, and technological change can lead to important changes in security, normative, and economic data attributes.

Data Context: Security and Privacy

Even among advocates of open data, it is generally agreed that access to data should be restricted when there are legitimate security or privacy concerns.[1] The challenge, of course, is determining what constitutes a legitimate concern. Many individuals recognize both instrumental arguments for security and privacy—ensuring data isn't used to cause harm—as well as an intrinsic right to privacy, and many nations have laws in place to protect sensitive data. In some cases these laws limit data access to government agency officials for use only in the specific cases for which the data was collected.[2] However, as mentioned above, lack of clarity in the interpretation of these laws has the potential to lead to the release of sensitive information or, conversely, to the placement of restrictions on data that could be made available.[3]

Privacy
Although the government collects a great deal of personal information, this data has generally not been the focus of efforts to increase the availability of government data. This has led some in the open data community

to argue that privacy is a "nonissue."[4] Indeed, much of the data held by the government, such as transit schedules, expenditure information, maps, and environmental data, does not contain, and never did contain, personal data, so the release of these datasets generally does not pose a privacy risk. The statement cannot be absolute, because nonpersonal information can sometimes be used to learn about an individual. For example, data about average house prices by zip code could be matched with an individual's address to estimate the value of their home.[5]

Still, most privacy issues relate to data that is, or once was, directly associated with an individual. Raw personal data, data that includes personally identifiable information such as names or social security numbers, is rarely considered appropriate for release. Exceptions are sometimes made in the case of public officials, where transparency may be viewed as more important than privacy. More often, data must be anonymized—all personally identifiable information has to be removed—before there can be any consideration of its release.[6] Access to anonymized personal information has the potential to provide significant benefits: data on travel into and around the United States can be used by the hospitality industry, standardized test scores could allow parents to better understand school performance, and sharing healthcare data could improve medical research.[7]

However, release of personal data, even after it has been anonymized, is controversial due to the risk of deanonymization, in which individuals can be reidentified, intentionally or unintentionally, through cross-linking of databases.[8] Borgesius et al. (2015) argue that irreversible anonymization may be impossible, so even anonymized datasets should not be released openly. One option some nations have used to address this is data licenses that expressly forbid any attempts to reidentify individuals in the dataset. Borgesius et al. suggest "restricted disclosure" or "managed access" as other potential compromises for data access. For example, anonymized medical data may be released to an academic research team for use in an approved research proposal, perhaps even requiring that researchers access the data using secure government systems. Another option would be to create systems that allow third parties to query the dataset, but which only return statistical data, rather than individual information. The appropriateness of access and/or reuse restrictions, or of the release of the data of any kind, must be determined on a case-by-case basis. Borgesius et al. suggest considering the goal that is being pursued by releasing the data and whether there is another way to achieve that goal. If there is not another option for achieving the goal, the risks of releasing data should be assessed, and decision-makers should determine how probable these cases are and how great the harm would be if they occurred.

Security

Just as confidential data is restricted, data that has the potential to pose a national security risk is generally not considered eligible for sharing outside the government. This includes classified information, as well as data that is unclassified, but sensitive. At one extreme, some might argue for restriction of any data that could be used for nefarious purposes. However, almost no data would be released under this rubric. For example, one could argue that access to public transit schedules should be restricted, because terrorists could use them to plan and coordinate attacks. While true, clearly this would be an overreaction. Transit schedules provide a great deal of benefit to the general public, and terrorists would have other ways of gathering this information if the timetables were no longer published.[9] Florini (2004) argues that officials must consider not only the costs of openness, but also the costs of secrecy.[10]

Focusing specifically on geospatial information systems, Salkin argues that a lack of access to environmental and public health data will harm public safety, rather than help it. Like Borgesius et al., he argues that release of these datasets should be considered on a case-by-case basis, pointing to guidelines developed in a report by the RAND Corporation as well as the US Federal Geographic Data Committee (FGDC) to direct evaluation. These guidelines focus on usefulness, uniqueness, and costs and benefits. They suggest that data holders examine (1) whether the data are useful for selecting potential targets or planning an attack on a target; (2) whether the data is unique: not easily observable and not available from alternative sources; and (3) whether the security costs of sharing the data outweigh the benefits to the public of doing so.[11]

With respect to geospatial information, the RAND report found that almost none of the publicly accessible datasets they examined met the first two criteria, as alternative information sources exist from nongovernment entities.[12] The FGDC guidelines closely match those proposed by RAND, adding that even if the data does need to be safeguarded, additional steps should be taken to develop policies that limit access in intelligent ways. In particular, the FGDC suggests examining options for making changes to the data—deletions, aggregations, or other adjustments—that would increase the safety of release. Limited restrictions—on users eligible for access or on redistribution—should also be considered.[13]

In some cases, access to data may significantly enhance national security or public safety. Timely information about severe weather developments, for example, can help to save lives and property. Good data about public infrastructure could be critical for first responders or individuals in preparing for, or responding to, a terrorist attack. Sharing information about the environment may allow scientists to improve the understanding of global challenges, allowing governments and individuals to better mitigate or adapt to changes. Sharing crime data may help individuals to make more informed decisions about where to live, work, or travel. Analysis of public health could lead to insights that save lives. As always, the benefits of data sharing must be considered with respect to the risks of doing so.

Data Context: Normative

Normative arguments involve identifying moral or ethical responsibilities related to data sharing. Rather than looking at which policies *can* be implemented, a normative framework asks which policies *should* be implemented. This framework helps to explicitly identify the values that underlie the goals these policies should aim to achieve, including those implicit in the economic and security sections in this chapter. Debates surround both the appropriateness of data sharing processes as well as the importance of the potential outcomes of data access and use. Those interested in normative issues are concerned with equity and the distributional effects of data sharing policies. The applications that are enabled or deterred via data sharing have the potential to impact individual's lives in important ways, and some argue that particular types of data uses create an imperative for greater data sharing or greater data protection.

National Security, Public Safety, and Privacy

It is a normative judgment that the government ought to consider national security, public safety, and privacy in the development of data sharing policies. There is widespread agreement that protection of national security and public safety is one of the key responsibilities of government and that citizens have a right to privacy. As discussed earlier, pursuing these goals may mean choosing to restrict access to data that could threaten national security or public safety if it were to fall into the wrong hands, or ensuring that data is only released if it has been reliably anonymized to protect privacy. Sharing information about environmental or other challenges may enhance national security and public safety, advancing scientific understanding of major global challenges or facilitating the development new safety applications.

Distributional Effects and Equity: Repository or Public Trust

In a classic economic view, the only concern is the maximization of net social benefit, regardless of who actually captures those benefits. In reality, many are concerned with equity and the distributional effects of data sharing policies. Do the benefits accrue to the producer of the data or the consumer? Do the government and general public shoulder the costs of data collection and sharing, or does the private sector share this burden? Are existing companies and noncommercial data users favored over those who have not yet discovered or used the data?

Distributional effects of data sharing policies most commonly manifest in debates about whether it is more appropriate for the government to operate as a repository for data, maintaining data and making it freely available, or as a public trust, restricting access to data and charging for its use. Those who argue for government as a data repository emphasize that citizens have already paid for the data once through taxation and shouldn't be asked to pay again to access the data themselves.[14] Advocates argue that when fees are imposed, it privileges companies that have the ability to pay. Researchers and others that could develop applications with broad social benefit will have more difficulty accessing the data. When a cost is imposed, access is especially challenging for the poorest individuals, including those in developing countries.[15] However, if the argument is that citizens should have free access because they pay taxes, then this argument doesn't actually apply to data sharing outside the boundaries of the district, state, or nation that collected it.[16] The argument would, however, apply to commercial users, as corporations do pay taxes.

The traditional argument in favor of government acting as a public trust is based on the idea that valuable public resources should be protected by the government, generally by restricting access, to ensure they're available for future generations. However, this logic does not apply well to data—any amount of data use today will not decrease the amount of data available in the future. If anything, it will have the opposite effect, with data use generating additional knowledge and applications that benefit future generations. In the case of data sharing, the concept of a public trust is sometimes interpreted to mean that the government should protect this valuable public asset by not allowing individuals or companies to derive private benefit, in the form of revenues, at the expense of the general taxpayer.[17] Commercial users—corporations and entrepreneurs—are most likely to have the time, expertise, and incentive to analyze large datasets. These users reap disproportionate personal benefits from data, while data collection, maintenance, and distribution are financed by taxpayers. By privileging these groups, open data policies would increase inequality and reinforce the digital divide.[18] Charging commercial users a fee would help to offset the costs of the data collection and sharing system to the public. Onsrud argues that the public trust model is more appropriate when the primary uses of the data are commercial, while the repository model is better when the data is useful for scientific research, transparency, or other applications that would primarily benefit the general public.[19]

Proponents of both models are concerned with the advantages of users with time and expertise, particularly commercial users, over the general public. Supporters of a public trust model and fee-based system argue that this returns at least some of the private gains to the public. Supporters of a repository and free data provision argue that fees would only exacerbate existing commercial advantages. Martin suggests that education and technology could be used to offset these inequalities. Agencies can advertise the availability and value of their datasets to the general public, or adopt visualization software that lowers barriers to use posed by a lack of statistical or analytical skills.[20]

In general, it is important to understand that any data sharing policy will have winners and losers. A decision to implement a free and open data sharing policy will benefit the value-added commercial sector, which will incorporate the free data into its products and business plans. Data sales will benefit companies that have developed applications that are sufficiently profitable to operate given data costs. Public-private partnerships, including government data purchases, benefit data collection companies that receive contract guarantees and revenues. A policy that provides exclusive use of data to scientists who helped develop a data collection system (i.e., principal investigators), benefits scientists with the resources, reputation, and expertise to apply for and win large grants. Even if these types of distributional issues were not always considered in the initial development of a policy, they often come to the fore when a change in policy is being contemplated. In these cases, existing users have a strong incentive to protect the status quo.[21] Hence, this issue will generally enter agency decision-making via inputs from interest groups, rather than as a normative consideration on its own. Just as there is no economic motivation to privilege one group or another, neither existing nor potential users generally have a stronger normative claim to disproportional benefit from a data sharing policy.

Increasing Government Transparency, Accountability, and the Right to Know

One of the most common arguments in favor of greater access to public information, and open data policies in particular, is the important role data access can play in increasing government transparency. Increased transparency is sometimes seen as an economic benefit, due to its instrumental role in increasing economic growth—discussed in more detail in the economic section below. Normative arguments instead focus on the intrinsic value of transparency—citizens' right to know what their government is doing or to have access to the information on which government officials are basing decisions.[22] Calls for transparency sometimes also reference its instrumental role in increasing government accountability, which is seen as an intrinsic good. Some effects of transparency, such as the enabling of "armchair auditors" to identify government fraud or abuse, have both economic and normative implications.[23]

Yu and Robinson argue that the term "open government" was originally used to refer to the release of politically sensitive government information. Its origins can be traced back to the freedom of information movement in the 1950s. Now, the term "open government data" is used to describe not only data directly relevant to government transparency and accountability, but more broadly to any government data shared freely over the Internet.[24] However, some of the broadening of this definition may be appropriate. Even data that are released with the primary goal of improving government services, rather than providing public accountability, provide insight into what the government is doing. The release of environmental data, which may seem politically innocuous, can provide citizens with an insight into the information on which policy-makers are basing important decisions affecting public health and safety. A number of scholars argue that insight into government decision-making is necessary in a democracy, helping to affirm the legitimacy of the government and maintain the trust of its citizens.[25]

Some argue that openness can undermine trust, as citizens uncover fraud or other abuse.[26] However, this is likely a short-term effect, as in the long run the increased ability to detect fraud would decrease its actual occurrence.[27] When information is shared with the public, the government is more accountable for its decisions.[28] Others counter that while this is true in theory, empirical evidence for an increase in transparency related to open data is limited. There is a concern among some public officials that trust can be harmed, because data is disproportionately used in negative stories and in "gotcha" journalism.[29] Others point again to the advantage of wealthier individuals and corporations, in terms of time and expertise, in accessing and using government data. These individuals and organizations can use the data to track government actions and decision-making and more effectively lobby for their own preferred policy solution, once again furthering inequality.[30] Like Martin, above, Halonen argues that even if open data does not empower the average citizen now, this may change as education increases and technology lowers the threshold for using public data.[31]

Efficiency as a Normative Goal

Before moving on to the next section, it's important to acknowledge that the pursuit of economic efficiency–maximizing net social benefit, implicit in the economic section below, is a normative goal. As noted in the next section, the decision to calculate these costs and benefits on a global, rather than a national or agency level, has normative implications, as well. Economic efficiency is only one of a number of goals that individuals argue the government should be attempting to achieve, and the relative importance of achieving economic efficiency, compared to security or transparency, for example, is debatable. Furthermore, due to the challenges of directly measuring utility, net social benefit is almost always measured in dollars, giving preference to a narrow focus on maximization of benefits that are easily monetized, rather than on effects that help drive economic growth, but whose exact effect is difficult to measure.

Data Context: Economic

Despite these limitations, economic arguments are at the heart of many debates about data sharing policy development and changes. From an economic perspective, the goal is to determine which data sharing policy will maximize net social benefits: total benefit to society minus total costs. Yet actually carrying out this calculation, and even determining which components should rightfully be included, can be quite complex. Thus, this section goes into significant detail on this important topic, discussing the economic attributes of data, the theoretical effects of data sharing policies, and practical limitations faced in implementation. Although the economic efficiency of a policy depends significantly on the specifics of the type of data and its potential uses, this section also develops a framework to determine the relative economic efficiency of open data sharing given the economic context of the data.

Economic Attributes of Data

Much of the debate regarding the appropriate choice of data sharing policy is related to whether data is most efficiently treated as a public good and shared openly or as a commodity to be sold. To understand this debate, it is important to first understand the economic attributes of data itself. Within economics, goods are typically classified based on two dimensions: excludability and rivalry. The first questions whether users can be excluded from accessing or benefiting from a good, and the second examines whether one person's use of a good reduces the amount of that good available for others.[32]

A hamburger is a classic example of a pure private good, a good that is both excludable and rivalrous: it is easy to limit someone's access to the hamburger, and when one person eats it, it is no longer available for anyone else. The traditional example of a pure public good is national defense: it is not possible to exclude individuals from the benefits of national defense, and the fact that one person is protected by national defense does not limit the amount of protection available for another citizen.

Data and information as a whole don't fit neatly into these categories. For example, data can sometimes be considered rival. Privileged access to data can give the holder an advantage that is lost once others gain access to that information. This is one reason that firms sometimes choose to keep key technological information secret rather than to pursue a patent. It also explains why scientists who help design a new instrument sometimes receive a period of exclusive use of the data, allowing them the first chance to publish new findings. After corporate secrets are released or initial scientific findings are published, the data is technically still available for use, but its strategic value has decreased. For the most part, however, data can be thought of as nonrival: the fact that one person uses the data does not reduce the amount available for others. Once the data has been collected or produced, it can be used over and over by many users with no additional production costs. In economic terms, the marginal cost of providing the data is zero.

The extent to which data and information are excludable is also debatable. In some cases, once information is made publicly available, it is difficult or impossible to exclude individuals from accessing and using it. This is particularly true for simple facts or single data points, which, once made known, cannot be made private again. However, for more extensive datasets or collections of information, exclusion is possible through the use of legal restrictions in the form of licenses governing access, use, and redistribution of the data. Individuals can also be excluded from accessing or using information if that information is kept in secure systems and not made available outside the originating organization.

Although there are some exceptions, for the most part data can be thought of as nonrival and excludable. These two attributes together suggest that information is best thought of as an impure public good. The fact that data is nonrival—the marginal cost of the data is zero—means that net social benefits are maximized when every user who places a positive value on the data can access and use it. The more a dataset is used, the more knowledge is created or applications are developed, and the more benefit is generated overall. Maximizing use will maximize total social benefits.[33]

One way to ensure that everyone who places a positive value on the data can access it is to make the data freely available. Since a commercial entity could not make a profit by giving its products away for free, this would require the government to collect and distribute the data. Because data is excludable, it is also possible to sell data. If a company were to sell data to every user at exactly the amount they were willing to pay, a concept called *perfect price discrimination*, this would also result in a situation in which everyone placing a positive value on the data would be able to access it. Therefore, in theory, it would be equally efficient to treat data as a public good and provide it for free or treat it as a commodity and sell it using perfect price discrimination. The only difference would be whether the benefits accrue primarily to the data users—the case in free government distribution—or to the data producer—in the case of commercial data sales. Questions of efficiency come down to the extent to which each of these two methods can be implemented in practice. A review of these practical issues below shows that both policies have drawbacks in practical implementation, but the relatively larger challenges in approximating perfect price discrimination mean that free and open data policies will almost always result in greater social benefit than data sales.

Free Data Provision in Practice

While in theory the marginal cost of providing data to an additional person is zero, in reality the cost of maintaining and sharing data can be significant.[34] Because of this, some agencies that claim to have a free data policy do charge a fee for access. While the data itself is free, and users are not responsible for the costs associated with data collection, the user pays the marginal cost of data provision or the cost of fulfilling a user request.

In determining the marginal cost of data provision, some agencies factor in salaries of the personnel needed to process, copy, and distribute data, or the costs of the data storage and processing systems. Initial processing of raw data into datasets that are error free and properly documented with metadata can require hours of expert work. Organizations also require advanced technology for data storage, which requires maintenance and updates over time. Enabling easy access to the data requires development of a user-friendly data portal. If agencies pass on the costs of these efforts to users, the fee for data access can become quite high.

However, in most cases, the inclusion of personnel, infrastructure, and maintenance costs in marginal cost fees is inappropriate. Typically, initial processing, development of metadata, and data storage are required even if government officials only expect to share the data within the agency or maintain the data for their own future use. In these cases, making the data

available to additional external users does not require significant investments in personnel, infrastructure, or maintenance.

More easily justifiable are marginal cost fees that include the cost of storage media (paper, DVDs, CDs, or hard drives, for example) and shipping. In
1962, Kenneth Arrow, an expert on the economics of information, noted
that the cost of transmitting data is frequently very low.[35] In some cases,
governments have found that imposing marginal cost pricing results in a
net cost to the government due to the need to develop and maintain a system for collecting and processing fees.[36] Further, the increasing capabilities
and decreasing costs of information technology, particularly the growth of
the Internet, have resulted in marginal costs that truly do approach zero.
Data can be provided online without any media or shipping costs. Further,
if agencies do not restrict redistribution of the data, it can be retransmitted
by others at no cost to the agency: truly zero marginal cost, at least from the
perspective of the agency.

Perfect Price Discrimination in Practice

Perfect price discrimination is not practical in reality, because it is nearly
impossible to determine each user's true willingness to pay. Attempts to
negotiate prices on a case-by-case basis would be time consuming and
entail transactions costs that on their own would exceed the resources of
some users. Instead, perfect price discrimination is typically approximated
by prices that differ depending on some attribute of the user or the data.
As noted in the discussion of definitions in chapter 2, data policies sometimes differ based on whether the data will be used for commercial, public,
research, or educational uses. Access conditions may differ depending on
the timeliness, accuracy, or other attribute of the data. The extent to which
these policies approximate the benefits of perfect price discrimination
depends on how well the tiers align with various groups' actual willingness
to pay. It is inefficient to the extent that users who place a positive value
on the data, cannot, or choose not to, use it because of the existence of fees
or restrictions. Additional practical issues suggest that the number of users
that fall into this group can be quite large.

Appropriability and externalities   One group of users negatively impacted
by fees and restrictions are those who do not derive personal financial
return from their use of the data. For example, a scientist who uses data

to improve understanding of climate change, a government analyst who uses data to improve the efficiency of Medicare, or a nonprofit organization that develops a tool to more easily monitor forest fires are all unlikely to generate any direct revenue from their activities. These users may be unable or unwilling to raise the funds needed to purchase the data, and the social benefits that would have been generated by their activities are lost. Moreover, the benefits of these activities extend well beyond the individuals involved in the particular project, resulting in positive externalities that benefit society in general.

A tiered policy that provides data for free for noncommercial uses may help to alleviate this problem, but data use will still be limited by restrictions on access and reuse needed to protect the ability to sell the data to others, and these restrictions can adversely affect data use.[37] This may be due to increased transaction costs—the need to read and sign license agreements, for example. Restrictions can also limit the ability to share data that underlie research, making collaboration and replication of studies difficult. In some cases, restrictions on redistribution prevent open sharing of the research or application produced using the data. It's worth reiterating the observation by Overpeck that about half of international environmental modeling groups were restricted from sharing digital climate model data beyond the research community because of intellectual property rights imposed by governments.[38] By contrast, empirical evidence suggests that the returns to science of open data policies are considerable; the accelerated pace of discovery in genomics is often credited to action taken by that community to support the rapid and open sharing of data.[39]

Evaluation challenges   Another complication is that for some, the value of the data is uncertain, and it is often difficult to evaluate information goods without actually working with them. An entrepreneur developing a new product may not know whether a particular dataset will prove useful in his or her application, or whether the finished product will be successful at all. Such a potential user may not be willing or able to purchase a dataset given this uncertainty. Data sellers may provide limited access to users on a trial basis to help mitigate this issue.

In practice, there are some indications that this negative effect on the value-added sector can be large. A 2000 report by Pira International entitled "Commercial Exploitation of Europe's Public Sector Information" noted that the market for public sector information in Europe, which often charged fees for access to government data, was significantly smaller than the corresponding market in the United States, where data was often provided on an open basis. The report found that cost recovery policies had resulted in a net financial loss for European governments, arguing that revenues generated from licensing fees were smaller than the taxation revenues that would have been generated from the value-added market had the data been given away for free.[40]

Data Sales, Cost Savings, and Distributional Effects

Given the relatively greater challenges of approximating perfect price discrimination compared to free data provision, and the existence of nonappropriable data uses, externalities, and evaluation challenges, a free and open data policy is much more likely to maximize data use and benefits than one that imposes fees and restrictions.[41] Because policies incorporating data sales and restrictions have lower total social benefits compared to free and open data policies, they will only be the more economically efficient option in cases in which they decrease the total social costs associated with data collection and provision. These cost savings can come about in two ways.

First, data sales enable the private sector to get involved in data provision. It is generally accepted that the private sector is more efficient than the government and has incentives to continually lower costs through innovation. If the private sector is able to develop and operate the data collection and distribution systems at a lower cost than the government, then this decrease in system cost would represent a decrease in total social costs, as well. For example, if a satellite system would have cost the government $100 million to build and operate, but a commercial company can do the same job for $90 million, this represents a savings of $10 million for society as a whole.

Second, to the extent that data collection is funded by resources raised through sales to the private sector rather than funds raised through general taxation, it is possible to avoid the deadweight loss associated with taxes, which microeconomic estimates suggest could be up to 30 percent.[42] These savings can occur when government agencies purchase data from commercial providers, decreasing the total amount of government funds needed to access data. They can also be secured if government data producers sell their data to private actors and use the funding to offset the cost of activities that would otherwise have been provided through regular budgeting procedures.

When considering this second type of cost savings, it is common for governments and others to conflate true societal cost savings with distributional effects. It is important to remember that only a portion (up to 30 percent) of the decrease in government spending represents true cost savings from a societal perspective. For example, if the government spends $100 million to build its own satellite and raises $10 million in revenue from data sales to commercial entities, this looks like $10 million in savings from the government agency's perspective. However, from a societal perspective, that $10 million cost has simply been transferred to another sector within society—commercial data users. True societal savings in this case would only be $3 million (or less), representing the up to 30 percent savings resulting from efficiencies associated with using funds that were not generated through general taxation.[43]

This also means that if the government agency in this example were to sell its data to other government agencies, or researchers funded by government grants, there would be no societal cost savings at all: the costs would only have been shifted around within the government and would still have been paid for with funds raised through general taxation. In fact, the need to negotiate the fees and exchanges among agencies would likely have added transaction costs.

If our definition of society includes the global community, this finding holds for data sales to foreign governments, as well: if a government agency sells data to a foreign government agency, costs are simply transferred from one national government to another. They are still dependent on general taxation and subject to the associated deadweight loss. There is no change in total social costs and no true cost savings. This means that data sales to a foreign government are equivalent—from an economic efficiency perspective—to other types of cost sharing arrangements: for example, cofunding the development of a new data collection system.

That is not to say that distributional effects are of no importance. A government agency may have a strong incentive to decrease its own costs, even if those costs are simply shifted to another agency in the same government. Some argue this provides a fairer distribution of costs and a more accurate picture of the distribution of data users. Similarly, a government may see significant benefits in shifting some of the costs of a system to a foreign government. In practice, particularly in the short run, agency budgets are usually relatively fixed. To the extent that cost sharing arrangements allow agencies to engage in the collection of data they would not otherwise be able to procure, both agencies and data users may favor these arrangements, regardless of the lack of net economic impact. However, it is important that decisions about these distributional effects be recognized as normative and practical, not economic, decisions.

Finding a Balance between Open Data and Data Sales

In determining the most economically efficient data sharing policy, a government agency should determine how alternative policy options would affect both the total costs of data collection and provision and the total benefits derived from the data.

To calculate the potential for cost savings, governments need to estimate the cost of developing a government-owned data collection system as well as the estimated revenues from data sales to commercial entities and/or the cost of purchasing equivalent data from the commercial sector. These can be difficult questions to answer, particularly if the government is developing a new type of data collection system for which costs are not well understood, or if a commercial market for data sales does not already exist in the area of interest. However, as discussed below, the potential for cost savings will typically vary depending on the size of the commercial market for data, thus making it possible to make informed data policy decisions based simply on general estimates about the market.

When calculating the total social benefits of a data policy, the value of data sold at market prices can be estimated by total revenues. However, quantifying data that is sold below market prices or given away for free is more difficult. Even if the data is used to produce a commercial value-added product or service, it is often difficult to determine the proportion of resulting revenue that can be attributed to a particular dataset. Applications developed by government agencies or nonprofits are even more difficult to value. Advancing scientific knowledge, improving the quality of public policy, and improving transparency are all common uses of government data that are associated with economic growth, but whose exact benefits are exceedingly difficult to quantify.[44]

Table 4.1
Scenarios for Relative Economic Efficiency of Open Data vs. Data Sales

	Narrow Noncommercial Uses *Less benefit to open data*	Broad Noncommercial Uses *Greater benefit to open data*
Nonviable Commercial Market (government funding required) *Less savings from data sales*	I Open data policy or Tiered data policy	II Open data policy
Viable Commercial Market (no government funding required) *Greater savings from data sales*	III Data sales	IV Open data policy or Tiered data policy

Further, estimating the relative effects of various levels of fees or types of restrictions is particularly difficult. Agencies must ask: When a price is put on the data, are many potential commercial users left out of the market, or very few? How much of a disincentive is data cost to new entrepreneurs uncertain about the potential of their product? To what extent are scientists and others able to raise funds to support research that requires the purchase of data? If the data is made freely available for noncommercial uses, to what extent do restrictions on access and redistribution limit its benefits? Luckily, as in the case of cost savings, it is possible to relate the relative social benefits of an open data policy to a single key attribute: the extent to which the data has broad, as opposed to narrow, noncommercial uses.

There is no simple formula to determine the most economically efficient data policy, but it is possible to say where free data provision and data sales each have a relative advantage, and which types of policy designs or public-private interactions may prove most economically efficient in various situations.

The first key attribute is the viability of a commercial market for the data. The size of this market determines the extent to which a commercial data collection entity would be likely to be successful, and the extent to which data sales can generate revenue that would decrease the reliance on general taxation, thus decreasing the total social costs of the program. Importantly, the definition of commercially viable here means that a private entity engaging in data collection and distribution is sustainable

without requiring funding from the government in the form of investment or data purchases. Since raw or minimally processed data is an intermediate good, this would mean that there were a sufficient number of
value-added companies willing and able to purchase the data for use in
value-added products or services to allow the data collection company
to recoup the costs of data collection and distribution. It's possible, even
likely, that a data collection company would also sell some value-added
products or services on its own. If it could generate a profit on the sale of
these products and services alone, or in combination with raw data sales,
this would also be considered a viable commercial market. Commercial
viability is presented as a dichotomous variable, but it can also be thought
of as a spectrum. To the extent that there is a large commercial market for
a particular type of data, even if it is not fully commercially viable, it may
present opportunities for economic efficiency similar to those of fully commercial viable situations.

Much of the benefit of free and open data policies comes from the ability
for the data to be used by a broad range of users, particularly those producing products and services that provide broad social benefits rather than
private financial benefit. (Open data policies also benefit the commercial
value-added sector, but data sales represent less of a barrier for these actors.)
The second of the two key attributes thus looks at the breadth of noncommercial uses of the data. Is the data useful primarily for fulfilling one narrow government need or addressing one specific scientific question? Or is
the data likely to be broadly useful for a number of government agencies,
nonprofits, and researchers? This can be a difficult question to answer, and
some open data advocates would argue that all data has the potential for
broad uses, even if these aren't obvious when data collection and sharing
policies are being planned. However, some distinction between these situations is typically possible, at least in the short term, and the concept is
useful as a general guide for decision-making.

The quadrant into which a particular type of data falls has implications
for the relative economic efficiency of free and open data policies versus
data sales. The quadrant also has important implications for the types of
public-private partnerships or interactions that are likely to be most successful and generate the greatest economic benefit overall. Below, each of
these four situations is examined and discussed.

Nonviable Commercial Market, Broad Noncommercial Uses: Open Data Policy

In cases in which the commercial market for data is not viable, but there are many noncommercial uses for the data—for research or to conduct government activities, for example—then it will be most efficient for the government to collect the data and provide it for free. This is because the small commercial market means that data sales do not offer significant opportunities for cost savings, and the broad noncommercial uses mean that free data provision will be particularly important for ensuring data use and maximizing data benefit.

This doesn't mean that there are no opportunities for interaction with the private sector. Many governments already contract with commercial firms to build data collection systems that will be owned and operated by the government. This allows governments to share the fixed costs of facilities and labor needed to build these assets with other agencies and entities, domestically and internationally. It also results in commercial competition that can drive innovation and advancements in the technology sector as a whole. Other agencies outsource data collection to the private sector, purchasing the data itself, rather than the data collection system. If the data is purchased with an open license, the government would benefit from commercial efficiencies, but still have the same ability to share the data as it would if the data collection system itself was government owned, and it would thus retain all of the benefits of open data sharing.

Nonviable Commercial Market, Narrow Noncommercial Uses: Open Data Policy or Tiered Data Policy

Some data collection programs are designed to meet a narrow government need, and do not have broad commercial or public applications. In these cases, the advantages of adopting an open data policy are relatively low, and data sales are not likely to result in significant savings. In this type of situation, almost any data policy can be justified. The government could collect the data and make it freely available, hoping to generate benefits from at least some additional uses of the data. There is always the possibility that new uses will be identified that weren't foreseen early in planning. Since the effect of restrictions on data use is likely to be low, the government may choose to sell some of the data itself, offsetting just a small portion of its costs. If the commercial market is large enough (but not fully viable), the government may choose to act as an anchor tenant for a commercial firm, again decreasing costs by taking advantage of commercial efficiencies and reducing the amount of government funding required.

Viable Commercial Market, Broad Noncommercial Uses: Open Data Policy or Tiered Data Policy

It's more difficult to determine the optimal data policy in the case of a viable commercial market and broad noncommercial uses. On the one hand, broad noncommercial uses suggest that an open data policy is particularly important for ensuring data use and maximizing benefits. On the other hand, the existence of a viable commercial market means that there is the potential for significant savings through a decreased reliance on funding generated through general taxation and efficiency advantages offered by the commercial sector.

To attempt to balance these two goals, an agency could pursue a tiered policy that maximized data availability for noncommercial uses, while allowing for some data sales by the commercial sector. For example, the user could purchase data from the commercial entity under the condition that the data could be shared freely for all noncommercial purposes. If the data used persistent identifiers and put the onus for compliance on data users, it would not need to restrict data access or redistribution. Value-added companies that ignored the licensing restrictions and used the data in a commercial product would risk having their company fined or shut down, if this was discovered.

Tiered policies based on some particular attribute of the data, rather than the type of user, could also allow for a balance between data sales and open data. For example, the government could purchase all data that is older than two weeks, or one year, or five years, depending on the relative commercial and noncommercial value of these types of data. The government may be able to purchase data that has been degraded or aggregated to serve noncommercial uses, while the highest-precision or most detailed data is sold to private actors.

Viable Commercial Market, Narrow Noncommercial Uses: Commercial Data Policy

When there is a viable commercial market, normal market incentives should result in the creation of private entities to fulfill this demand. If the data is also useful for fulfilling a narrow government need, and does not seem to have broad noncommercial uses, there is not a strong incentive for the government to make the data widely available under an open data policy. Instead, the government can purchase this "commercial off-the-shelf" data and expect to have significant savings compared to a situation in which it built the full data collection system itself.

It's worth reiterating here that there is a good deal of uncertainty involved in estimating whether the noncommercial uses of the data are broad or narrow, and there are benefits even for the commercial value-added sector of free and open data policies. For these reasons, to the extent that the government can negotiate agreements that increase the ability to share data (using the types of arrangements in the viable commercial market, broad noncommercial uses scenario, for example) without significantly increasing cost, this would be a worthwhile option to pursue.

Data Context: Technical

Technological developments have had, and continue to have, the ability to fundamentally change the landscape in which data sharing decisions are made. Technology affects the types and volume of data that can be collected; it affects how much data can be stored and how easily it can be accessed and distributed. Technology and technical standards affect the extent to which data can be processed, analyzed, linked, and otherwise manipulated to enable the many applications that have been discussed previously. These possibilities in turn affect the security or vulnerability of the data, and they affect the economic costs and benefits associated with sharing. Technological change can create new challenges and new opportunities, and new problems and new solutions.

Large-Scale Technology Developments

It's impossible to explain changes in data sharing policies over time without understanding changes in the underlying technologies that enable data collection and sharing. Many of these technological changes apply broadly across sectors, particularly the transformations in electronics and information technology. Others may be driven by the development of specialized equipment, new techniques, or new algorithms for collecting, analyzing, or using data in a particular field.

Before the advent and spread of computers and the Internet, the volume of data collected was often smaller, and the physical space it occupied was larger. The methods for copying and distributing data could be time consuming and expensive, limiting the incentives to share data on the part of the producer. It was more difficult to discover the existence of new datasets, particularly outside your own field. This decreased the number of requests for data on the part of consumers.

With the advent of computers and their accompanying storage devices, both the physical space required and the cost of distributing data steeply declined, sometimes by orders of magnitude. Evans and Wurster noted that in the late 1990s, printing, binding, and shipping a set of encyclopedias cost about $200, while producing an encyclopedia on CD-ROM was about $1.50.[45] As access to the Internet spread, it became easier to discover what datasets existed and where they could be found. While the creation of a well-designed, user-friendly portal is a nontrivial activity, once it is created, the marginal cost of providing data to an additional user is nearly zero. Data users can search, select, and download data on their own.[46]

Emerging technology developments have the potential to create both challenges and opportunities. For example, in some cases the volume of data being collected exceeds the ability to transmit or process it with existing technologies. Big data, for example, poses new opportunities for analysis, but also challenges to widespread access.[47] The emergence of cloud computing is one potential solution to this challenge. In the future, users may not actually copy or move the data to their own systems at all, rather accessing it and manipulating it in the cloud. Other technological advances, such as the development of data visualization tools, will also help to improve accessibility of data, continuing the trend of reaching new users without expertise in the given field, or even without any advanced data analysis skills at all.[48]

Technical Opportunities and Challenges That Aren't Really Technical

A number of scholars list technical issues as one of the key barriers or enablers of data sharing.[49] While it is true that developing a robust, user-friendly data portal or data sales interface is a technical challenge, it is one that can be met using existing technology and knowledge.[50] Therefore, with regard to policy-making, I argue that the development of such a system is more correctly thought of as an economic or political challenge. The cost of developing a portal should be taken into account in an agency's calculations of net social benefit. The availability of funding to support these activities is an important input from national-level policy-makers. If the agency is interested in developing a data portal and the funds are provided to implement it, then technology will not pose a significant barrier.

Similarly, ensuring high data quality, including relevant metadata, and documenting and sharing algorithms are important tasks that undoubtedly affect the success of data sharing efforts and require technical expertise.[51] However, if the funds are available to support the hiring of appropriately trained personnel to complete these tasks, a government agency would not find the technical challenges of this activity to be a major barrier.

Many authors point to the importance of international technical standards, and significant data supports the idea that the existence of standards can greatly enhance effective data sharing.[52] However, agreeing on technical standards is often more of an international relations challenge than a technical one, and in this model would be considered in the context of nongovernmental and intergovernmental organizational activities.

Finally, Zuiderwijk and others point to the importance of building feedback mechanisms into data sharing infrastructures to better understand how the data is used and improve user experiences.[53] This feedback is important, but, as implied by Zuiderwijk, the primary reasons are related to improved interactions with user groups and a better ability to explain the value of data sharing to national-level policy-makers and others.[54] These issues are captured in the external actor sections of this model, focusing on the concept of this interaction as the key factor, rather than the technical means by which this interaction is achieved.

Feedback Loops and Other Dynamics

It is important to note that this model is dynamic, not static. The right choice of data sharing policy at one time may no longer be the best option later on. Almost every element of the model is subject to change over time. Bureaucratic mission or preferences may change, although this process is generally slow. National-level policy-makers, their directives, and budgetary support can change much more rapidly. Nongovernmental and intergovernmental groups may form or disband, increase or decrease their activity level, or even change the nature of their activities.

Changes in the security, economic, normative, or technical attributes of data and data systems can also occur, and in many cases these changes are interrelated. In particular, new technology developments may change not only the technological possibilities, but also the nature of the security or privacy issues faced or the economics of data sharing. New types of security threats may increase concern about existing practices. Commercial innovations can change the economic effects of data sharing. The rise of norms in society, calling for government transparency, for example, or concerned with public-private interactions, may raise the profile of normative issues in policy development.

Sometimes these changes come not from external developments, but from improved understanding of the existing situation. Sometimes the answers to the questions raised in this chapter, and posed explicitly in the table below, are partially or completely unknown. As this information becomes available, through study, or through trial and error, data sharing policies can be reevaluated and updated as necessary. For example, a study examining the economic value of downstream applications developed with free government data can provide better insight into whether the data sharing policy is resulting in positive net social value. Experimenting with a data sales model after implementing an open data policy, or vice versa, may provide useful data with regard to the types of users most interested in data use or the willingness of the data users to pay. Actively seeking to answer these questions can improve our understanding of data policy development, or improve development of the policies themselves.

Data Sharing Policy Development Model Framework

Each of the elements of the model presented above presents key questions that have implications for the success of a chosen data sharing policy. The following table presents these key questions along with their implications. This table is designed to be used as a tool to examine and understand an existing policy or as a checklist to guide the successful development of a new data sharing policy.

Table 4.2

Key Questions and Implications in Data Sharing Policy Development

---

Government Agency

What are the agency's goals? Does the agency view data collection, use, and sharing as central to its activities or as a secondary issue/ byproduct?	Achieving the mission or goals of the agency is often a high priority for officials developing a new policy.
What is the dominant professional culture within the agency? (For example, does the agency have a culture of secrecy or of openness?)	Agency culture impacts the type of data sharing policy the agency views as appropriate or natural in their circumstances. Agency culture is often affected by the dominant professional culture within the agency and agency history.

National-Level Actors

Are there national-level initiatives or laws supporting particular data sharing policies?	Laws or initiatives that encourage or require the release of data generally result in some agency action on the issue, but can often be subverted in implementation (e.g., slow or partial compliance, poor technical infrastructure and/or data quality, little or no user support, etc.).
Are there national-level initiatives or laws forbidding the release of some types of data?	Laws forbidding the release of data are generally successful in meeting their objectives. However, if the interpretation is not clear, these laws may dampen data sharing more than intended, as agency officials worried about inadvertently breaking the law tend to err on the side of restricting access.
Is there budgetary support to cover the costs of data collection, maintenance, and distribution? Is there budgetary support to offset losses from data sales, if applicable?	Sufficient budgetary support (or lack of support) is a key enabler for agencies wishing to implement their data sharing policies. More open data sharing generally requires more budgetary support.
Is there an understanding among national policy-makers of the potential revenues and/or nonquantifiable benefits of data use?	The ability to communicate both the commercial and noncommercial benefits of data use is key to gaining and maintaining political support.

Nongovernmental Actors

Are there existing companies, researchers, or others who benefit from maintaining the current policy?	Existing data users will generally have an incentive to lobby to maintain the existing policy.

Table 4.2 (continued)

Are there companies, researchers, or others who would benefit from a change in the data sharing policy? Are these potential users aware of the data's availability and potential benefits?	These users have an incentive to lobby for a change in data sharing policy. However, these users may not be aware that the data exists or that is relevant to them.
Are there nongovernmental organizations (e.g., professional associations, nonprofit organizations) active in this area? What activities (if any) have these NGOs taken to enable data sharing and/or raise awareness of data sharing issues and activities?	Nongovernmental actors can be instrumental in developing technical standards and building and reinforcing professional norms related to sharing. They can also increase the visibility of existing challenges or opportunities.

Intergovernmental Actors

Are there intergovernmental organizations active in this area? What activities (if any) have these IGOs taken to enable data sharing and/or raise awareness of data sharing issues and activities?	Intergovernmental organizations may present a forum for developing technical standards. They can play an important role in building and reinforcing international norms related to sharing. They can also increase the visibility of existing challenges or opportunities.

Security and Privacy Attributes

Are there legitimate privacy, public safety, or national security concerns with respect to sharing the data?	Releasing data has the potential to improve or harm privacy, public safety, and national security depending on the circumstances. Careful analysis of the risks and opportunities posed by various data sharing policies is necessary.

Economic Attributes

Are there broad noncommercial uses for the data? To what extent would data use be inhibited under the data policy being considered?	Free and open provision of data is particularly important for data that has broad noncommercial uses. Policies that restrict access or redistribution are likely to decrease total societal benefits.
Is there a viable commercial market for the data? To what extent can total costs of the program be decreased through data sales?	If there is a viable commercial market for data, or the size of the commercial market is relatively large, there may be opportunities for significant cost savings through data sales to the private sector. If these cost savings are larger than the decrease in total benefits (above), data sales and tiered policies may be efficient.

Table 4.2 (continued)

Normative Attributes

What is the relative value the agency places on public safety, national security, privacy, efficiency, equity, and transparency concerns?	Each of these can be a legitimate goal in developing a data sharing policy, and agencies will often have to make trade-offs among them. Their relative importance is often related to agency norms.
Is the data primarily useful for commercial applications?	If the data is primarily useful for commercial purposes, data sales may decrease the extent to which the general public pays for a good that results in private gains for commercial users (though some argue that commercial users also pay taxes and thus have a right to benefit from the data without paying for access).
Does the data have the potential to significantly improve government transparency and accountability? Do citizens have a "right to know" this information?	Free and open policies are more appropriate when the data provides significant transparency benefits. This may be especially true for politically sensitive data about the government, but also applies to data upon which the government relies to make policy decisions.
To what extent can technology be used to lower the barriers to data access and use for average citizens?	Some concerns with regard to equity of access can be alleviated by implementing technologies, such as visualization software, that make the data accessible to those without significant time or advanced statistical skills.

Technical Context

What technological limitations (in data collection, processing, distribution, use, etc.) may impact data sharing in this area? How might these change in the future?	Technology can affect the limits of what is possible in data sharing as well as the calculations of security, economic, and normative risks and benefits.
Does the organization have the technology and expertise to implement their data sharing policy (e.g., build a robust data sharing infrastructure) while following technical best practices?	This should be seen primarily as a political challenge, requiring adequate political support and funding to support these activities (discussed above under national-level actors). Lack of attention to good technical implementation may also reflect agency incentives and priorities related to data sharing (i.e., lack of emphasis on technical implementation is one way agencies may subvert national directives).

Table 4.2 (continued)

Are there adequately defined international technical standards, including relevant metadata, in this field?	This should be seen as an issue related to nongovernmental and intergovernmental actors. Are there collective action challenges that have limited the development of these standards?
Does the agency have a robust infrastructure to allow user feedback with regard to the data?	This should be seen as an issue of agency interaction with nongovernmental, intergovernmental, and national-level actors. A robust technical system is important, but the key challenges are in the engagement with these organizations on a political level.

Part II Sharing Satellite Data Case Studies

This part applies the model developed in part I to the issue of satellite data sharing by examining a series of historical case studies. It begins with case studies of two international organizations that have been active in this area: the World Meteorological Organization (WMO) and the Group on Earth Observations (GEO). These case studies focus on international dynamics particularly important in the area of satellite Earth observations and data sharing, examining how the organizations both influenced, and were influenced by, their member nations. The case studies are designed to provide the reader with an understanding of the larger international trends and debates that affected the individual satellite-operating agencies.

The part then moves into the seven core case studies, each focusing on a space or meteorological organization in the United States, Europe, or Japan. This part also includes a short review of data sharing activities related to US military, intelligence, and commercial satellites, due to their numerous interactions with their civil counterparts. The core case studies are followed by summaries of data sharing policy development in the BRICS countries: Brazil, Russia, India, China, and South Africa.

The case studies and summaries trace the process of data sharing policy development within each agency from the launch of its first Earth observation satellites to the present. The focus is on the development of data sharing policy, looking at the way the policy issue was viewed and discussed within the agency, particularly in the context of its goals and culture. The case studies and summaries include attention to how agency developments were affected by the actions of external actors, including national-level government actors, non-governmental actors, and international organizations, and they highlight discussions and developments related to the contextual elements of the data, including security, economic, normative, and technical issues.

5   World Meteorological Organization

The World Meteorological Organization (WMO) has a history that stretches back almost 150 years, and international data sharing has been central to its activities from the very beginning. The organization was formed to facilitate international cooperation in the collection and exchange of meteorological data and products, and it has continued to emphasize the importance of free and unrestricted data exchange throughout its existence. When member nations began to explore commercial meteorological activities that required restrictions on data sharing, the WMO was the venue for international debates on the issue. The organization itself had an interest in maintaining free and open exchange, as this principle was at the foundation of its own key functions. The formal international compromise on this issue was laid out in 1995 in WMO Resolution 40, which ensured that free and unrestricted exchange of data essential for saving lives and property would continue, while restrictions on other, "nonessential," data would be allowed.

WMO and its affiliated bodies were also actively involved in international debates and understandings related to climate change. WMO helped to form the Global Climate Observing System (GCOS), which over the years has identified the key observations needed to adequately monitor climate change and advocated for greater international sharing of climate-relevant data. In 2015, WMO passed a new resolution—Resolution 60—officially calling on its members to support free and open exchange of essential climate data.

Before the Space Age: From IMO to WMO

The study of climate change, free and open international data sharing, and even ideas about the environmental applications of satellites, predate the space age. The first large-scale international data sharing efforts began in the nineteenth century, focused on the issue of meteorology and enabled by one of the first innovations in communication technology: the telegraph. It had long been realized that information about weather in one country could be used to anticipate weather soon to occur in another. With the development and spread of telegraph technology in the mid-1800s, this knowledge could finally be used operationally.

In 1873, meteorologists around the world held the first International Meteorological Congress, and formed the nongovernmental International Meteorological Organization (IMO) to coordinate and facilitate data sharing. In addition to its efforts to coordinate day-to-day meteorological activities and data sharing, the IMO organized a number of major international endeavors to improve the state of knowledge in the field of meteorology. It coordinated the First International Polar Year (IPY) in 1882–1883 and the second IPY fifty years later from 1932 to 1933. These programs brought together nations from around the world to coordinate research and data collection regarding the Arctic and Antarctic. They led to the construction of new observing systems and important new discoveries.

In the 75 years following its formation, weather forecasting capabilities improved and meteorology grew in importance. In 1950, the IMO was replaced by the World Meteorological Organization (WMO). Nongovernmental representatives were replaced with government officials as the agency gained status as an intergovernmental organization under the auspices of the United Nations.[1] The new organization was given five purposes, including facilitating worldwide cooperation in meteorological observations and promoting the establishment of systems for the rapid exchange of weather information.

The organization's main decision-making body was the WMO Congress, made up of the director of the meteorological service of each member nation. This group would meet every four years to determine regulations and policies for the WMO, and members were expected to "do their utmost to implement the decisions of the Congress." WMO had more than 75 members when it began operations in 1950, and eventually grew to include nearly every country in the world.[2]

Satellites, Data Sharing, and the World Weather Watch

One of WMO's early activities was the third IPY, also known as the International Geophysical Year (IGY), organized together with the International Council of Scientific Unions. The IGY took place during 1957 and 1958 and involved nearly 70 countries.[3] Data exchange was a key issue for the IGY. Recognizing the difficulties experienced in accessing data collected during the second IPY, the Special Committee for the IGY implemented a new system for global data sharing. This system was based on the creation of three World Data Centers (WDCs) located across the globe. WDC-A was located in the United States, WDC-B was located in the USSR, and WDC-C was comprised of specialized data centers in a variety of nations. All IGY observations were sent to at least one of these centers, which then provided copies to the others. According to the IGY data exchange agreements, data and information at the centers was to be made available to all countries and scientific bodies—this was perhaps one of the earliest open data policies, although it wasn't referred to as such.[4] Among the many achievements of the IGY were the first systematic measurements of atmospheric carbon dioxide, carried out by Charles David Keeling in Mauna Loa, Hawaii, and the first spacecraft, with both the United States and the Soviet Union launching their first civilian satellites in support of the IGY.[5]

The fact that the first satellites were launched in relation to an environmental and meteorological effort was no surprise; it had been clear for years that weather monitoring would be one of the first civil applications of satellites. In 1958, less than a year after the launch of Sputnik, the WMO established a panel of experts on artificial satellites, with members from the USSR and the United States, to explore the issue.[6] Just two years later, NASA launched the first weather satellite, the Television Infrared Observing Satellite 1 (TIROS I), in April, 1960.[7]

While the space race and superpower competition was front page news, the international community was eager to find opportunities for peaceful cooperation in space, and meteorology seemed a promising option. In December, 1961, months after the first men were lofted into space, the United Nations General Assembly adopted Resolution 1721 on international cooperation in the peaceful uses of outer space. This resolution included a request that the WMO investigate opportunities for international cooperation related to meteorological satellites.[8] The WMO eagerly responded, drafting a report proposing the creation of the World Weather Watch (WWW), an ambitious cooperative global system to assist the meteorological services of the world. The proposed system included a global observing system to monitor the weather; a global data processing system to turn data into useful information products, including forecasts; and a global telecommunications system to distribute the data and information.[9] The UN General Assembly endorsed these plans in November 1962, and in 1963, the World Weather Watch was officially established.[10] Although the design of the program and its approval occurred rapidly—at least in international political terms—there was still significant work to be done in the detailed planning and implementation of the new program.

Satellites were expected to play a central role. A 1966 publication about progress toward the WWW stated, "It is quite clear that the advent of meteorological satellites marks a turning point in the science of meteorology and that their use will figure prominently in the World Weather Watch now being developed."[11] From the perspective of the meteorological community, satellites had two advantages over other techniques: satellites could provide a means of monitoring the weather from outside the atmosphere, and they could provide data promptly on a global scale. Even though the United States and Russia were the only countries capable of lofting these satellites at the time, the ability and willingness to share the data had already been demonstrated. The Automatic Picture Transmission (APT) system, developed by NASA and placed on its weather satellites beginning in the early 1960s, transmitted satellite imagery automatically to any location with the appropriate ground receiving equipment. This meant that any nation in the world that chose to acquire this equipment could access satellite data.

The WMO was not content to be a passive spectator in the development of these new technologies. In the 1970s, the WMO planned another large-scale endeavor, an alphabet soup of acronyms, alternatively referred to as the First Global Atmospheric Research Program (GARP) Global Experiment (FGGE) or as the Global Weather Experiment (GWE). The goal of the one-year experiment was to determine the requirements of a global observing system that would allow routine, operational, long-range weather prediction.[12] Satellites, including the relatively new geostationary weather satellites, first launched by the United States in 1966, were a major component of this effort. Europe and Japan developed and launched their first Earth observation satellites—both geostationary weather satellites—in support of the GWE. The geostationary satellites were located in a much higher orbit than the polar-orbiting weather satellites that had been developed earlier. This higher orbit allowed them to circle the globe at such a rate that they appeared to remain stationary over one area of the Earth. This meant that with just five geostationary satellites, it would be possible to have constant, global coverage of the entire world.

To coordinate their efforts to provide this full global coverage, officials from the United States, Japan, Europe, and the Soviet Union formed the Coordination Group for Meteorological Satellites (CGMS) in 1972. When Russia announced in 1977 that it could not provide its planned satellite, the United States was able to adjust the position of one of its geostationary satellites to cover the gap, so that in the end, three of the five geostationary satellites were furnished by the United States.[13] Russia contributed two polar-orbiting satellites, complementing two others provided by the United States. As expected, there was a surge of global meteorological observing activities during the GWE, which produced a massive amount of data. Datasets generated for the GWE were held at the World Data Centers in the United States and the Soviet Union and provided to any user that requested them at the cost of duplicating the data and postage.[14]

It's interesting to note that the CGMS, originally created for the GWE, proved important in later years as well. Working through this forum, operators of geostationary satellites provided backup support to each other on numerous occasions. When the European geostationary satellite, Meteosat-2, failed in 1984, the United States moved GOES-4 further east over the Atlantic to cover the area. When a Soviet geostationary satellite failed unexpectedly in 1989, Europe placed its satellite over the Indian Ocean to replace it. Europe also aided the United States when it was left with only one operational geostationary satellite in 1991, shifting its Meteosat-3 satellite west to cover the US east coast. In 2003, a Japanese satellite stopped operating, and the United States moved one of its satellites to cover the western Pacific.[15] A policy of free and open data exchange among meteorological agencies enabled these types of arrangements—without such a policy, coverage would not have been useful or would have required significantly more complex negotiations.

Free and Open Exchange of Data under Threat

Despite the fact that by the 1980s, free and open exchange of data had been the norm within the international meteorological community for more than 100 years, the practice was not destined to remain unchallenged. Some nations began to consider alternative policies, prompted largely by the rise of commercial meteorological organizations and national initiatives that aimed to reduce government spending by implementing market mechanisms and engaging in commercial activities within government agencies.

Facing tightening budgets and significant political pressure, a number of National Meteorological Services (NMSs), particularly in Europe, developed commercial branches within their government agencies and began selling weather data and products to help recover agency costs. These efforts were not consistent with free and open data sharing: data given away for free could not also be sold. This trend caused some alarm within the WMO, and participants at the Tenth World Meteorological Congress, held in 1987, determined that the issue needed careful consideration. In the meantime, the Congress released a resolution reiterating its long-standing belief that "the principle of free and unrestricted exchange of meteorological data between National Meteorological Services should be maintained."[16]

In the following years, national activities continued to clash with international data sharing efforts. Agencies under pressure to recover costs through sales of meteorological data and products began to limit their own free distribution of some items, but would find their commercial efforts undermined when those same data or products were made freely available from another country. Government agencies engaging in commercial activities expressed frustration with what they viewed as unfair competition from transnational companies that received free data from one nation and used it to develop products sold in another nation. Agencies argued that these companies could undercut government prices, because government agencies were required to adopt prices high enough to recover the costs of expensive observing infrastructure, while the commercial entities didn't face these costs. Some developing countries voiced the concern that private meteorological companies could undercut and undermine the efforts of their national meteorological agencies altogether. Private-sector companies, in return, argued that government agencies that restricted data access were undermining private competitors by placing unreasonably high prices on data of interest to the private sector, or by not making data available at all.

It was clear that the issue of data sales and commercialization was not going away, and in the debates between government agencies and private-sector meteorological organizations, WMO saw a threat to its core functions. The Eleventh World Meteorological Congress, held in 1991, "noted with concern that commercial meteorological activities had the potential to undermine the free exchange of meteorological data and products between National Meteorological Services." The Congress was determined to find an internationally agreeable solution that would ensure the continued operation of the international system. It recommended the establishment of a working group to study the issue in more depth and make proposals on future policies.[17]

A Compromise: WMO Resolution 40

In 1995, a compromise was finally reached. The Twelfth World Meteorological Congress passed Resolution 40, the "WMO policy and practice for the exchange of meteorological and related data and products including guidelines on relationships in commercial meteorological activities." The resolution recognized the trend toward commercialization of meteorological activities and the pressure some meteorological agencies were under from their national governments, including "the requirement by some Members that their NMSs initiate or increase their commercial activities." It recognized that national governments had the right to choose the extent to which they made data available for international exchange, and laid out a compromise: a tiered system for future data sharing. The new system required free and open sharing of "essential data" needed to support safety and security, but allowed members to restrict access to "nonessential" data. The nonessential data would be available for official use by National Meteorological Services and for noncommercial use by research and education communities, but could not be redistributed to third parties. Satellite data was included in the list of essential data, but only "those data and products agreed between WMO and satellite operators," including data and products necessary for operations regarding severe weather warnings and tropical cyclone warnings.[18]

Despite the fact that the key element of Resolution 40 was the formal acceptance of some limitations on data sharing, WMO also attempted to use the resolution to encourage nations to continue the tradition of free and unrestricted data exchange, emphasizing this point multiple times and in multiple ways. It reminded Members that promoting the exchange of meteorological and related information was part of their obligation under Article 2 of the WMO Convention, the WMO's founding document. The resolution stated that the exchange of meteorological data and products among the elements of the World Weather Watch system was fundamental for the provision of meteorological services in all countries, and that these services provide safety, security, and economic benefits for the citizens of Member nations. It explained that the research and educational communities depend on access to meteorological data and products and noted that data exchanged with WMO was also important to WMO programs dealing with climate, the oceans, and other issues. It reminded Members that Parties to the UNFCCC had committed to promoting and cooperating in the full, open, and prompt exchange of climate information, and that world leaders at the 1992 UN Conference on Environment and Development had called for increasing the commitment to exchanging scientific data and analysis.[19]

The Resolution went so far as to adopt a formal policy on international exchange of data that didn't mention restrictions at all, stating, "As a fundamental principle of the World Meteorological Organization (WMO), and in consonance with the expanding requirements for its scientific and technical expertise, WMO commits itself to broadening and enhancing the free and unrestricted international exchange of meteorological and related data and products." The Congress urged members to "strengthen their commitment to the free and unrestricted exchange of meteorological and related data and products," and to "increase the volume of data and products exchanged."[20] In 1999, WMO passed a similar policy covering hydrological data.[21]

Satellites and Climate Change

While the WMO was struggling with the threats to free and open exchange of meteorological data in the late 1980s and early 1990s, it was also recognizing and examining ways to address the growing need to collect and share data related to climate change.

WMO involvement on this issue stretched back at least to the GWE in the late 1970s. While the GWE had largely been seen as a success with respect to weather, some argued that it had not contributed sufficiently to the objective of studying the physical basis of climate.[22] This issue was of growing importance for the WMO, which along with other international bodies, had sponsored the first World Climate Conference in 1979. The conference was attended by more than 300 scientists from 50 countries. Attendees acknowledged the importance of understanding and addressing climate change and called on nations to support the creation of the World Climate Program (WCP) within the WMO. The WCP included the World Climate Research Program (WCRP), and encouraged the use of in situ and satellite-based observations for climate research.[23]

Nearly a decade later, in 1988, the Intergovernmental Panel on Climate Change (IPCC) was formed to collate and assess the evidence on global warming. In 1990, WMO and other international organizations held the Second World Climate Conference, and the results of the first IPCC Assessment Report were presented. The report stated that there was a greenhouse effect and that human activities were resulting in greenhouse gas emissions that were increasing that effect, leading to warming of the average temperature of the Earth. Two of their five key recommendations included the need to improve the systematic observation of climate variables with both satellite and surface-based instruments on a global basis and the need to facilitate international exchange of climate data.[24]

The Second World Climate Conference led to the creation of the Global Climate Observing System (GCOS), a joint program of the WMO and other international organizations. GCOS was to coordinate among national and international entities and organizations to implement an observing system that would meet the monitoring needs of the World Climate Program— essentially aiming to duplicate in the area of climate the success of the WMO and its World Weather Watch.[25] It also directly supported article five of the United Nations Framework Convention on Climate Change (UNFCCC), which required parties to support international and intergovernmental efforts to strengthen systematic observation and promote access to, and exchange of, data and analysis.[26]

Many meteorological services were receptive to this need. The European Organization for the Exploitation of Meteorological Satellites (EUMETSAT) proposed an amendment to its convention in 1991 that explicitly added "operational monitoring of the climate and detection of global climatic changes" as part of its primary objective.[27] The United States National Polar-orbiting Operational Environmental Satellite System (NPOESS), announced in 1994, was expected to include new climate-monitoring instruments in addition to traditional meteorological instruments.[28]

In 1995, the same year the WMO Congress passed Resolution 40, GCOS also adopted a set of data principles. The principles stated that global environmental concerns were an overriding justification for the unrestricted international exchange of data and called for full and open sharing at the lowest possible cost to users. Recognizing the same challenges discussed with regard to meteorological data, the principles clarified that unrestricted access would apply only to noncommercial scientific and applications uses. They also acknowledged the GCOS policy was nonbinding and that each country would develop its own data policy.[29]

Development of the Essential Climate Variables

Despite the growing recognition of the global importance of climate change and the creation of a new organization, GCOS, to coordinate global observations, progress in this area remained slow. The IPCC released its second assessment report in 1995 and its third in 2001. In both reports, the IPCC called attention to the need for improved systematic observations. The third assessment report warned that observational networks in many parts of the world were in decline and stated that international cooperation, including free exchange of data among scientists, was crucial to better use of scientific, computational, and observational resources.[30]

Gaps in the satellite record were a particular concern. It is necessary to ensure new satellites are launched before existing satellites cease functioning to ensure continuity of the very precise measurements needed for climate. This overlap in satellite operation allows the instruments to be cross calibrated, measuring the same phenomenon at the same time. If this is not done, and the new satellite provides different readings than its predecessor, it is difficult to determine whether this change is due to differences in instrument functionality or the result of real changes in the environment. Potential gaps in satellite solar irradiance measurements were particularly worrisome, because this variable is critical to understanding climate change, requires very accurate measurement, and can only be observed with space-based sensors. GCOS stated that overlap between these spacecraft was more important than for any other environmental variable observed by satellites.[31]

To focus and facilitate climate data collection and sharing, in 2003 GCOS developed a set of 50 essential climate variables (ECVs). These variables were based on analysis of the underlying needs of the parties to the UNFCCC and the IPCC, and included data needed to characterize the state of the global climate system and its variability, attribute causes of climate change, and support prediction of climate change. The final set of essential variables chosen comprised those that would have a high impact on UNFCCC requirements and were dubbed technically and financially feasible for global implementation. It was expected that additional ECVs would be added as knowledge of climate change and technical capabilities advanced.[32]

GCOS stated that free and unrestricted exchange of all ECVs was urgently required, but this task was even more difficult than it appeared at first blush. In some ways, the list of essential climate variables oversimplified the requirements of climate monitoring. The ECVs were defined at a high level; adequate collection of a single essential climate variable often required collection of many types of specific measurements. For example, measuring the "cloud properties" ECV requires measurement of cloud cover, height, temperature, and composition. In all, collection of the ECVs required about 150 different types of measurements to be taken.[33]

The task was made even more challenging due to the fact that requirements and standards for climate data collection are generally much more demanding than those for other disciplines, such as weather. Data must be collected continuously over long time periods and large geographic areas, and high accuracy is required in order to separate small climatic changes from larger short-term dynamics. These requirements make global climate data collection a larger, more complex, and more expensive undertaking than data collection efforts for other needs.[34]

GCOS identified about half of the ECVs as largely dependent on satellite observations. The organization acknowledged that Earth observation from satellites is a costly activity to which only a small number of UNFCCC parties are able to contribute, but argued that the information derived from these satellites was a global utility that required global access. They also noted that every nation could contribute to the ground-based measurements that are critical for validating satellite records.[35] The Committee on Earth Observation Satellites (CEOS), an organization that had been formed in 1984 to coordinate international Earth observation satellites, responded to the GCOS report in 2006. CEOS validated the satellite component of the plan and providing detailed analyses of required actions.[36]

The CEOS report also called attention to the fact that most satellite systems that could contribute to the GCOS plan were not intended or optimized for climate purposes. GCOS had included satellites from both research agencies, such as NASA, ESA, and JAXA, and operational agencies, such as NOAA, USGS, EUMETSAT, and JMA. Each type of agency posed particular challenges for climate monitoring. Research agencies, which develop the majority of the climate-relevant satellites, typically focus on cutting edge research and development activities, developing one-off Earth observing missions to help answer specific scientific questions. In most cases, they do not have a mandate or funding to support the long-term, continuous data collection, satellite overlap, or cross calibration needed for climate monitoring. Operational agencies are expected to maintain these types of continuous satellite observations. However, the satellite programs within these organizations are typically smaller than in research-oriented space agencies, and the satellites they operate are generally optimized for meteorological uses, not climate purposes, so they do not cover the full

Table 5.1
Essential Climate Variables Largely Dependent on Satellites

Domain	Essential Climate Variables
Atmospheric (over land, sea, and ice)	Precipitation, Earth radiation budget (including solar irradiance), upper-air temperature, wind speed and direction, water vapor, cloud properties, carbon dioxide, ozone, aerosol properties
Oceanic	Sea-surface temperature, sea level, sea ice, ocean color (for biological activity), sea state, ocean salinity
Terrestrial	Lakes, snow cover, glaciers and ice caps, albedo, land cover (including vegetation type), fraction of absorbed photosynthetically active radiation (FAPAR), leaf area index (LAI), biomass, fire disturbance, soil moisture

spectrum of climate needs and often do not meet accuracy or precision requirements that would be best for climate change uses.[37]

WMO on Climate Data Sharing: WMO Resolution 60

Following the fourth IPCC Assessment Report in 2007 and the Third World Climate Conference (WCC-3) in 2009, GCOS updated its implementation plan. The 2010 update noted that while much progress had been made, the Global Climate Observing System still fell short of the information needs of the UNFCCC and broader user communities. Yet the ambitions of the international community, and the corresponding need for data, continued to expand. The WCC-3 had created the Global Framework for Climate Services (GFCS) within the UN framework, recognizing the need to provide not just observational data, but climate information and services, as well. These information products and services would need to be based on high-quality observations across the climate system. GCOS added support of this new activity to its plan. The list of ECVs had evolved, with new additions and changing designations reflecting new requirements and new technological developments.[38] In 2011, GCOS updated the satellite supplement to its implementation plan, and CEOS issued a response in 2012.[39] These organizations were working hard to identify what was needed, but there was still much work to be done in actually fulfilling these needs.

The GFCS released an implementation plan in 2014, identifying eight principles important to fulfilling its goals. Among these, it asserted that "climate information is primarily an international public good provided by governments, which will have a central role in its management." The GFCS planned to "promote the free and open exchange of climate-relevant data, tools, and scientifically based methods while respecting national and international policies." The implementation plan directly addressed data policy issues, stating that full and open access to climate data was an important requirement for implementing the Framework. It noted that WMO Resolution 40 did not sufficiently cover climate needs, and suggested that a WMO policy on the exchange of climate data and products could help promote the issue within the United Nations system.[40]

The WMO Congress obliged, passing Resolution 60, the "WMO policy for the international exchange of climate data and products to support the

implementation of the Global Framework for Climate Services (GFCS),"
the following year. The resolution stated that all climate data and products covered by Resolution 40 and Resolution 25 (which addressed hydrological data sharing) should continue to be governed by those resolutions,
and that additional data, including the GCOS ECVs, would constitute an
essential contribution to the GFCS and should therefore be made available through the GFCS Climate Services Information System on a free and
unrestricted basis. It urged members to strengthen their commitment to
the free and unrestricted exchange of GFCS-relevant data and products and
to increase the volume of this data made accessible. As in Resolution 40,
passed 20 years earlier, it identified a set of data considered essential for
the implementation of GFCS, including "climate-relevant satellite data and
products."[41]

Resolution 60 reiterated a number of statements from Resolution 40,
emphasizing the ethical importance of data sharing, existing international
interdependencies, and the fundamental role of data exchange in enabling
global awareness of environmental issues. It stated that NMSs "provide universal services in support of safety, security, and economic benefits for the
peoples of their countries," and that NMSs are dependent on "cooperative
international exchange of meteorological and related data and products
for discharging their responsibilities." It noted that free and unrestricted
exchange of GFCS-relevant data was of fundamental importance.

Like Resolution 40, Resolution 60 also recognized that reality did not
match up with the organization's ideals. It acknowledged that different
NMSs have different business models, including some that require cost
recovery, and noted the right of governments to choose the manner by,
and the extent to which, they share data and products. It reiterated that any
conditions or restrictions on use should be respected by Members using the
data. The resolution gave some indication that the WMO plans to continue
pushing on this issue, however. The Congress requested that the WMO
Secretary-General undertake a global survey of data policies, including cost
recovery and public services models, to identify successful strategies and
best practices that can assist meteorological agencies in making the case
to their governments for new and continued funding for global climate
monitoring.[42]

Summary

International data sharing underlies nearly everything the WMO does. Coordinating and growing this practice was a core element of IMO and WMO activities from the beginning. Major WMO programs, particularly the World Weather Watch, were predicated on the common provision of meteorological data. In the 1980s, the WMO saw commercial activities and the subsequent restrictions on data sharing as a threat to its core activities. Economic arguments about the mutual benefit to all members of free and unrestricted data sharing did not win over those who were convinced of the economic efficiency of commercialization and market mechanisms. Normative arguments about the long-standing practice of international data sharing and the role sharing plays in saving lives and property were also insufficient to prevent these policy changes.

The WMO and its member agencies did not have the political strength to reverse the trend toward cost recovery favored by national-level policy-makers. NMSs believed that failure to successfully sell their products would likely lead to a decrease in their budgets, and subsequently to a decrease in their ability to collect data and provide data, under any conditions, to the WMO. The WMO was put in a position in which it had to accept some limitations on data sharing in order to halt a trend that could otherwise undermine its ability to carry out its core functions. This pressure, and the desire to achieve this balance, can be seen in Resolution 40. Although the main purpose of the resolution is to formally accept limitations on data sharing, much of the preamble and even the core policy statements are arguments for the importance of maintaining free and unrestricted sharing.

Passed 20 years later, WMO's objective with regard to climate data in Resolution 60 was quite different. Rather than trying to protect a long-standing norm that was being threatened, WMO was trying to extend its existing policy to an area not previously explicitly included. The resolution provides WMO with a framework on which to build additional climate-related programs and services and gives member agencies a further tool for demonstrating the international importance of sharing climate data.

6 Group on Earth Observations

The Group on Earth Observations (GEO) was formed in 2005, with promotion of full and open data sharing as one of its primary goals. GEO grew out of a desire to improve international cooperation on collection and sharing of environmental data for a wide range of environmental issues, rather than focusing on one domain, such as weather or climate. While it requires high-level political participation, the organization was formed as a stand-alone international body, not affiliated with the United Nations or another group, and its policies and plans are considered nonbinding on its members. This structure has given GEO the flexibility to continually advocate for forward-looking data sharing principles and to push its members to adopt increasingly open policies. In its first ten years, GEO has been an important venue for international debates on data sharing and for exchanging information about the impact of various data policy choices.

UN Principles on Remote Sensing

The space age began with the launch of Sputnik in 1957, and the first civil satellite capable of imaging the Earth launched in 1960, but it wasn't until the 1970s and 1980s that the number of countries involved in Earth observing satellite activities really began to grow. As the United States and Russia continued to develop remote sensing satellites for reconnaissance, weather, climate, and land cover research, Europe, Japan, India, and China all began their own remote sensing satellite programs.[1]

The increasing number and capabilities of civil remote sensing satellites and the associated wide dissemination of data spurred international discussion of new opportunities. France proposed the creation of an International

Remote Sensing Agency (ISMA) that would use remote sensing satellite data to monitor arms control agreements.[2] The idea was studied in detail by the United Nations Conference on Disarmament, but was eventually abandoned due to opposition from the United States.[3] Other reports and discussions focused on the value of the newly available satellite data for agriculture, mining, news, and a host of other civil applications. There was even some discussion of creating a privately funded "Mediasat" to be operated by the US news media, though development of such a system was never undertaken.[4] Perhaps the greatest area of interest was the development of commercial remote sensing, actively pursued in the United States, Russia, Europe, and India.

Unsurprisingly, these new developments also raised concerns, many of which were debated during the development of the UN Principles Related to Remote Sensing of the Earth from Outer Space, eventually adopted in 1986. Some nations, particularly less-developed countries without satellite capabilities of their own, argued that nations should be required to get permission from any nation that would be sensed before conducting remote sensing activities over their territory. Others argued that remote sensing could be carried out without the prior consent of the sensed nation, but that permission for dissemination should be required before those images could be shared or sold to a third party. Many wanted free access to data of their own state or region. Nations without space programs worried that companies or other nations could gain knowledge of their nation's natural resources, putting them at a disadvantage in the development of contracts or international agreements, or causing national security issues. Some argued that nations have sovereignty over information collected regarding their nation's land and natural resources, just as they have sovereignty over the land itself.[5]

The United States and others argued that requiring permission from sensed states was impractical and would greatly diminish the ability to research and address global environmental challenges such as desertification or flooding that do not follow political borders. Consulting with states before dissemination would be similarly impractical, greatly limiting the timeliness of data availability. Officials in the United States and Europe were particularly sensitive to the concerns of the emerging commercial remote sensing sector, and argued that the principles had to be consistent with the ability to generate revenue—meaning that nations could not demand free access to data or information, even if it was about their own state.[6]

In the end, the principles reflected a compromise mostly favoring the position of the United States and Europe. The principles did not include a requirement to gain the consent of a sensed state before collecting or disseminating data, and there was no requirement to provide data to a sensed state for free. Instead, the sensed state was to have access to the data of its territory on a nondiscriminatory basis and on reasonable cost terms. The principles stated that remote sensing activities should be carried out for the benefit and in the interests of all countries, and should not be conducted in a way that would be detrimental to the rights and interests of the sensed state. International cooperation and data sharing was encouraged, particularly in cases related to protection of the Earth's natural environment and protection from natural disasters.[7]

Data Sharing in the Committee on Earth Observation Satellites

Adoption of the UN Principles was an important step, but it only provided a very basic framework of what was required in terms of data sharing, and it was a static document that didn't require further interaction and negotiation as time went on. Another early effort at international cooperation in Earth observations that allowed more dynamic interaction emerged from the Group of Seven (G7) Economic Summit in 1984. Based on the recommendation of this group, the Committee on Earth Observation Satellites (CEOS) was formed, bringing together space agency officials to discuss in detail the plans and policies of their respective agencies. CEOS's goal was to coordinate international civil Earth observing satellite missions, and address planning, development, and interoperability of satellite missions and their related data.[8]

CEOS members recognized that a key element of international cooperation in Earth observations would be the ability to easily exchange satellite data. In 1991, the organization developed a set of data sharing principles. Recognizing the potential sensitivity of data sharing in some areas, including the contentious debates ongoing in the meteorological community, the organization distinguished between four categories: (1) global change and environmental research; (2) other research not necessarily related to global environmental change, such as that focused on the regional or local level;

(3) operational environmental monitoring, including weather; and (4) other, including commercial. The first set of CEOS data exchange principles only addressed the first group: global change research. The principles advocated for a relatively open data policy, calling out maximization of satellite data use as a fundamental objective. They promoted the development of a sharing mechanism among CEOS members and called for nondiscriminatory access to data for non-CEOS members. They noted that data should be released within three months after the start of routine data acquisitions, with no period of exclusive use reserved for government scientists or principal investigators.[9]

However, this open policy was short-lived, and in 1992 these data principles were amended, reflecting the ongoing struggle between the desire to promote research and the desire to protect commercial opportunities. This divide was written into the new policy, which recognized "various policy aims such as maximizing the use of data from all sources" and "shifting the funding responsibility for certain remote sensing systems to users or other sources." It explicitly called out the goal of providing data to global change researchers at a price reflecting only the cost of fulfilling the user request, but in a concession to those nations wishing to sell data, the policy also acknowledged that constraints of mission operations and available resources may require other mechanisms for data exchange.[10]

In 1994, CEOS adopted a set of satellite data exchange principles in support of operational environmental use, including weather. These were similar to the 1992 principles, reflecting the conflicting desires to encourage free and open data sharing and awareness of the political and budget realities faced by many agencies that precluded open sharing. The policy noted that free and unrestricted exchange of basic meteorological data had occurred for more than 100 years in the international meteorological community and stated that providing data for operational environmental use for the public benefit was a common goal. Yet CEOS also noted the economic value of the data, the major investments made in support of operational environmental efforts, and the constraints that may require different mechanisms for data provision. The policy further acknowledged that various legal regimes for data provision already existed, with different policies for data pricing and ownership. Even with these caveats, and the nonbinding nature of the principles, adoption of these principles by CEOS members was not unanimous: EUMETSAT abstained, demonstrating the extent of ongoing sensitivities and uncertainties with regard to national data sharing policies.[11]

Creation of the Group on Earth Observations

Despite CEOS's efforts to encourage international cooperation in satellite Earth observations, and the World Meteorological Organization's extensive work in coordinating meteorological observations, there was a feeling that cooperation was still too limited, particularly given the rise of global challenges, such as climate change.

In 2002, the United Nations held the World Summit on Sustainable Development (WSSD), a follow-up to the Rio Earth Summit held 10 years earlier, at which the UNFCCC had been adopted.[12] Jose Achache, the head of CEOS, spoke at the event, reminding attendees of the fundamental importance of observations. "There is no sustainable development without adequate information about the state of the Earth and its environment," he said.[13] The US NOAA Administrator, retired Navy Vice Admiral Conrad Lautenbacher, went further in his statement, calling for the creation of an international global observing system for climate. He argued that such a system could provide the tools needed to take the pulse of the planet, and that the information collected would be essential to sustainable development throughout the world.[14] The message was well received. The WSSD Plan of Implementation included numerous mentions of the need for improved collection and international sharing of data, including data from Earth observation satellites.[15]

In 2003, the G8 met in Evian, France and adopted a science and technology action plan to support the implementation of the WSSD. The plan focused attention on three areas, the first of which was coordination of global observation strategies. This plan called on nations to minimize gaps, and improve reporting, archiving, and reciprocal sharing of data. The G8 planned to develop an implementation plan for these activities by the following year.[16]

To support this effort, the United States hosted the first Earth Observation Summit, bringing together ministerial-level representatives from 35 developing and developed countries as well as key international organizations, including CEOS, WMO, and GCOS. It was hoped that requiring high-level participation would raise the visibility of the conference and make it

possible to achieve political commitments on a level not possible in existing organizations.[17]

The stated purpose of the summit was to promote the development of
a comprehensive, coordinated, and sustained Earth observation system to
understand and address global environmental and economic challenges.[18]
The summit focused on the broad, multidisciplinary and interdisciplinary
uses of Earth observation data. This was a much broader goal than any of
the existing organizations. In a speech to the WMO in the months leading up to the summit, US NOAA Administrator Lautenbacher recognized
the contributions of the WMO, the WWW, GCOS, and others, but argued
that there was much more to be done. He stated that existing observing
systems needed to be brought "to the next level of Earth Observations."
What WMO did for weather, the new system would do for climate, water,
ecosystem, natural resources, sustainable development, and other environmental challenges.[19]

The summit directly addressed the issue of data sharing. Participants
developed a declaration that stated that a system was needed to ensure
exchange of observations "in a full and open manner with minimum time
delay and minimum cost, recognizing relevant international instruments
and national policies and legislation." Participants in the summit also
called for the preparation of a 10-year implementation plan, and created
an ad hoc intergovernmental working group, referred to as the Group on
Earth Observations (GEO) to achieve this objective.[20] GEO met for two days
immediately following the summit. Even this working group had high-level
participation: the US representatives to the ad hoc GEO included the NOAA
administrator and the director of USGS.[21]

The organization took a major step forward at the Second Earth Observation Summit in Tokyo, Japan in 2004, with the approval of the 10-year
implementation plan for a Global Earth Observation System of Systems
(GEOSS). GEOSS would be a distributed system of systems building on current cooperation efforts while encouraging and accommodating new components. Participants were clear in acknowledging the efforts of existing
organizations, noting the success of the WMO in achieving coordinated
and sustained global cooperation through the World Weather Watch, and
the promising work of groups like GCOS. In fact, GCOS would officially
be in charge of the climate component of GEOSS and WMO would lead
on the issue of weather, illustrating GEOSS's plan to build on and stitch

together existing efforts. GEOSS would also push for cooperation in the areas of land, water, ice, and ocean observation, where existing international efforts were less advanced.[22]

GEOSS's efforts were organized around nine areas of socioeconomic benefit expected to result from comprehensive, coordinated, and sustained Earth observations, including: (1) reducing loss of life and property from disasters; (2) understanding environmental factors affecting human health; (3) improving management of energy resources; (4) understanding, assessing, predicting, mitigating, and adapting to climate variability and change; (5) improving management of water resources; (6) improving weather information, forecasting, and warning; (7) improving management and protection of territorial, coastal, and marine ecosystems; (8) supporting sustainable agriculture and combating desertification; and (9) understanding, monitoring, and conserving biodiversity.[23] Like the WMO's World Weather Watch, the GEOSS architecture would include an observation component, a data processing and archiving component, and a data exchange and dissemination component. The document reiterated the 2003 call for full and open exchange of observations with minimum time delay and minimum costs.[24]

Speaking at an event in Washington, DC, in 2004, NOAA Administrator Lautenbacher explained the logic behind the key elements of GEOSS's design. The "system of systems" terminology was important, he explained; it reflected the fact that GEOSS would not be one integrated system run from a centralized area. Instead, GEOSS would connect and coordinate many existing and planned systems in a distributed effort. He identified data policy as one of the four big challenges for GEOSS implementation, along with integration, data management, and engaging developing countries. He acknowledged the challenges posed by the different business models in each country, particularly those involving efforts at cost recovery. He also explained that the focus on socioeconomic benefit areas was meant to emphasize areas of mutual interest and concern, such as natural and manmade disasters. "Even enemies will help each other in these situations," Admiral Lautenbacher noted, due to the ethical responsibility to share data when it can save lives.[25]

The 10-Year Implementation Plan was finalized at the Third Earth Observation Summit, held in Brussels, Belgium in 2005, by which time GEOSS had grown to 60 members.[26] The plan laid out an ambitious vision for

GEOSS: "to realize a future wherein decisions and actions for the benefit of humankind are informed by coordinated, comprehensive, and sustained Earth observations and information." GEOSS would encompass all areas of the world; cover ground-, air-, and space-based observations; and build on existing and future national, regional, and international observational systems. It included not only collection and management of data, but also the transformation of that data into information and products to aid decision-making.[27]

Data sharing was a major component of the plan, which stated that "despite laudable efforts, the current situation with respect to the availability of Earth observations is not optimal," and added, "This situation is particularly true with respect to coordination and data sharing among countries, organizations, and disciplines." The GEOSS implementation plan echoed the sentiment of Jose Achache at the WSSD in 2002, stating, "The societal benefits of Earth observations cannot be achieved without data sharing." The plan laid out the GEOSS data sharing principles, which called for full and open exchange of data, metadata, and products with minimum time delay and at minimum cost, recognizing relevant international instruments and national policies and legislation. It encouraged provision of all shared data, metadata, and products free of charge or no more than the cost of reproduction for research and education.

To implement GEOSS, participants made the ad hoc intergovernmental Group on Earth Observations (GEO) permanent. Membership in GEO was open to all UN Member States and the European Commission. International, intergovernmental, and regional organizations could also join as participating organizations, subject to approval of GEO Members, and others could join as observers. GEO would oversee implementation of the GEOSS plan, monitoring and evaluating progress. It would act as a forum for dialogue and resolution of issues at levels varying from ministerial to scientific and technical. It would coordinate with national agencies and international organizations and engage in advocacy within and across existing systems. To carry out the work identified by Members, GEO established an elected executive committee, a secretariat, and subsidiary bodies, including science and technical advisory committees. The organization began meeting in plenary at least once a year with representatives at the senior-official level, and also held meetings on the ministerial level every two to three years.[28]

Data Sharing Implementation in GEOSS

The GEOSS data sharing principles called for a level of openness beyond the current practice of many of its Members, and the organization recognized that further effort would be needed to implement and refine these principles. In 2006, GEO established a data principles task force focused on "furthering the practical applications of the agreed GEOSS data sharing principles." A special workshop was organized by the Committee on Data for Science and Technology (CODATA) of the International Council for Scientific Unions (ICSU) to exchange ideas related to this task.[29] Workshop participants provided a number of recommendations for GEO. Recognizing the nonbinding nature of GEO's recommendations, they emphasized the need to develop incentives for compliance, rather than enforcement mechanisms, and to give credit to data providers. They felt that peer pressure could be effective in promoting sharing. They argued that in cases where restrictions are in place, the costs and benefits should be reviewed carefully, and suggested that more work be done to demonstrate the value of open access data.[30]

GEO took these recommendations to heart, issuing a "call to action" on data sharing at the fourth Earth Observation Summit in late 2007. Participants agreed that to fully realize the benefits of GEOSS, it was imperative to support the GEO principle of free and open exchange of data.[31] Many nations used their individual statements to call attention to the importance of data sharing and describe their own efforts in this area, ensuring their nation or region would gain greater recognition for their efforts and applying some subtle peer pressure on members not yet openly sharing their data. The European Commission stated that data sharing was one of the greatest challenges for the future of GEOSS, and highlighted its own efforts at increasing data sharing in Europe through the Infrastructure for Spatial Information in Europe (INSPIRE) directive and the Global Monitoring for Environment and Security (GMES) initiative.[32] Egypt noted that reducing costs for satellite images was one of the keys to ensuring that developing nations benefit from GEOSS.[33] Finland stated that it endorsed the GEOSS Data Sharing Principles and had already decided to share environmental data owned by its national Environmental Administration free of charge.[34] Germany rallied members, arguing, "we must forcefully advance the implementation of the GEO principles for data sharing."[35] One of the efforts highlighted at this event, and mentioned by numerous members and participating organizations in speeches, was the announcement by China and Brazil that they would provide data from the China-Brazil Environmental Resource Satellite (CBERS) to users in China, South America, and Africa on a free and open basis.[36]

The United States followed up with a similar announcement at the 2008 GEO Plenary, stating that GEO efforts had played a role in its decision to make all Landsat data freely available online. EUMETSAT noted that all of its archived data was now available through its Earth observations portal, illustrating the willingness of the meteorological community to embrace the data sharing principles. China announced that data from its newly launched weather satellites would be universally available free of charge. The European Commission voiced its support of full and open data sharing, and suggested that promoting the benefits of full and open access would help to engage additional providers and users.[37] It seemed that the efforts to provide credit to data providers and exert peer pressure to promote sharing were bearing fruit.

The GEO data principles task force saw this progress and continued to press further, proposing a set of implementation guidelines for the GEOSS data sharing principles that emphasized the importance of open data sharing. The guidelines reiterated that promoting the full and open exchange of data was essential to ensuring GEOSS reached its full potential. They reminded members that the value of data lies in its use, and emphasized the importance of avoiding restrictions on reuse and redissemination, as these types of restraints could "significantly diminish the societal benefits intended." The implementation guidelines reminded members that data should be provided at no more than the cost of reproduction and distribution, and preferably for free. Increasing the amount of data made available was also a key goal, and the guidelines stated that efforts should be made to make available products, or a subset of products, not only from government-owned systems, but also from public-private partnerships or even private-sector organizations.[38]

GEO Members discussed the proposed implementation guidelines at the 2009 Plenary. ESA agreed with the guidelines and noted that its own policy supported full and open access to satellite data. The United States also strongly supported acceptance of the guidelines, reiterating the "fundamental importance of the principles of free and open access to data and minimum time delays and costs." Brazil noted the progress already made by itself, China, the United States, and ESA in promoting full and open access to satellite data. Canada, which maintained some restrictive data policies to allow for public-private partnerships and data sales, asked for clarification on the implications for a country wanting to contribute data to GEOSS, but not adhering to the data sharing guidelines. Despite some concerns, the final version of the implementation guidelines were accepted at the meeting.[39] The nonbinding, best efforts nature of GEO allowed the organization to continue to push its members to adopt open data policies, beyond what many currently had in place.

One of the early goals of the GEO was to gain high-level recognition of the importance of global observations and data sharing. This goal was also being realized, as G8 leaders continued to voice their support for GEOSS, specifically mentioning it in the 2008 Declaration on Environment and the 2009 Declaration of Responsible Leadership for a Sustainable Future. The 2008 statement noted, "To respond to the growing demand for Earth observation data, we will accelerate efforts within the Global Earth Observation System of Systems (GEOSS)... by strengthening observation, prediction and data sharing."[40] The 2009 statement emphasized the importance of supporting GEOSS as part of the commitment to address the increased threats of natural disasters and extreme weather phenomena caused by climate change.[41]

Data Sharing Innovations: Data Democracy and Data-CORE

The 2010 GEO Plenary marked the halfway point in GEOSS implementation, and as part of a midterm evaluation process, GEO surveyed the Earth science community for their opinions on its efforts. A large number of survey respondents identified the GEOSS data sharing principles as both the most important accomplishment of GEOSS to date and as one of the major challenges still facing GEOSS. Community members felt that GEO had facilitated discussion and some consensus on a historically contentious issue, but that the discussion needed to be transitioned to action.[42]

GEO members agreed. The Data Sharing Action Plan, designed to continue moving GEO forward toward implementation of the principles, had already been in development and was adopted at the 2010 GEO Plenary. The document laid out arguments for sharing data on a full and open basis.

They called out, for example, the need for policy-makers and managers to have access to relevant information to make good decisions on issues that have the potential to affect citizens' safety and well-being. They noted that numerous studies had shown that the whole economy benefits from open access to data. It pointed to examples in which adoption of open data policies led to dramatic increases in data use: when CBERS data became available free of charge, access rose from 1,000 scenes a year to 10,000 a month. In the year after Landsat data was made freely available online, more than a million images were downloaded. That was more than the total number of images distributed in Landsat's entire 38-year history prior to the policy change.[43]

The Action Plan noted that restrictions on redistribution, charges, and other challenges in accessing data still in place in some member nations had limited full implementation of the GEOSS data sharing principles. To help highlight and address this issue, the GEO Data Sharing Task Force proposed the GEOSS Data Collection of Open Resources for Everyone (the GEOSS Data-CORE). While all data was welcome within GEOSS, the CORE included only data that fully adhered to the GEOSS data sharing principles of full and open access. This made freely available data easier for users to find and also provided special recognition for data providers that followed the GEO principles. The task force also supported data democracy efforts, originally developed within CEOS, which emphasized data sharing and capacity building specifically with respect to developing countries.[44]

Responding to these new concepts at the plenary, many countries, including the United States, Germany, Italy, Japan, and South Africa, voiced their support for data sharing, including the establishment of the Data-CORE and the concept of data democracy. Other countries, particularly those with more restrictive policies, proved more reticent with respect to the new system, and noted the importance of accepting a range of definitions of "minimum cost" and "restrictions" in defining the Data-CORE system. For example, France suggested that it is important to respect the view that data pricing policy is not necessarily a restriction. The European Space Agency (ESA) suggested that the word "conditions" should be used rather than "restrictions," because it has a more positive connotation, and argued that it is reasonable to view prices as a condition rather than a restriction. Canada welcomed the plan's recognition that attribution and registration requirements were not to be considered restrictions.[45]

Despite these misgivings among some member nations, the GEOSS Data-CORE and Data Democracy concepts were endorsed at the 2010 GEO Ministerial Summit that followed the plenary. The importance of data sharing was again highlighted at the summit, with multiple countries calling out the issue in their official statements. The United States provided a list of datasets that it planned to contribute to the Data CORE, and shared greetings from President Obama praising GEO's efforts to ensure open access to data and information.[46] Estonia noted that full and open exchange of data was one of the biggest achievements of GEO and argued that it had caused a complete change in philosophy.[47] The European Commission called the adoption of the Data Sharing Action Plan "the beginning of a new era of international cooperation in Earth observation," and reminded members of the policy of open access for GMES data.[48] ESA noted that it had put in place a new policy of free and open access to its data, including past ERS, Envisat, and Earth Explorer missions.[49]

The 2011 GEO Plenary focused on implementing the GEOSS Data-CORE concept.[50] The 2012–2015 work plan called for maximizing the number of datasets available within the GEOSS Data-CORE, developing metrics to show the impact of open data, and encouraging countries to identify specific institutional, legal, and technical barriers to full and open exchange of data.[51] The plenary established the Data Sharing Working Group (DSWG) to support further efforts to put the GEOSS Data Sharing Principles into practice.[52]

GEO and the Open Data Movement

GEO's efforts to implement its data principles coincided with a wider surge in political interest in open government data initiatives that called on government agencies to create a culture of open government and publish their data online.[53] The United States Open Government Directive, issued in 2009, was followed by a wave of open data movements around the globe, with more than 44 countries developing open data platforms by 2013.[54] The G8 developed an Open Data Charter laying out a set of principles for access to, and release and reuse of, government data. The principles stated that data should be open by default, available free of charge to all users without bureaucratic or administrative barriers, including registration requirements. Earth observation was one of the fourteen data categories identified as high value for this effort.[55]

GEO's fourth Ministerial Summit, held in Mexico City in November 2015, marked the end of its first 10-Year Implementation Plan. GEO Members recognized that while significant progress had been made, there was much that still needed to be accomplished to achieve the GEO vision.[56] As part of its new strategic plan, the GEOSS Data Sharing Principles were completely revised, reflecting the global open data trend and GEOSS's strategy of adopting forward-looking data sharing principles. Drawing partially on language from the G8 Open Data Charter, the new principles stated that "data, metadata, and products will be shared as Open Data by default, making them available as part of the GEOSS Data Collection of Open Resources for Everyone (Data-CORE) without charge or restrictions on reuse, subject to the conditions of registration and attribution when the data are reused." It went on to say that in cases where open data provision was not possible, due to international or national policy or law, the data should be made available with minimal restrictions on use and at no more than the cost of reproduction and distribution. All data should be made available with minimum time delay.[57]

At the same time GEO approved the new data sharing principles, it also released a document titled, "The Value of Open Data Sharing" laying out the rationale for why it is important for GEO to make the data in its GEOSS portal available on a free and unrestricted basis. The report argued that public data made openly available had been shown to provide significantly greater economic returns than restrictive, proprietary approaches. Research and private innovation also increase with a policy of openness, it stated, particularly with the growth of big data and data mining techniques. Open data helps to facilitate improved educational opportunities. It argued that providing data on an open basis was viewed by society as the appropriate, ethical way to manage public digital resources, providing improved transparency and helping to build trust in governance.[58] The new data sharing principles and the report on the value of data sharing demonstrated GEO's continued commitment to remain on the leading edge of international debates about environmental data sharing, actively pushing for wider application of open data policies through both peer pressure and provision of supporting information. With a membership of more than 104 Member nations and 106 participating organizations as of the beginning of 2017, GEO is poised to remain an important actor in increasing international sharing of Earth observation data.

Summary

GEO was created because, despite the existence of multiple efforts to facilitate international cooperation on environmental observation, progress had not been sufficient. GEO was designed to overcome the challenges of these earlier organizations. As a voluntary organization formed outside the UN Framework, GEO had greater freedom in its actions. Emphasizing high-level political membership ensured decisions made in GEO would have the level of support needed to result in implementation at the national level. GEO is all inclusive: global space and in situ measurements across a broad range of societal benefit areas and communities of interest are included, but GEO doesn't reinvent the wheel: WMO, GCOS, CEOS, and others are members of GEO, and their efforts are linked into the larger GEOSS vision.

Although deficiencies still exist in global monitoring capabilities, it is undeniable that the organization has made some impact on the Earth observation scene. In particular, GEO has become an important forum for discussing international cooperation on Earth observations of all kinds, looking at issues and challenges that cut across typical data provider or user silos. This is particularly true with respect to the issue of data sharing. Examples can be found in every GEO Plenary and Ministerial meeting of nations promoting open data policies and providing recognition for efforts in this area. The benefits of open sharing and challenges caused by restrictions have been reiterated in many ways. GEO has been successful in keeping the issue of data sharing high on the agenda year after year.

The GEOSS Data-CORE is one example of an innovative idea developed within GEO that helps to clearly identify the state of data sharing and to make concrete arguments related to data sharing outcomes. With the Data-CORE, it is possible to clearly see which nations or agencies comply with the GEOSS Data Sharing Principles and which do not. This makes it easier to exert peer pressure, and to use political and economic arguments to influence those nations with noncompliant data sharing policies. The existence of the Data-CORE also works as a demonstration project to show the benefits of easily accessible, freely available data.

The voluntary, best efforts nature of GEO could be seen as a limitation, as the organization has no way to enforce compliance with its data sharing principles, but it can also be an asset. Because compliance with its data sharing principles is voluntary, GEO has been able to set aspirational goals, consistently pushing the boundaries in promoting increased data sharing. GEO's most recent effort to do so, adopting language calling for provision of environmental data on a free and open basis, continues this trend.

7 US National Oceanic and Atmospheric Administration

The US National Oceanic and Atmospheric Association (NOAA) operates the constellation of US weather satellites. The organization has one of the longest histories of satellite activity among all space and meteorological agencies, and data sharing policies have been a constant issue of debate throughout this time. Although it has been buffeted by trends and arguments by actors on the international and national level, NOAA's policy has remained relatively consistent over time. Its story is one of debate and constancy.

The United States enthusiastically engaged in free and open international sharing of data from its early weather satellites and contributed significantly to establishing international systems as part of the WMO. As part of broader trends toward commercialization in the 1980s, Congress authorized NOAA to charge commercial entities fees to access its data, officially transitioning the agency to a tiered policy—data was still provided without charge to research and other users, but commercial entities were sometimes required to pay fees for data access. Despite this official change in its policy, NOAA remained a strong proponent of full and open sharing of data on both the international stage and domestically, throughout many policy debates and innovations, and in practice the agency typically provides its satellite data at no cost to all users.

Space and Weather

NOAA, as an agency, was not formed until 1970, but a closely related predecessor, the Weather Bureau, worked with NASA in the earliest planning for space activities. The US Weather Bureau traced its beginnings to the 1870s, when it was created to take meteorological observations across the continent and provide warnings of approaching storms.[1] Monitoring weather conditions, let alone predicting them, was a significant challenge for the United States, whose territory spans a vast area and is surrounded on two sides by oceans. The ability of satellites to contribute to this mission had been clear since before the first satellites were launched. US meteorologists had written scholarly articles about the potential of satellites to contribute to their work, and were heavily involved in planning for the International Geophysical Year, which resulted in the launch of the first satellites. Some of the earliest sounding rocket test flights carried cameras to take high-altitude images of clouds. With promising results, meteorologists were eager to begin applying satellite technology to their field.

NOAA's International Beginnings

Given the interest and involvement of the meteorological community, it is perhaps no surprise that the first satellite dedicated to observing the Earth was a weather satellite, the US Television Infrared Observing Satellite 1 (TIROS I), launched in April, 1960.[2] The satellite carried two cameras that provided black and white images somewhat distorted by the spinning of the satellite. At its best, the spatial resolution of these images was 300 meters, in some cases, the resolution was only 3,000 meters. Collecting data from this new machine was not easy. TIROS I had a magnetic tape capable of storing 32 images. For the few minutes a day the satellite was in range of the two US ground stations, operators had to decide between transmitting stored images—typically recordings of non-US territory—or transmitting images of the United States collected in real time. In either case, pictures transmitted to the stations were simultaneously displayed on a television screen and recorded onto magnetic tape. A 35-mm camera photographed the television screen.[3] By today's standards, the technology was very limited, but the thousands of images taken by TIROS I during its 78 days of operation were unlike anything that had been seen before. TIROS I confirmed that satellites could provide valuable contributions to global weather observation.

As engineers and scientists demonstrated the technical and scientific potential of weather satellites, political leaders hoped to use them to achieve political goals as well. By the time the first satellites launched, the

World Meteorological Organization (WMO), together with its predecessor, the nongovernmental International Meteorological Organization (IMO), already had a history of more than 75 years of successful international cooperation, including free and unrestricted exchange of meteorological data. Many in the international community, including the United States, were eager to use this existing framework to promote international cooperation in the peaceful uses of outer space.[4] In 1958, two years before TIROS I launched, the Director of Research at the US Weather Bureau and the head of the Soviet Central Institute of Forecasts, among others, formed a panel of experts on artificial satellites within the World Meteorological Organization.[5]

In his first State of the Union Address, in January, 1961, President Kennedy invited all nations, including the Soviet Union, to join the United States in developing a weather prediction program—one of many projects "to invoke the wonders of science instead of its terrors."[6] President Kennedy reiterated this idea in his September, 1961, address at the UN General Assembly, stressing the importance of reserving space for peaceful use.[7] The UN General Assembly responded in December, 1961, with Resolution 1721, which recognized "the common interest of mankind in furthering the peaceful uses of outer space and the urgent need to strengthen international cooperation in this field," and requested that the WMO investigate opportunities for international cooperation in light of the developments in meteorological satellites.[8]

In light of these goals, and the inherent potential of the information to help save lives and property, data from even the earliest satellites was shared freely with the international community. Henry Wexler, director of research at the US Weather Bureau wrote, "Since some of the phenomena observed by meteorological satellites may have rapid and serious consequences to the safety of the population and to the economy of a nation, it is imperative that meteorological satellite information be conveyed speedily to all the nations."[9] An official US Weather Bureau report on satellite programs concurred, stating "weather knows no boundaries," and "requires close cooperation between nations... for protection of life and property."[10] Although limitations in communication technology posed challenges to doing so, satellite images from the experimental satellites were shared widely to support timely warnings. By 1962, TIROS images had been sent to Australia to warn of a storm in the south Indian Ocean, to Mexico regarding a tropical storm off its borders, and to many other countries for similar warnings.[11]

NASA rapidly developed more advanced data sharing technologies to help make this data available on a more routine basis. Beginning with TIROS IV, launched in 1962, the United States initiated an international fax transmission network to regularly share cloud images with meteorological services around the world.[12] By the end of 1963, NASA flew the first prototype of the Automatic Picture Transmission (APT) system. The APT was a television camera designed to automatically transmit a live stream of images whenever it was within range of a ground station. This allowed local meteorological services in any nation that purchased such a station to access the real-time data needed for useful weather forecasting.[13]

Other NASA and Weather Bureau experts emphasized the importance of international data sharing for advancing meteorological science and prediction capabilities. NASA aimed for its program to be completely known and understood by the scientific world community. Dr. Morris Tepper, Director of Meteorological Systems at NASA, stated that sharing satellite data with other countries had been an active part of the program from the beginning, "reflecting the spirit of the very Act which gave birth to NASA."[14] The Director of the Weather Bureau, Dr. Reichelderfer, wrote that the immense amount of data made it suitable for independent studies, and thus it would be distributed widely for research in universities, private institutions, and government agencies.[15]

Archiving data for research use was also a priority from the beginning. This issue was addressed at an international workshop on meteorological satellites organized by the Weather Bureau and NASA in 1961 and attended by 27 participants from 25 nations. The Weather Bureau explained that the data had already been proven valuable, but its full potential would only be realized if scientists from all parts of the world were able to work with it, discovering and developing new uses and applications. To enable this, data would be made readily available to any interested party efficiently and promptly.[16] Accordingly, NASA released a complete listing of cloud images obtained by the TIROS I satellite accompanied by maps showing their geographical coverage, as well as a guide to how to make use of the images. Copies of the 35-mm images could be ordered from the National Weather Records Center in Asheville, North Carolina at a price of $4.00 per 100-foot reel (about $30 in 2016 dollars), a price primarily reflecting the cost of reproduction and distribution. A full set of TIROS I images occupied 50 reels.[17] Cloud images from follow-on systems were dealt with in a similar manner. Data from radiation monitors that were included on TIROS II and later satellites was coded onto magnetic tape, archived, and made available as well.[18]

Meteorologists saw satellites, computers, and other technological developments as a turning point for meteorology. Although international cooperation in meteorology had already been occurring for nearly a century by this time, experts felt that the possibilities offered by these new technologies required a redesign of the entire international data exchange system. The proposal was well received by the international community. The idea for the WMO World Weather Watch (WWW) was proposed by the WMO panel of experts on artificial satellites in 1962, endorsed by the UN General Assembly that same year, and officially established in 1963. The new system included three interconnected components: a global observation system, a global telecommunication system, and a global data processing system. Although adoption of the concept was rapid, it took many more years to implement the full system.[19]

The United States was a leader in the development of the WWW, which took advantage of satellites, along with a vast network of surface stations and other in situ sensors, to achieve truly global observations, the lack of which had previously been the single largest obstacle to improving weather forecasting. Washington, DC, was home to one of three World Meteorological Centers (along with Moscow and Melbourne), housing the complex and costly technology needed to receive, process, and disseminate meteorological data from the global observing system. A high-throughput circuit was installed to connect these centers, forming the trunk of a global telecommunications system. Secondary connections linked the WMCs to regional and national meteorological centers where data could be used by local meteorologists.[20] Free and open data sharing was the policy foundation on which this technical system was built.

In the midst of the Cold War and with the space race in full swing, the United States also sought to establish a bilateral arrangement for sharing satellite weather data with the only other nation capable of lofting spacecraft at the time: the USSR. In a March, 1962, letter to Khrushchev, President Kennedy proposed a number of cooperative space activities, writing, "Perhaps we could render no greater service to mankind through our space programs than by the joint establishment of an early operational weather satellite system. Such a system would be designed to provide global weather data for prompt use by any nation."[21] Khrushchev replied positively, and by the end of that year, the United States and the Soviet Union had agreed to establish a communications link to allow two-way transfer of cloud-cover imagery and radiation data collected from meteorological satellites.[22] This connection was referred to as the "cold line" to distinguish it from the nuclear "hot line" also being developed at the same time. The two countries began exchanging meteorological data from ground-based sensors in 1964, but satellite data was not exchanged until 1966, when Russia launched Cosmos 122, its first official meteorological satellite.[23]

Going Operational

Officially, the TIROS satellites were experimental research satellites, but the Weather Bureau had been providing forecasters with cloud analysis based on satellite images since TIROS I. Discussions began almost immediately on plans to develop a National Operational Meteorological Satellite System (NOMSS). In 1960, officials from NASA, the Weather Bureau, the Federal Aviation Agency (FAA), and the Department of Defense (DOD) formed the Panel on Operational Meteorological Satellites (POMS), and recommended in April, 1961, that development of an operational system be undertaken as soon as possible. The program had high-level support across the government: the President requested funds for the program a month after the report was released, and Congress passed a supplemental appropriations act providing the funds in September.[24]

A 1961 report by the US House of Representatives Committee on Science and Astronautics explicitly addressed the multitude of reasons for developing such a system and sharing the resulting data with the rest of the world, particularly noting the strategic benefits for the United States. The report stated, "International cooperation in the exchange of weather observations is of long standing and represents one of the outstanding examples of amicable international relationships. It would be to the advantage of the United States to take the lead in extending such international meteorological cooperation to include satellite data." The report noted that a system designed to meet national needs would simultaneously meet the needs of many in the international meteorological community, with no additional burden. It stated that data should be shared via fax in the earliest phases of the program, and expressed approval of the fact that for a minimum expense, any nation could obtain a ground station to obtain cloud images directly through the APT system. The committee recognized that weather satellite systems would improve storm warning, saving lives and reducing property damage in all areas of the world, and meteorological data would provide a significant contribution to the welfare and economic growth of underdeveloped nations. These benefits were not only intrinsically worthwhile, but would "be an excellent opportunity for the United States to win many friends in many lands" and be one of "the greatest peaceful weapons against communism."[25]

Unlike the research program, the operational system provided global coverage on an operational basis without interruption. This required a steady cadence of launches to ensure there were no gaps in data availability. The Weather Bureau took on a much larger role: financing, managing, and operating the meteorological satellite system.[26] The Bureau was also in charge of processing, dissemination, and archiving of data for operational and research uses.[27] NASA was still involved. The space agency designed, built, and launched the satellites to meet Weather Bureau requirements, transferring control to the Weather Bureau after an initial checkout of the spacecraft in orbit.[28] In concert with this transition in responsibilities, President Johnson created the Environmental Science Services Administration (ESSA) within the Department of Commerce in 1965, giving it the functions and employees of the Weather Bureau and the Coast and Geodetic Survey.[29]

The first operational weather satellites were launched and began providing daily global coverage in 1966. In the same year, the United States launched the first geostationary weather satellite. Located in a much higher orbit, this system orbited at a rate that allowed it to appear stationary over one area of the globe. The satellite captured an image every 20 minutes, allowing the United States to continuously monitor weather developments over its territory. This was particularly valuable for monitoring the formation of storms off the coasts that could evolve into hurricanes or other severe weather, and that were difficult or impossible to detect before the availability of satellite imagery. Data from both the polar-orbiting and geostationary satellites was made available worldwide. Additional APT stations were installed and fax transmission networks developed.[30] Archived data was cataloged and made available at marginal cost through the National Weather Records Center in Ashville, North Carolina.[31]

In 1970, President Nixon created the National Oceanic and Atmospheric Administration (NOAA), stating, "We face immediate and compelling needs for better protection of life and property from natural hazards, and for a better understanding of the total environment—an understanding which will enable us more effectively to monitor and predict its actions." ESSA, which by this time had expanded to include not only the Weather Bureau and the Coast and Geodetic Survey, but also the Environmental Data Service, the National Environmental Satellite Center, and a number of research agencies, would be incorporated into the new agency. ESSA would be the largest component of NOAA, its 10,000 employees making up 70 percent of NOAA personnel.[32]

By the time NOAA was created in 1970, the United States had launched more than 20 low Earth orbit weather satellites, and five geostationary weather satellites.[33] Satellite technology continued to advance and new instruments were added to the systems, improving the quality and quantity of data available. NOAA began referring to its satellite series as the Polar-Orbiting Environmental Satellite (POES) series, with each POES satellite renamed "NOAA" and given a number designation after its successful launch, and the Geostationary-Orbiting Environmental Satellites (GOES) series, also given a number designation after launch. APT receiving stations had been installed in 500 locations across more than 50 countries, and the United States continued to exchange information with Russia via the "cold line" connection. Archived data in manuscript, microfilm, photographic, or magnetic tape form was available to all users at the cost of production from the National Climatic Center in Ashville, North Carolina.[34] Operational weather monitoring and international data sharing had become routine.

From December, 1978, to November, 1979, the United States participated in the Global Weather Experiment (GWE) led by the WMO and ICSU in support of the World Weather Watch Program.[35] A major component of the experiment was the establishment of a network of geostationary weather satellites providing full global coverage of the Earth. Since 1972, the United States had been working with officials from Japan, Europe, and the Soviet Union to coordinate the planned five-satellite system. Russia was unable to provide its planned satellite on time, so the United States ended up contributing three of the five geostationary satellites in the system.[36] The United States also provided two polar-orbiting satellites, which were complemented by two launched by Russia. The United States maintained many of the datasets generated for the GWE at the World Data Center in Washington, DC, providing data to any user who requested it at the cost of duplicating the data and postage.[37]

By 1979, interest in satellite data was booming, and more than 120 countries had APT systems for directly receiving satellite imagery. Another 16 countries had also installed more advanced ground systems that could receive high-resolution imagery and data from the new sounding instruments placed on US weather satellites, which could be used to produce three-dimensional profiles of atmospheric temperature and humidity. This data was becoming increasingly useful due to advancements in computing technology and numerical weather forecasting techniques. The satellite data was being used not only for meteorological purposes, but also to support hydrological, oceanographic, agricultural, and other operations and research.[38] Archived satellite products, particularly useful for scientists and researchers, continued to be available at the cost of reproduction.[39]

Commercialization Considered and Rejected

After 15 years, the operational weather satellite system had become indispensable, yet in the early 1980s, the program began to experience budgetary challenges. In some ways, the weather satellite program was the victim of its own success. The increase in uses and users had created demand for more and better observations and data distribution capabilities. Sounding instruments on polar-orbiting satellites had helped to make numerical weather forecasting feasible, and this new activity was highly dependent on continued data provision. Similarly, geostationary satellites provided imagery that became critical for severe weather and flash flood warnings, and users could not go without it. NOAA satellite data was being used on a daily basis to monitor crop production, crop disease, drought, and other important environmental conditions. Maintaining and improving the system to support these users required that NOAA sustain a high level of budgetary support.[40]

Over the same period, support of NOAA missions within NASA, which was facing its own budgetary challenges, decreased significantly. After decades of focusing its Earth observation efforts primarily on experimental weather systems, NASA's interest in Earth science had begun to expand beyond meteorology. Its program of developing meteorological instruments to support NOAA under the Operational Satellite Improvement Program was cut back and eventually terminated. As a result, NOAA had to take over responsibility for funding for new developments and improvements in its weather satellite sensors.[41] A NOAA report showed that total weather program expenditures, considering both NASA and NOAA efforts, had increased only a few percent from 1971 to 1985. However, because NASA funding had decreased significantly, the off-setting increases in the NOAA budget appeared dramatic. NOAA spending on its satellite program increased from about $40 million in 1971 ($240 million in 2016 dollars) to $250 million in 1985 ($560 million in 2016 dollars), more than doubling its budget in real terms.[42]

The Reagan administration, which promoted policies of fiscal constraint and privatization across the government, considered multiple options for decreasing weather satellite costs, including reducing the number of satellites, combining civil and military weather satellite programs, and transferring government civil remote sensing activities, including both Landsat and weather satellites, to the private sector.[43] All three plans faced opposition from the meteorological community, and commercialization of weather satellites faced bipartisan opposition from Congress.[44]

In a March, 1983, Congressional Hearing on the issue, Dr. David Johnson, former Assistant Administrator of NOAA for Satellites, stressed the US role in the existing international system of meteorological satellites, noting that countries around the world relied on data from US systems. He also expressed concern about the possibility of weather data being subject to copyright, which he argued "flies squarely in the face of a long tradition of free international exchange of weather data."[45] The DOD argued that US data must continue to be freely shared with international partners to ensure that foreign governments do not limit the data they provide to the United States.[46]

In November, 1983, the House of Representatives passed a resolution stating that it was the sense of Congress that it is not appropriate to transfer ownership or management of the civil meteorological satellite system to the private sector. The resolution called out normative, economic, and security problems with such a transfer. It stated that the government was the principal user of the data and had the responsibility to provide forecasts and severe weather warnings in order to protect property and public safety. It argued that a transfer to a private system would result in a government subsidized monopoly, and that security could be an issue, as the civil system provides data important for national security purposes. The resolution also noted that the United States had engaged in free international exchange of weather data for over 100 years.[47]

Congress followed up the House resolution by addressing the issue in law. The Land Remote Sensing Act of 1984, passed in July of that year, primarily dealt with plans to commercialize the Landsat system, but it also included two strongly worded provisions prohibiting the commercialization of weather satellites. Section 701 of the Act stated that no effort can be made to "lease, sell, or transfer to the private sector... any portion of the weather satellite systems operated by the Department of Commerce." Section 702 reiterated this, stating, "Regardless of any change in circumstances subsequent to the enactment of this Act, even if such changes make it appear to be in the national interest to commercialize weather satellites, neither the President nor any official shall take any action prohibited by section 701 unless this title has first been repealed."[48]

Commercialization's (Partial) Comeback

While privatization of the weather satellite system itself was quickly and soundly rejected by Congress, discussions about the role of private industry in providing value-added weather products and services continued both in the United States and abroad. In the United States, new companies had emerged specializing in the provision of services aimed at radio or TV stations reporting the weather, and at shipping companies, oil companies, commercial fisheries, or others with special needs related to meteorology and hydrology. In Europe, some NMSs had developed commercial branches within their agencies to sell meteorological products and services. These developments generated concern about competition being created between government and private weather product and service providers, both domestically and internationally.[49]

These trends also called attention to the need for international coordination, and the problems of attempting cost recovery in some nations while others maintained free and open sharing. NMSs in Europe, in particular, argued that data they provided freely to other nations was shared with private companies that then used that data to develop and sell products in competition with NMS commercial branches. It viewed this competition as unfair, since the companies received the data for free while the NMS had to pay the large cost of the observation systems. These developments also raised concerns about the role of data collected by the agencies: should agencies continue to provide this data at the cost of reproduction, or should companies have to contribute something to the system, and if so, how much?[50] These debates gained attention on the international level, and the Tenth WMO Congress passed a resolution in 1987 reiterating the importance of its long-standing open data tradition, stating "The principle of free and unrestricted exchange of meteorological data between National Meteorological Services should be maintained."

Despite its opposition to privatizing the weather satellites, and NOAA's advocacy for continued free and unrestricted data sharing, the US Congress showed an interest in exploring the types of commercial activities being undertaken by European meteorological agencies. In 1988, it passed a law authorizing the Secretary of Commerce to assess fees based on fair market value for access to environmental data archived by NOAA's National Environmental Satellite, Data, and Information Service (NESDIS). Revenues from data sales would be retained by NOAA to support its data archive centers. The fees were only to be applied to commercial users of data. The Act stipulated that data should remain available at the cost of reproduction and transmission to government, university, and nonprofit users for research use. It also stated that fees should be waived to continue to provide data to foreign governments and international institutions on a data exchange basis or under international agreements.[51] The Omnibus Budget Reconciliation Act of 1990 extended the provisions of the bill to include all data, information, and products collected and/or archived by NOAA, rather than just archived data.[52]

As it examined whether and how to implement the new ability to assess fees and engage in its own commercial activities, NOAA was also determining how to react to the growing number of private-sector meteorological companies. In January, 1991, NOAA released a policy statement elaborating the US position on public- and private-sector roles in weather product and service provision. The policy, titled, "The National Weather Service and the Private Weather Industry: A Public-Private Partnership," was jointly prepared by NOAA and the Privatization Branch of the Office of Management and Budget (OMB). The policy stated that the primary mission of the National Weather Service (NWS) was the protection of life and property and the enhancement of the national economy, and identified the NWS as the one official voice for issuing warnings for life-threatening situations. Collection, exchange, and distribution of meteorological, hydrological, climatic, and oceanographic data and information were listed as some of its basic functions. The role of the private sector would be to use the basic data and information from the NWS to provide general and tailored hydrometeorological forecasts and value-added products. The policy stated that the NWS would not compete with the private sector when a service was currently provided or could be provided by commercial enterprises, unless otherwise directed by law.[53]

The policy specifically addressed the issue of free and open international exchange of data, stating, "The private weather industry and the NWS will work together to protect the free and open international exchange of meteorologic, hydrologic, and oceanographic data provided by the NWS by ensuring that the data are not used to compete directly with, or to interfere with, internal policies of national meteorological agencies in those countries where they also provide commercial weather services."[54] This wording reinforced NOAA's long-standing support for free and unrestricted data exchange—a policy that benefited US companies that relied on free access to data—while also directly addressing some of the concerns raised by foreign nations in the ongoing WMO data sharing discussions.

Climate Change and Open Data Sharing

While interest in government participation in commercial activities in the 1980s had led to some restrictions in data sharing—specifically the ability to assess fees for access to US weather data—growing awareness of the challenge of climate change in the late 1980s and early 1990s created a wave of support for more open environmental data sharing. In 1989, President Bush created the US Global Change Research Program (USGCRP), and in the Global Change Research Act of 1990, Congress codified its membership and purpose. NOAA was identified as one of 14 US agency members, and one of the missions of the new group was to make recommendations regarding "bilateral and multilateral proposals for improving worldwide access to scientific data and information."[55] These national efforts mirrored growing awareness on an international level. The results of the first Intergovernmental Panel on Climate Change (IPCC) Assessment report were presented in 1990, laying out the evidence of global warming and highlighting the need to improve systematic observation of climate variables and facilitate international exchange of climate data.[56]

In 1991, the USGCRP released a policy statement on "Data Management for Global Change Research." Sometimes referred to as the "Bromley Principles," after the President's Science Advisor, Dr. Allan Bromley, who led the group, the principles were an important step in supporting free and open data sharing. The policy called for full and open access to data for global change research. It explained that data should be available at the lowest possible cost to global change researchers, and this cost should be no more than the marginal cost of filling a specified user request. The focus was clearly on making data easily available for researchers, but an annex did address commercial users, noting simply, "as required by appropriate public law, global change research agencies will develop plans for commercial access to the global change databases."

NOAA promoted these principles on the international level. As the chair of the working group on data within the Committee on Earth Observation Satellites (CEOS), the agency proposed adoption of a set of data sharing principles calling for the full and open sharing of global change research data. Adopted in December, 1991, the principles called out maximization of satellite data use as a fundamental objective, promoted the development of a sharing mechanism among CEOS members, and called for nondiscriminatory access to data for non-CEOS members.[57] In 1992 these data principles were amended, reflecting the ongoing international debates regarding commercial data sales. Among other changes, CEOS added a statement acknowledging that constraints of mission operations and available resources may require alternative mechanisms for data exchange.[58]

NOAA also referenced the Bromley Principles when updating its own data policy. In 1992, the agency released a report detailing its strategic plan for the Earth System Data and Information Management (ESDIM) Program, including a new draft data policy dealing with the cost, availability, and location of NOAA's data holdings. The policy stated that all environmental data collected, purchased, or otherwise obtained by NOAA were public property, and reiterated its support of free, full, and open exchange of environmental data with the international community. The data policy did not specifically mention commercial access.[59]

Further national-level support for data sharing came in 1993, in the form of Office of Management and Budget (OMB) Circular A-130. This document provided guidance on data dissemination by federal agencies, stating that government data and information products should be priced "at a level sufficient to recover the cost of dissemination, but no higher." It also allowed that agencies could establish user fees at less than the cost of dissemination if higher fees would constitute a significant barrier to fulfilling its information dissemination responsibilities, and also allowed exceptions in cases where laws required agencies to follow a different procedure.[60]

NOAA saw this new directive as an opportunity to move even further toward its preferred policy of free and open data exchange. In 1994, NOAA updated its schedule of fees—originally created in response to Congressional action in 1988—stating that it "strongly supports" the policy directives of OMB A-130, and that NOAA "believes it is in the Nation's best interest to maximize the distribution of government scientific data at the minimum cost." Accordingly, NOAA stated that it would begin charging no more than the cost of disseminating its data, except where otherwise directed by statute.[61]

The changes in policy were accompanied by changes in technology that increased the impact of NOAA's data sharing efforts. In July, 1994, NOAA premiered its Satellite Active Archive (SAA), a system for online access to satellite data. SAA allowed users to discover data by electronically searching satellite metadata, determine its applicability by viewing low-resolution images, and order the desired data through online menus for electronic delivery or through the mail.[62] More users requested NOAA online in 1994 than had requested data offline in any previous year that decade. The number of users requesting data and information online increased dramatically in subsequent years, from less than half a million users in 1994 to more than two million in 1997.[63]

WMO Resolution 40: Acceptable Restrictions on Data Sharing

Although NOAA and the US executive branch had moved to embrace increasingly open data sharing policies, European and other nations continued to explore commercial sales and public-private partnerships that necessitated the adoption of more restrictive policies. The debate between these two policies continued on the international stage within the WMO, until a compromise was finally reached at the Twelfth World Meteorological Congress in 1995.[64] This compromise took the form of WMO Resolution 40, which required free and open sharing of "essential data" needed to support safety and security, but allowed members to restrict access to "nonessential" data. The nonessential data would still be made available for official use by National Meteorological Services, but the producing agencies could require that redistribution to others, including private companies, was restricted, thus enabling commercial meteorological activities.[65] At its next meeting, in 1999, the WMO Congress passed a similar policy covering hydrological data.[66]

Not everyone was happy with the international acceptance of restrictions on data sharing. Some researchers expressed concern that the new restrictions placed on data and the fees aimed at commercial users would inhibit the wide dissemination of data and interfere with both research and value-added product development.[67] The US National Academies argued that restrictions on redistribution aimed at preventing unauthorized commercial use were already resulting in an inability to publish data that underlies scientific results, harming credibility in science. Restrictions on redistribution also prevented sharing data with scientists in foreign countries and restricted international collaboration. They noted that delays in access to satellite data, in particular, could be problematic, as it may take the scientific community much longer to detect issues in data quality that require quick attention to maintain the integrity of climate data series. They also noted that copyrighted or restricted data was not eligible for sharing within the World Data Center network, which would decrease the value of this long-standing international system.[68]

Data Restrictions at NOAA

In the late 1990s, NOAA's policy of free and unrestricted data sharing was once again threatened by budget realities. In 1997, NOAA estimated that updating its computer hardware and software systems—the very systems that had allowed it to greatly increase data distribution in recent years—required a $30 million investment, and announced that the cost of this upgrade would be recovered by reinstating above marginal cost pricing for commercial users. Prices for noncommercial users in government, universities, nonprofits, and other organizations were not increased.

Interestingly, the cost for online access to satellite data listed in the Federal Register was the same for commercial and noncommercial users, with limited access to satellite data costing $30 and unlimited access $200.[69] The number of online users continued to grow, doubling to nearly 4 million by 1999.[70] Over the same period, revenues from data sales decreased—likely because users were opting to access data online, which entailed lower costs—and this led to shortfalls in funding for NOAA's data centers.[71]

Security concerns with respect to sharing of meteorological satellite data were not often raised, but did make some impact at this time in bilateral negotiations between NOAA and EUMETSAT. In 1998, the two organizations signed an agreement to develop a joint polar-orbiting system, in which the agencies would coordinate the launch and orbits of their satellites and in which satellites would carry instruments developed by each of the two agencies. All data would be available to both agencies for their official use. NOAA was able to ensure that all instruments on the satellites that it operated, even those developed by EUMETSAT, as well as the NOAA instruments on the EUMETSAT-operated craft, would all be subject to NOAA's data policy. The agreement also included a clause regarding data denial for US instruments on the EUMETSAT satellite. It stated that during a crisis or war, a US Cabinet-level authority could require that data from US instruments not be provided to some or all users for a specified period of time.[72]

Challenges of Cost Recovery

As NOAA debated and changed its data policy throughout the 1990s, the US meteorological community as a whole had also continued to wrestle with the issue of appropriate government data sharing practices. In the first decade of the new century both NOAA and other organizations in the community released a string of reports, statements, and policies on this issue, largely decrying the problems of cost recovery efforts and advocating for a return to free and open data sharing.

In 2001, the National Research Council published, "Resolving Conflicts Arising from the Privatization of Environmental Data." The report acknowledged the fundamental difference in goals between public-sector users, who generally rely on open access to data at the cost of reproduction, and private-sector organizations, which generally restrict access to data in order to generate a financial return. The report recommended that environmental information systems created by the US government to serve a public purpose should provide data without restriction for all purposes at no more than the marginal cost of reproduction. It argued that private-sector contributions should focus primarily on value-added distribution and specific observation systems.

It likened this model to a tree, in which the roots represent a set of government data on which the government and private industry can build to create core products (the trunk) and a multitude of value-added products (the branches) serving a variety of user communities. It argued that if the government did decide to transfer government data collection to the private sector, it should only do so after ensuring that it would not provide any firms with significant monopoly power, that full and open access to core data and products would be preserved, and that high-quality information could be ensured.[73]

In 2001, NOAA released "The Nation's Environmental Data: Treasures at Risk," which stated that collecting, archiving, and disseminating environmental data were critical to its mission, and highlighted growing challenges in these areas. These challenges included decreasing funding for data management tasks, data on aging media that was inaccessible and at risk of being lost forever, and the inability of observing systems to keep up with industry demands. The report also specifically addressed challenges caused by NOAA's cost recovery data sharing policy.

It argued that the cost recovery policy had alienated some in the university research community, with whom NOAA must work closely to achieve its mission. NOAA also reported that the agency had discovered that its data was being redistributed by a multitude of websites, including those of other federal government agencies. The agency did not see a ready solution: US law prohibits the copyright of government-produced data, which means there was no way to legally restrict this redistribution. Perhaps related to this was the fact that revenues were lower than expected and declining more each year, so that NOAA was nearing a point at which it would cost more to recover fees than it did to provide a free service.

Advocating for these challenges to be addressed, the report stated, "The accessibility of data is key to the value of data. A data set has no value if no one knows it exists, if it is on a media or in a format that makes it unusable, or if the cost associated with being given access to the data is higher than the user can afford to pay."[74] Despite these concerns, cost recovery policies were not changed, and a 2003 follow-up report noted that revenues continued to fall, primarily because users were choosing to access data online, which involved lower fees.[75]

In 2002, The American Meteorological Society (AMS) reinforced NOAA's own concerns, publishing a statement in support of free and open exchange of environmental data. The AMS argued that weather forecasts rely on free international exchange of data, and that monitoring and predicting climate change requires access to global datasets. Research and education depend on free or low-cost environmental data to improve knowledge of these phenomena. The organization argued that free and open provision of data was one of the most important ways that the developed world assisted underdeveloped nations, and that cost recovery disproportionately affected the nations with the least resources. They argued that cost recovery was ineffective, resulting largely in a shuffling of funds among public agencies and support for government subsidized private contractors.[76]

Public-Private Debates Continue

The debate about appropriate data sharing policies was always intricately tied to the relationship between the public and private sectors, and a number of reports in the early 2000s focused on this issue in particular. In April, 2003, the White House released a new US Commercial Remote Sensing Policy. Though developed primarily with land-monitoring systems in mind, the policy was relevant to weather and other remote sensing satellite systems as well. The policy directed US government agencies to rely on privately owned and operated US commercial remote sensing capabilities to the maximum practical extent, developing government systems only when the data was not available from the commercial sector or in cases that required collection, production, or dissemination by the government "due to unique scientific or technological considerations."[77] This suggested that if private companies expressed an interest and ability to develop weather satellites, NOAA may be obliged to purchase these services, rather than continuing to launch its own systems. Many worried that such an arrangement would jeopardize NOAA's ability to provide data on a free and unrestricted basis.

The same year, the National Research Council released "Fair Weather: Effective Partnerships in Weather and Climate Services." The report noted that there was no easy way to delineate the proper responsibilities of the government, academic, and commercial industries, which would always have overlapping interests and activities related to the provision and use of weather, climate, and other environmental information. The report recommended that NOAA update its 1991 Public-Private Partnership policy. The 1991 policy had stated that NOAA would not provide services that the private industry was currently providing or could provide. However, with technological developments over the past 12 years, this had become untenable—the private industry was now technically capable of taking on many of the publicly provided meteorological services, but there were other good public policy reasons for them to remain under government control.

The report stated that the NWS should continue to carry out activities essential to its mission of protecting life and property and enhancing the national economy. This included, among other things, collecting data and providing unrestricted access to those publicly funded observations at the lowest cost possible to all users. Recalling the 2001 NRC report on conflicts arising from privatization of environmental data, it suggested that purchase of commercial data or privatization of data collection activities may be appropriate in cases in which commercial entities can provide sustainable, high-quality sources of data, but only if the government could purchase data rights that allowed full and open access to support public-sector needs and if a commercial market for the data existed that would not be compromised by full and open access provided to the government.[78]

Following the publication of this report, in 2007, NOAA released a new Private-Public Partnership Policy to replace its 1991 policy. It stated that the policy was "based on the premise that government information is a valuable national resource and the benefits to society are maximized when government information is available in a timely and equitable manner to all." NOAA would continue providing open and unrestricted access to publicly funded observations at the lowest possible cost to users to contribute to its mission. It reiterated its commitment to open and unrestricted international exchange of environmental data. NOAA recognized that the private sector uses NOAA information and develops its own observation, communication, and prediction infrastructure to create commercial products. The agency pledged that it would not institute significant changes in its existing information dissemination activities without first considering the full range of views and capabilities of all parties, as well as the public's interest.[79]

NOAA Leadership in Creating GEOSS

In response to an initiative called for by the Group of Eight (G8), NOAA, together with NASA and other US agencies, hosted the first Earth Observation Summit in July, 2003. The Summit brought together representatives from 35 developing and developed countries as well as key international organizations, including CEOS and WMO. The purpose of the summit was to promote the development of a comprehensive, coordinated, and sustained Earth observation system to understand and address global environmental and economic challenges. It would have a broader focus than existing international efforts. In a speech to the WMO in the months leading up to the summit, the NOAA Administrator, Conrad Lautenbacher, explained that what the WMO did for weather, the new system would do for climate, water, ecosystem, natural resources, sustainable development, and other environmental challenges.[80]

The summit resulted in the creation of the intergovernmental Group on Earth Observations (GEO) and an internationally agreed-upon plan to develop the Global Earth Observation System of Systems (GEOSS). Data sharing was a key element of GEOSS. The implementation plan stated that without data sharing, the societal benefits of Earth observations could not be achieved. It laid out three GEOSS data sharing principles, including (1) full and open exchange of data within GEOSS, recognizing national policies and legislation; (2) provision of all data, metadata, and products with minimum time delay and at minimum cost; and (3) sharing of data, metadata, and products free of charge or at no more than the cost of reproduction for research and education.[81]

In late 2004, the United States created the US Group on Earth Observations (US GEO), cochaired by officials from NOAA, NASA, and the White House Office of Science and Technology Policy, to examine national contributions to GEOSS. In 2005, the group released the Strategic Plan for the US Integrated Earth Observation System. One of the goals expressed by the group was to "establish US policies for Earth observations and data management and continue US policies of open access to observations, encouraging other countries to do likewise." It clarified that this included availability for all users, including commercial and operational users, with minimum time delay and at minimum cost.

With international cooperation and technological progress, the importance of meteorological satellites continued to increase. In 2006, a cooperative project between the United States and Taiwan resulted in the launch of the first satellite constellation dedicated to the global positioning system radio occultation (GPS-RO) technique, which analyzes bending angles and delays in GPS signals that have traveled through the Earth's atmosphere to derive three-dimensional information about pressure and temperature. Originally designed as an experimental research program, the data proved so useful that only a year after its launch, data was being used operationally to improve numerical weather forecasting.[82] The majority of the funding for the program came from Taiwan, and permission from that country was required for access to the data.[83] In practice, access was granted to any users, regardless of whether they were engaged in research, operations, or commercial work.[84]

Proliferating Policies and Free Satellite Data

As NOAA headed into the 2010s, it released a series of new data sharing policies, although none represented a significant change from NOAA's existing practice of providing the vast majority of its data for free, but retaining the legal right to charge. The first of these new policies was the "NOAA National Data Centers Free Data Distribution Policy," released in March, 2009. The policy listed 11 instances in which data may be provided at no cost. These included access by a number of federal agencies, educational uses, and valid data agreements, including WMO data exchange agreements. Anyone with a .gov, .edu, .k12, .us, or .mil domain could also access data online for free. Commercial users were not listed as one of the groups eligible for free data, and the document included a note reiterating existing policy that data would be provided to government agencies, universities, and nonprofit institutions at the cost of reproduction and transmission if the data were to be used for research and not for commercial purposes.[85]

Two years later, in February, 2011, NOAA released a policy recognizing the need for full and open exchange of data and products collected by environmental satellites. It listed the GEOSS Data Sharing Principles and WMO Resolution 40 among relevant reference documents, and noted the importance of NOAA satellite data to achieving the GEOSS goal of freely sharing environmental data used to benefit society.[86] The vast majority of NOAA satellite data continued to be made freely available online through the Comprehensive Large Array-Data Stewardship System (CLASS), the follow-on to the Satellite Active Archive developed in the 1990s. Although it is not formally specified anywhere on the NOAA or CLASS website, online access to data on CLASS has been made available free of charge to all users, including commercial users, and was not subject to any restrictions on use or redistribution. Only requests for data too large to be accommodated by ftp included service fees, estimated on a case-by-case basis.[87]

Throughout this period, between 2009 and 2016, the NOAA schedule of fees published in the Federal Register has included a price on the order of $100 for a "satellite image product," and a price on the order of $1000 for "offline satellite, radar, and model digital data (average size is 1 terabyte)."[88] Both are physical products that require some effort by NOAA personnel to create and ship. While the data itself is provided for free, the fees are designed to cover the costs of government labor, physical media, and equipment use.[89]

Ongoing Debates about Public-Private Interaction in Meteorology

More than 30 years after commercialization of weather satellites and other weather products was first raised, the debate about the interaction between the government and private sector in meteorology is still not resolved. For example, multiple companies have expressed an interest in developing their own privately owned and operated GPS-RO satellite constellations, which would collect the same type of data currently provided by the aging COSMIC constellation. Congress has enthusiastically supported the concept of commercial data buys from these companies, while NOAA has been more hesitant, raising issues related to data quality and reliability and stressing the importance of maintaining free and open international data sharing, particularly its data obligations under WMO Resolution 40.[90] In 2015, Congress mandated that NOAA conduct a pilot program to test these potential data sources and provided funding for this purpose.

In 2015, NOAA released a draft Commercial Space Policy focused on NOAA's space-based observations. The policy stated that the services that NOAA provides, including severe weather warnings and other information critical to protection of lives, property, and the US economy, are public goods: they benefit the society as a whole and all individuals have equal access to equal quality products and services. The linchpin of this forecasting capability is timely access to global environmental data from satellites and other sources, which is built on a system of reciprocity for global data. The policy stated that "the US policy of full, open, and free data" also supports the commercial weather sector, which relies on NOAA data to produce value-added products and services. The policy stated that NOAA was interested in leveraging commercially available space capabilities to complement government-owned assets and international partnerships, but that the data must meet the highest standards of quality and allow NOAA to uphold its international data sharing commitments.[91]

The final version of the policy, released in 2016, begins by noting that NOAA is "a science-based services agency charged with the responsibility of understanding and predicting changes in Earth systems." This critical environmental intelligence depends on reliable, timely access to global observations from satellites and other sources. The policy explains that such a large volume of data is needed that no single entity can collect it on its own. As a result, an international data exchange regime exists in which "all nations share essential Earth observations as global public goods, on a full and open basis." Participation in this system provides the United States with much of the environmental data that it uses. The policy expresses NOAA's interest in exploring new business models to complement current public and international data supply arrangements. However, it notes that ensuring continued access to global observations, upholding international conventions for the full, open, and timely sharing of data, and ensuring continued access for the research community remain guiding principles.[92]

One last example demonstrates this ongoing struggle between open data and commercialization. NOAA has long been seen as a leader in the

US open data efforts: weather data was listed in President Obama's 2013 Open Data Policy as an example of the US government's historic commitment to free and open data sharing and of the benefits that such a policy can produce.[93] Recently, however, NOAA has stated that technical challenges have limited the extent to which it can make data available. With more than 20 terabytes of data produced each day, full access to NOAA data online would overwhelm its servers. As of 2016, users only have access to about 10 percent of NOAA data online. One solution NOAA is exploring is the use of cloud computing for provision of data and data analysis capabilities, a project being developed through a Cooperative Research & Development Agreement with private industry. Under the agreement, the private companies involved would pay the marginal cost of accessing data and would be able to charge a fee to users to access the data using their services. One NOAA official describing the project explained, "it may be open, but it's not free," and another stated "freely available doesn't mean there's not a cost associated with it."[94] It has yet to be seen how this new development may affect existing NOAA data sharing practices.

NOAA Data Sharing Summary

Multiple factors aligned in the earliest days of weather satellite development to support open data sharing. Agency officials from the Weather Bureau were part of the meteorological community, which had been sharing data internationally for more than 75 years before the first satellites launched. In their view, continuing to share data from new technology seemed natural, and collection and dissemination of data were part of NOAA's core functions from the beginning. National-level government actors took advantage of this situation to promote their own foreign policy agendas, sharing weather satellite data to demonstrate US technical capabilities, and to gain soft power and prestige. By the time the first meteorological satellites began operating, the user community was already well established, with meteorologists around the world connected and active through the World Meteorological Organization, national meteorological services, and academic communities. Although meteorological systems have military value, sharing of weather data was not viewed as a national security risk because of their broad civil value. The global meteorological community emphasized normative and public safety benefits of data sharing, noting the global nature of weather, and the potential of new weather data to improve forecasts that would save lives and property.

After 25 years of free and open sharing of meteorological data, the situation began to change. The Cold War was winding down, prices of satellite systems were rising, national budgets were tight, and economic issues came to the fore. The value of weather satellites had been proven, but the meaning of this fact was interpreted differently by different groups. NOAA and the meteorological community saw it as proof that these systems were successful and worthy of continued financial support from the government. National-level policy-makers interpreted this value as a sign that NOAA could engage in data sales to recover costs and reduce budget pressures. In response to pressure to commercialize or recover costs, NOAA raised concerns about the importance of maintaining unrestricted international sharing of meteorological data and supporting the research community. This resulted in a compromise, tiered policy, in which NOAA was required to charge market prices to commercial users, but could continue to provide data to international and research users at the marginal cost of reproduction.

Despite the tiered policy mandated by Congress, NOAA remained an active proponent of data sharing internationally, promulgating data sharing principles that would ensure international exchange of data among national meteorological services, through WMO, and space agencies, through CEOS. It took the lead in forming a new international body, GEO, to promote cooperation in collecting and sharing Earth observations. Domestically, NOAA cost recovery efforts were failing to generate significant revenues. Given the flexibility provided by vague and sometimes contradictory laws and directives, paired with low marginal cost distribution made possible by the Internet, NOAA transitioned to a practice of providing the vast majority of satellite data to all users at no cost. Still, NOAA's official, written policy remains one of tiered access, in which free data is only assured for noncommercial users, and national-level policy-makers continue to pressure the agency to engage in commercial arrangements that could jeopardize unrestricted international sharing.

Current NOAA officials and policies continue to defend existing practice, emphasizing the normative, public safety, and economic value of data sharing. They state that data underpins the provision of forecasts that save lives and property. They note that the current international data sharing regime results in economic benefits to the US, providing the country with much more data than it could afford to collect on its own, and that unrestricted data access promotes research and development, resulting in new scientific understanding and new value-added products and services, both of which lead to increased economic growth. Still, debates about the proper roles of NOAA and the private sector continue, with the potential to affect data sharing policies and practices in the future.

Even though it is only one satellite series—made up of a total of eight satellites as of 2016—the Landsat series has proven to be one of the most interesting cases for understanding data sharing policies. Over its 45-year history, the program has transitioned from a free and open policy, to full privatization and commercial sales, to marginal cost pricing, and eventually back to free and open provision.

In the early years of the program, free provision of data was seen as a logical and useful extension of the US data sharing policy with regard to weather satellites. However, as global interest in commercial remote sensing began to increase, US policy-makers saw Landsat as the best candidate for engaging in this trend. Privatization of the system resulted in high prices and reduced data use, and ultimately it was determined that the system was not commercially viable. The Landsat program transitioned back to government control and marginal cost pricing was imposed to help recover the costs of maintaining data dissemination systems. In 2008, USGS adopted an open data policy for Landsat, and use of Landsat data skyrocketed. The program is now one of the most commonly cited examples in the space industry for illustrating the potential benefits of adopting an open data policy.

Experimental Origins and International Engagement

In December, 1968, the Apollo 8 astronauts captured the famous "Earthrise" image from the far side of the Moon. This image, together with the "Blue Marble" picture taken by the Apollo 17 astronauts in 1972 showing the full disk of the Earth, is synonymous with the environmental movement that took hold in the 1970s, and heralded the growing connection between space and the environment.[1] In September, 1969, just months after the first men walked on the Moon, President Nixon gave a speech at the UN General Assembly that brought the focus of the space program back to the Earth. The next step in the US space program, he explained, would be examining the benefits that space technology can have on Earth. He announced that the United States planned to develop an experimental Earth resource survey satellite that would produce information not only for the United States, but also for the world community.[2]

The USGS, as the agency responsible for monitoring and managing the nation's natural resources, had been advocating for such a program for years, going so far as to publicly announce plans for an Earth Resource Observation Satellite (EROS) program in 1966, before funding was approved.[3] The advocacy paid off, and the Earth Resources Technology Satellite (ERTS), later renamed Landsat 1, was developed by NASA and launched in July 1972. In addition to a TV camera, Landsat carried the first civil multispectral scanner (MSS). This instrument allowed Landsat to transmit data in a digital format that was more easily processed and analyzed.[4]

The Landsat program focused on collection of cloud-free images, rather than the cloud-cover pictures of interest to the meteorological community, and the scenes had much higher resolution than those collected by weather satellites, making it possible to identify cities or other man-made developments. This new type of imagery generated a significant amount of interest in a variety of communities and among the general public. To deal with this situation, the distribution of data was split between NASA and USGS. NASA provided the data without cost to foreign and domestic scientists and researchers who could help to develop and improve the system.[5] More than 106 Landsat-1 investigations were carried out by scientists in 37 countries.[6] Other users, including the general public, could purchase processed images from the USGS Earth Resources Observation and Science (EROS) Data Center in South Dakota at a "nominal cost." Nations could also purchase Landsat ground stations, allowing them to receive data of their region directly from the satellite as it passed overhead. Both Brazil and Canada had invested in Landsat ground stations by 1973.[7]

Under the Memoranda of Understanding with NASA, countries paid for construction and operation of ground station equipment, but paid no fee to NASA to receive the data. Each ground station operator agreed to provide data to NASA on a cost-free basis. It also stated that agencies participating in the program must pursue an "open data policy" comparable to that of NASA, which included making data catalogs and the data itself publicly available to the domestic and international community as soon as practicable.[8] It was also interpreted to mean that data would be provided to any interested user at a reasonable price.[9] The agreement was subject to change, and shortly after the launch of Landsat 2 in 1975, NASA announced that countries operating or planning to operate Landsat ground stations would be required to share in the costs of operating the Landsat space segment through an annual $200,000 fee per ground station.[10]

In recognition of this new, potentially sensitive, technical capability and expected future developments, in 1973, the United Nations called for a review of the legal implications of the Earth resources survey by remote sensing satellites.[11] The Outer Space Treaty (formally, the Treaty on Principles Governing the Activities of States in the Exploration and Use of Outer Space, including the Moon and Other Celestial Bodies) had entered into force in October 1967, formally establishing the principle of freedom of use of outer space and freedom of scientific investigation in outer space.[12] The freedom to acquire data from space, as was being carried out in US and Soviet reconnaissance programs, had tacit support. However, dissemination of land remote sensing data was a new issue. Some countries were concerned about their lack of control over the dissemination of information about their own national resources. In particular, France, Brazil, and the Soviet Union proposed that states should not acquire or disseminate data without prior consent from the countries being sensed.[13]

A 1974 US State Department report on this issue noted that the United States should retain its "open skies" position on the freedom to acquire satellite data without prior permission, which was necessary for national security. With regard to dissemination, the report noted that the US policy across most of its space programs, and particularly the weather program, had supported free international exchange of data. However, by 1974, a few companies, with substantial support from government research and development contracts, were already working on creating commercial satellite photo interpretation services, and the potential for commercial remote sensing was being considered. The State Department noted that these activities would require the power to restrict dissemination and place proprietary rights on primary data.[14]

NASA's response to the State Department report suggested making a distinction between the data, which the United States makes freely available, and analysis of the data, which required advanced technology and expertise, and could be held proprietary. NASA also argued that the most effective means of enhancing international acceptance of the "open skies" concept would be the establishment of additional Landsat ground stations, which would require countries to sign a memorandum of understanding that ensured open dissemination. In addition to existing Brazilian and Canadian arrangements, by 1974, NASA had signed an agreement with Italy and was in discussions with Iran, Venezuela, and West Germany.[15]

Transition to Operational Status and Consideration of Commercial Operations

The early concepts for land remote sensing satellites had always envisioned transition to an operational, and potentially commercial, system. Buoyed by the success of the commercial communications satellite industry in the 1960s and 1970s, and spurred by plans in Europe, Japan, and Canada to build domestic remote sensing satellites, many in the United States were anxious to make this transition. A 1976 GAO report suggested four organizational alternatives for management of an operational land remote sensing system: (1) a UN-affiliated agency; (2) a new international organization, following the model of the International Telecommunications Satellite organization; (3) a national program controlled by a US agency; or (4) a private corporation governed by corporate goals.[16]

France's 1978 proposal to create an International Remote Sensing Agency (ISMA) demonstrated some international interest in pursuing the first or second option, although the focus of France's proposal was limited to use of remote sensing satellite data to monitor arms control agreements.[17] The United States ultimately rejected the idea, arguing that an international organization would not be impartial.[18] Congress was in favor of an operational, and eventually commercial, system, putting forward bills to this effect in 1976 and 1977, both of which generated significant discussion, but were never enacted.[19] In November, 1979, the Carter Administration directed that management of the Landsat program be transitioned from NASA to NOAA to reflect a more operational focus for the program. NOAA was to continue to coordinate with USGS, the Department of Agriculture, and other relevant agencies. As part of the transition, NOAA was to adopt system financing that included cost sharing by data users. The agency was told to establish plans for private and international participation with the eventual goal of transitioning Landsat operation to the private sector.[20]

NASA and other agencies had warned for years against premature transition to operational and commercial activities. A 1974 State Department analysis argued that private investment in land remote sensing was unlikely, because most applications of the Landsat data, such as land-use planning, flood control, or pollution monitoring, provided general public benefit, rather than benefits that could be captured by an individual. Given this situation, the report concluded that "it is certain that governments will be the primary customers."[21] An internal NASA memo written in 1976 expressed a belief that private industry would eventually play a major role in land remote sensing, but that the federal government would remain the major user for the next 5–10 years. Given the high risks of an uncertain commercial market, NASA reasoned, industry would not be likely to be interested in the project.[22]

The General Accounting Office (GAO) chimed in on the issue in 1977, arguing that the large magnitude of investment, the long period of time before there would be a return on that investment, and the risks involved, meant private-sector operation of Landsat was unlikely.[23] A thorough review of the program by NOAA in 1980 provided specifics. With the annual $200,000 ground station access fee, computer tapes sold at $200 apiece, and images for between $8 and $50 each, total revenues amounted to about $6 million a year. This accounted for 1.5–5 percent of the estimated annual costs of the operational system. While noting that the future market was still unknown and difficult to project, they concluded that Landsat would probably "not be self-financing before the end of the century."[24]

President Reagan, who became president in 1981, disagreed. In his first budget proposal, the president stated his intention to transition operation of Landsat to the private sector as soon as possible, and withdrew funding for Landsat satellites beyond the two that were currently under development at NASA with the understanding that future systems would be built by commercial operators.[25] Congress was enthusiastic about the plan. In the same discussions in which Congress members vehemently spoke against privatization of the weather satellites, they spoke with equal enthusiasm

for commercializing Landsat. One senator suggested that the most compelling reason to officially reject the possibility of commercializing weather
satellites was because the issue was delaying action on the decision to commercialize land remote sensing satellites. The senator worried that with foreign governments developing their own systems, these delays would result
in the United States losing its lead in this area.[26] In preparation for the
transition to commercial operation, NOAA increased annual ground station fees to $600,000. The prices for imagery were updated to a range of
$650–$2,800 for a computer tape and $26–$235 for a photographic image.
The new fees were designed to recover program operating costs, but not the
costs of building or operating the satellite.[27]

Transition to the private sector was meant to improve efficiency, reduce
pressure on the national budget, and promote development of US companies in the emerging global remote sensing market. Many also hoped that
it would lead the program to focus more on the data user requirements
than on technology development, which was seen as a weak point of the
program while under NASA management. However, by 1982, despite the
launch of Landsat 4, which included a new, more capable imaging sensor
and an improvement in resolution from 80 to 30 meters, increases in cost
and uncertainties about the program's future had already led to decreased
use among federal agencies.

The number of Landsat scenes distributed to the federal government
dropped from 40,000 in FY 1978 to 17,000 in FY 1982.[28] Data sales to non-
federal users, such as universities, state and local governments, and industry also decreased, from more than 50,000 images and 2,000 digital items
in FY 1981 to fewer than 10,000 images and 600 digital items in FY 1983.[29]
When OMB required that all federal agencies, some of whom had previously received Landsat data for free, include funds for Landsat data purchases in 1983, some chose to decrease the amount of imagery they used,
or failed to receive the full funding requested. Because the Landsat program
had planned on these data purchases as part of its budget, these changes left
it with a multimillion dollar budget shortfall.[30] In FY 1983, revenues from
the sale of Landsat products amounted to about $7 million, less than a third
of Landsat operating costs, estimated at $22 million.[31]

In February, 1984, the GAO released a series of reports on the potential
impact of commercializing Landsat based on interviews with a range of
user groups. Although a few of those interviewed raised the possibility for

improvements in service or cost savings for the government, GAO reported that "the majority of users could not see any benefit to selling Landsat." On the contrary, there was widespread concern about the impacts of creating a commercial monopoly for satellite data. Many federal users believed that the types of price increases necessary to cover costs would lead them to restrict or end their use of Landsat data. They were also concerned that a private operator may discontinue less popular, specialized products on which some agencies depended, cut back on archiving of data, or decrease investment in research and development, as these activities were unlikely to be profitable.

International purchasers of Landsat data were concerned about the potential for a private company to end transmission to foreign stations or to increase annual fees beyond what they could afford. They also expressed concerns that exclusive data purchases may allow a company to gain information about a country's natural resources not available to the nation itself, putting the country at an economic disadvantage and, in the opinion of some states, impinging on its national sovereignty. Combined with the development of more sophisticated sensors, private data sales would also pose national security concerns. Value-added data companies expressed worries that a commercial Landsat operator would develop its own data products and use its monopoly power to limit distribution to existing value-added companies in that sector.[32]

A report by the Congressional Office of Technology Assessment (OTA) also released in 1984 cautioned the government to carefully consider economic, foreign policy, public interest, and national security issues related to privatization. It warned that until the commercial market for data sales expanded considerably and more efficient spacecraft were developed, it could cost the government as much to subsidize a private owner as to operate the system itself. The OTA report argued that nondiscriminatory international sharing of Landsat data was a powerful demonstration of the US commitment to open exchange of ideas and information. It noted that the United States Agency for International Development (USAID), the World Bank, and other organizations were using Landsat data in international aid programs, which could be damaged by high data prices. It called attention to security issues, as well, stating that the intelligence and military agencies together made up the largest federal consumer of Landsat data, noting that coordination with civil developments, technical oversight of the program, and the potential for preemption by the military in emergency situations would need to be addressed as part of any commercial transition.[33]

Commercialization of Landsat

Despite these issues, Congress passed the Land Remote Sensing Commercialization Act of 1984 in July of that year to direct phased commercialization of land remote sensing. It acknowledged that there was doubt that the private sector alone could develop a total land remote sensing system, but argued that competitive, market-driven private-sector involvement was in the national interest and that it was the appropriate time to begin cooperation between the government and private industry in this area. Congress provided assurances that data would remain available to federal government agencies, that the United States would remain committed to nondiscriminatory access to data, and that the United States would honor existing international agreements.[34]

Under the Act, the government would contract a private-sector entity to market the minimally processed Landsat data, including both newly acquired and archived data. The chosen company would set prices and engage in data sales. It would pay the United States a fee or percentage of revenues, under an agreement to be negotiated with the Secretary of Commerce, and it would pay the government the full market price of any data it retained itself for use in developing value-added products. The government would also solicit proposals for private development of a follow-on Landsat system to ensure data continuity for at least six years.[35] In 1983, the solar panels on Landsat 4 failed prematurely, leaving it only partially functional. Landsat 5 was launched in March 1984, with a design life of three years. This meant that to provide continuity, the new commercial system would need to be completed by 1987.

The proposals were to include explanation of any need for Federal funding to help pay for development of the follow-on Landsat system and outline plans to provide the government with a percentage of sales receipts or other returns for this investment. The act provided for government loans or loan guarantees to assist in covering the capital costs of the system. The company would be in charge of data sales, including sales to the US government, but would not receive any advance guarantee of government data purchases. The goal was full transition to private financing, ownership, and operation of the Landsat system.[36]

The Act stated that in addition to any data collected and maintained by the commercial operator, the United States government would continue to maintain its own archive of land remote sensing data for historical, scientific, and technical purposes, particularly long-term global environmental monitoring. The archive would include existing Landsat data and could add data purchased from the privately run Landsat program or obtained from foreign ground stations and foreign remote sensing space systems. The private Landsat operator would retain exclusive rights to sell data provided to the government for its archive for up to 10 years, after which time the data would be considered in the public domain and be provided at prices reflecting only the cost of reproduction and transmittal.[37]

In September, 1985, NOAA selected the newly formed Earth Observing Satellite Company (EOSAT) to manage the Landsat program. Under the agreement, EOSAT would develop and operate two satellites and an associated ground system. The US government would provide $250 million toward these development costs and provide the launch. The government would also continue to pay the costs of operating Landsat 4 and 5 for the duration of their lives. EOSAT had responsibility for marketing data products from existing and future Landsat systems in the United States and overseas, and retained revenues from those sales. It also received the annual fees paid by the existing Landsat foreign ground stations.[38] In preparation for the transition, NOAA raised Landsat data prices to $4,400 a scene.[39]

UN Remote Sensing Principles

International debates about legal issues of satellite remote sensing had been ongoing since the early 1970s and had gained even greater importance with the onset of commercial remote sensing activities. Some nations worried that with the availability of commercial imagery, companies or other nations could gain knowledge of their nation's natural resources, putting them at an economic disadvantage in negotiations or creating national security risks. They suggested that nations should be required to gain permission from a state before it was sensed, or at least get consent before distributing imagery to third parties. Others suggested that rather than restricting data collection or dissemination, nations should be provided access to data of their own state or region at no cost.[40] The United States and others argued that any attempt to gain prior consent before collecting or disseminating data would be impractical and could greatly diminish the ability to use satellites to address global environmental challenges such as desertification or flooding. They also rejected the idea of guaranteeing free access to data, as this could undermine the ability of the emerging commercial remote sensing sector to generate revenue.[41]

In the end, the UN Principles Related to Remote Sensing of the Earth from Outer Space, adopted in 1986, reflected a compromise. Permission would not be required for data collection or dissemination, and states would not be guaranteed free access to data of their own nation. However, the sensed state was to have access to the data of its territory on a nondiscriminatory basis and on reasonable cost terms. The principles also stated that remote sensing activities should be carried out for the benefit and in the interests of all countries, and should not be conducted in a way that would be detrimental to the rights and interests of the sensed state. Just as in the Outer Space Treaty, nations were responsible for ensuring that both government and nongovernment entities abided by the principles, ensuring that commercial remote sensing entities would be bound by the same requirements as states.[42]

Competition and Other Challenges

Over the next few years, Landsat began to face foreign competition. France launched the first System Probatoir d'Observation de la Terre (SPOT) satellite in 1986, with the SPOT Image Corporation in charge of commercial distribution of its imagery. Unlike the Landsat program, which had begun as an experimental research effort, SPOT had been designed from the beginning with commercial users in mind. It provided 10-meter-resolution data, the best on the market. Russia began marketing limited amounts of imagery from its reconnaissance satellites in 1987, and the Indian Remote Sensing (IRS) satellite was launched in 1988 as a direct competitor to Landsat. In addition to sales of individual images, more than 10 nations had purchased and installed ground stations capable of receiving data of their region directly from the Landsat, SPOT, or IRS satellites as they passed overhead.[43]

Despite the competition, EOSAT managed to increase revenues from 1986 to 1989, from $19 million to $25 million.[44] Increases were largely due to higher prices and more targeted marketing. The number of orders actually dropped from 35,000 in 1984 to less than 8,000 in 1990.[45] Many government and research users struggled to secure funding and were ultimately priced out of the system.[46] Academic users were particularly hard hit, going from tens of thousands of scenes accessed in the 1970s to only 450 purchased in 1990.[47] One researcher expressed his frustration with the government subsidized data sales, stating, "We have the worst of all possible worlds: we are both spending the money and making sure that we get nothing out of it."[48] There were problems on the government side of the partnership, as well, with Congress reconsidering plans to provide $250 million in subsidies. EOSAT received only $27.5 million in 1986 and $5 million in 1987. Funds for developing the Landsat archive mandated in the commercial remote sensing bill were also lacking.[49]

In 1988, a new contract stipulated that the US government would provide a one-time payment of $220 million for hardware development, and EOSAT would only be responsible for the development of one satellite, Landsat 6. Facing tight budgets, and knowing that Landsat 4 and 5 were both beyond their expected life-span, NOAA only requested funds to operate the satellites through March, 1989. When the satellites continued to operate beyond expectations, NOAA announced that it would have to shut them down and terminate operations due to lack of funds.[50] The system was saved at the eleventh hour by presidential intervention, with a commitment to continue operating the satellites while the administration conducted a review of the program by the National Space Council, chaired by the Vice President.[51]

The results of that review were announced two and a half months later. The Council recommended that funds be provided to continue operating Landsat 4 and 5 and to complete development of Landsat 6. In explaining their reasoning, the administration stated that Landsat provided imagery important for global change research, environmental monitoring, national security, and a variety of other uses. It also demonstrated US commitment to "the use of space for the common good." Echoing the results of a review completed almost a decade earlier, the council concluded "commercializing the entire Landsat program would not be feasible until at least the end of the century."[52]

The potential contributions of the Landsat system to understanding global climate change were especially timely. The issue of global warming had been identified as a presidential priority by President Bush when he entered office in 1989. A 1990 report by USGS highlighted the potential of Landsat to contribute to this issue and to the work of the US Global Change Research Program (USGCRP). The report concluded that the 18-year continuous global dataset provided by Landsat was uniquely valuable for studying global change. It also noted that high prices driven by commercialization of the system were impeding the use of large quantities of data needed for global change research. The report recommended finding ways to provide data to users reflecting only the costs of reproduction and dissemination, without use restrictions. It also highlighted the importance of ensuring data continuity and adequately maintaining the Landsat data archive.[53]

After the USGCRP released the Bromley principles, stating that data should be available on a full and open basis, at no more than the marginal cost of fulfilling a user request, the National Research Council published a report noting that despite Landsat's value to global change research, utilization of the data by researchers was seriously compromised by their high cost. The report argued that the US government should procure domestic and foreign Landsat data for use by researchers, through data purchases from EOSAT and foreign operators, if necessary.[54]

A 1992 OTA report on the Landsat program found that the prices charged for imagery were a key factor in determining how widely the data were applied by public and private sectors. It argued that low-cost access was critical for research use, and suggested that in the long run focusing on promoting privatization in the value-added sector may be more beneficial for the United States than privatization of raw data sales.[55]

Military utility of the system was also demonstrated during the Persian Gulf War when the US military used Landsat and SPOT data to develop maps of the region that could be shared with the coalition forces to support operations.[56] Any concerns about the national security risks of sharing or selling Landsat data had been overcome by the availability of higher-resolution imagery from commercial satellite systems operated by France, Russia, and India, over which the United States had much less control.[57]

In 1991, the president released a National Space Policy Directive providing guidelines for US commercial space activities. The policy reiterated the US commitment to encouraging commercial space activities, consistent with national security and foreign policy interests. It stated that US government agencies should utilize commercially available space products and services to the fullest extent feasible. As a rule, the government would pursue these objectives without the use of direct federal subsidies; however, anchor tenancy and guarantees of future purchases could be considered. The policy also provided a definition of commercial that included instances in which (1) private capital is at risk, (2) there are existing or potential nongovernmental customers, (3) the commercial market ultimately determines the validity of the activity, and (4) primary responsibility and management resides with the private sector.[58] Landsat did not meet these criteria, but there were hopes that future commercial efforts would be more successful.

President Bush addressed Landsat directly in another directive, released in 1992. The policy stated that continuation of Landsat was in the civil and national interests of the United States, and committed to providing continuity for Landsat-type data. Government development and launch of a seventh Landsat system would ensure that data was available for national security, global change research, and other federal users. The government would also seek to minimize cost of the data consistent with the Bromley principles while also not precluding private-sector commercial opportunities.[59]

Landsat Returned to Government Ownership

The Land Remote Sensing Policy Act of 1992 built on the presidential directive and addressed many of the issues raised by user communities. It acknowledged that satellite remote sensing was useful for studying human impacts on the global environment, managing natural resources, and carrying out national security functions. It also recognized that the cost of Landsat data had impeded its use for scientific purposes, particularly global environmental change research. Congress concluded that full commercialization of Landsat was not possible in the foreseeable future, but did not give up on commercialization completely, stating that commercialization of land remote sensing should remain a long-term goal of US policy.[60]

The Act directly addressed the data policy for the Landsat system, putting in place a tiered policy designed to meet the needs of researchers while also protecting EOSAT's right to market commercial data and services. It mandated a change in EOSAT's contract that would require the company to provide Landsat 4, 5, and 6 data to US government agencies, global environmental change researchers, researchers funded by the US government, and selected educational users for noncommercial use at no more than the cost of fulfilling user requests. These users were permitted to redistribute the data to others within these approved user groups, as long as the data was used for noncommercial purposes. Nonprofit and public interest groups would have access to the data via vouchers or data grants, also on the condition of noncommercial use. Commercial data sales remained the responsibility of EOSAT, and the company continued to retain revenues from these sales.[61]

The Landsat 7 system, which would be fully government owned and operated, was to follow an open data policy. Minimally processed data would be made available to all users, including civilian, national security, commercial, and foreign users, at the cost of fulfilling user requests. The value-added product and services sector would remain exclusively the function of the private sector. Private operation of ground stations, subject to a fee, could also be considered.[62] In addition to the policy changes, technology was also being used to help make access to data easier. In 1993, USGS debuted its Global Land Information System (GLIS), an online interface which allowed users to view inventories of data, directories with information about each dataset, and user guides with detailed information. Due to a lack of high-speed, high-capacity transmission lines, users could not access data directly through the system, but they could place order requests online to receive imagery on magnetic tapes or other media.[63]

Transitioning to an Open Data Policy

With plans for the future of the Landsat program laid out, there was great anticipation for the launch of Landsat 6, scheduled for October 1993. Landsat 5 had already been functioning three times longer than its original design life, so Landsat 6 was key to allaying concerns about a lack of data continuity. EOSAT also expected the launch of the new satellite, which would provide better coverage and higher-quality products, to result in a dramatic jump in sales.[64] The failure of the satellite to reach orbit, due to a problem with onboard engines, was a devastating blow to company officials and the user communities alike.[65]

Because of Landsat's importance for national security and global environmental change, the Land Remote Sensing Policy Act of 1992 had directed it to be run as a joint program of the DoD and NASA. However, almost immediately a conflict arose about the requirements of the program and the division of funding, and DoD withdrew from the program at the end of 1993, leaving NASA to manage the Landsat 7 program on its own. In May 1994, President Clinton directed that the program be transferred again, this time to a triagency coordinated structure. Under the new arrangement, NOAA would manage and operate the satellite, NASA would procure the satellite, and USGS would process, archive, and distribute data.[66]

In accordance with the Land Remote Sensing Policy Act of 1992, EOSAT's contract was renegotiated and extended. Under the new terms, EOSAT would pay the operational costs of Landsat 4 and 5, but as of 1996, would be required to sell data to US government agencies and other affiliated users, such as researchers working with the government on global change studies and education users, at prices below market cost. For example, each scene from the Landsat 4 and 5 would be $2,500 for the government and affiliated users and $4,400 for others. Images already residing in the EROS archive would be available for $425 per scene. As directed in the Act, once purchased, data could be freely copied and distributed among the government and affiliated users for noncommercial purposes.[67] In 1997, USGS announced the planned prices for Landsat 7, which would be launched in 1999. Raw data for each scene would cost $475, while minimally processed (Level 1) scenes would be $600 each. Value-added products would be produced by commercial vendors.[68] While these prices were significantly lower than prices under EOSAT, they reflected an interpretation of marginal cost that included some personnel and capital costs, rather than simply the costs of copying and postage. Foreign ground station fees remained at $250,000 per year.[69]

Although efforts to commercialize Landsat had failed, regulations put in place to allow licensing of commercial remote sensing were starting to see some success. By 1999, three companies—Space Imaging, Earth Watch (previously known as WorldView), and Orbital Image—had requested and received licenses to operate commercial remote sensing satellite systems. This was largely seen as a positive development, but it created tension with respect to government plans for Landsat. The companies were beginning to market data from their own soon-to-launch land remote sensing systems, and they were concerned about competition from Landsat 7. In June 1999, they sent a letter to USGS voicing their concerns. According to the companies, the prices set for Landsat 7 were less than half the price of any comparable commercially available data, and much lower than would be necessary for commercial firms to recover costs from their systems. They recognized that their systems would collect data with much higher resolution than the 30-meter Landsat 7 imagery, but still felt that Landsat 7 pricing was "creating unrealistic expectations in the international marketplace." They requested that the government reinstate a tiered policy in which data sales and international ground station fees return to commercial rates.[70]

This request was flatly rejected. Dr. Neal Lane, director of the White House Office of Science and Technology Policy responded that the government felt a tiered policy would not serve the best interests of the research or commercial sectors and that there were no plans to pursue such a policy change. He explained, "Obtaining the widest possible dissemination of Landsat data and the greatest possible participation from foreign ground stations will increase the scientific return on the US taxpayer investment in Landsat." He added that the White House was a strong supporter of the commercial value-added industry, and that provision of Landsat data to all users on the same low-cost terms was the best way to support these research and commercial objectives.[71]

Commercialization Explored and Open Data Advocated

Through these transitions, Congress continued to promote commercialization in the satellite remote sensing sector. In July 1999, just three months after the successful launch of Landsat 7, NASA and USGS issued a Request for Information (RFI) to gauge private-sector interest in providing the next Landsat system. The RFI noted that the data would support the USGCRP as well as international scientific efforts, and "thus, these data must be shared with the international community and readily exchanged from user to user without undue restrictions."[72] A letter from the USGS regarding the RFI recognized that Congress had recommended giving a fully commercial approach first priority, but expressed hope that there would also be careful consideration of alternative arrangements, such as a public-private partnership, a government-built and -operated system, or development by an international consortium.[73]

Ownership of the Landsat program changed hands again in October, 2000. NOAA, which had experienced budgetary challenges and didn't have a mission directly related to land remote sensing, would no longer manage the Landsat system, and USGS would take over this role. The presidential directive announcing the change stated that one of USGS's responsibilities was to seek to offset costs of US operations through arrangements with foreign ground stations and/or reimbursements for fulfilling user requests.[74] In July, 2001, EOSAT determined it was no longer economical to operate Landsat 4 and 5, and these satellites were transitioned back to government control, allowing data from these satellites to be distributed according to USGS pricing policies.[75]

Shortly after the Landsat program was transitioned to its new home agency, some USGS officials advocated removal of fees and transition to free and unrestricted data sharing. These officials argued that making data freely available would increase data use and result in greater benefits for society. They also argued that since the systems had been developed at taxpayer expense, all taxpayers should have equal access. Changes in technology were important, as well. It was now possible to provide data to users online, and advocates of free provision argued that if this transition were made, the cost of fulfilling user requests would essentially be zero.[76] Others in the agency argued against this change. Even with data prices well below commercial levels, USGS was able to bring in nearly $10 million in annual revenue from the sale of Landsat 7 products in FY 2001. Although this was only about 1 percent of USGS's annual budget, it was sufficient to support half of the annual costs of operating the satellite, and it was seen by Congress as evidence of the success of the program.[77]

One of the risks of cost recovery models, the dependence of the program on market forces, was demonstrated in May 2003, when Landsat 7 experienced a failure of one of its imaging systems. While useful information could still be collected, about 25 percent of the data was missing from each scene. USGS reduced the prices for scenes with data gaps from $600 to $250, but even with the discount, revenue from data sales decreased significantly. US government agencies, including USDA and DoD, cut their data purchases by about 40 percent.[78] Estimated sales for FY 2004 dropped under

$5 million, and emergency reprogramming of funds was necessary to keep the system operating.[79]

Concurrent with these challenges, USGS was engaged in the NOAA-led efforts to create the Global Earth Observation System of Systems, which promoted international data sharing. Secretary of the Interior Gale Norton was a featured speaker at the first Earth Observation Summit held in Washington, DC, in 2003, and highlighted the importance of Landsat to the global community.[80]

Further impetus for a free and open data sharing policy came from a 2004 announcement by the Brazilian space agency that it would make data from its China-Brazil Earth Resources Satellite (CBERS) freely available to Brazilian citizens via the Internet.[81] Proponents of a free and open policy argued that if Brazil could make its land remote sensing data freely available online, surely the United States was capable of doing the same.[82]

Of course, there was still the question of commercial involvement in the Landsat 8 project. The 2001 RFI aimed at commercial development and operation did not generate significant interest from industry, which argued that the market was insufficient to support private-sector development. The government also investigated the option of a public-private partnership, but failed to reach a satisfactory agreement with potential industry partners.[83] In 2004, the government briefly considered flying Landsat sensors on the joint NASA-NOAA-DOD National Polar-orbiting Operational Environmental Satellite System (NPOESS), until a review of this option showed that implementing it would be exceedingly complex.[84] (A dodged bullet for the Landsat system, as even without the added complexity of including Landsat sensors, NPOESS experienced years of delays and cost overruns before eventually being canceled altogether in 2010.[85])

Finally, in December, 2005, the White House announced that the Landsat 8 satellite would be procured by NASA and then turned over to USGS, following essentially the same model as the fully government-owned Landsat 7 system.[86] A 2007 follow-up report by an interagency working group confirmed that there was no viable commercial market for Landsat-type imagery in the United States, but stated that the societal benefits for land management, climate change monitoring, national security, and other uses justified continued US government investment in Landsat data collection.[87] The decision to develop Landsat 8 as a fully government funded mission helped to ensure continued support of its below-market-cost data policy.

The Landsat 8 decision was too little, too late for some customers, however, including some within the US government. The degradation of Landsat 7 data, combined with uncertainty regarding continuity of the Landsat
system and relatively high data costs, led the USDA to investigate other
sources of medium-resolution remote sensing data. The agency determined
that India's ResourceSat-1 could provide comparable data to Landsat at a
cost lower than the fees charged by USGS. In 2006, the agency ended its
use of Landsat imagery for operational applications. In doing so, USDA
noted that it was a "price-sensitive purchaser of satellite imagery" and concluded that Landsat was "not the best value for USDA."[88] USDA became
the largest commercial purchaser of ResourceSat-1 wide field sensor data
in 2006 and 2007. This put the United States in a position in which it
paid the full cost of Landsat development and operation, and also funded
data purchases for similar, redundant data from India, essentially subsidiz-
ing a foreign effort rather than its own. This also meant that the effort to
recover costs for USGS—by requiring other US government agencies to pay
data access fees—had resulted in higher costs for the US government as a
whole.[89]

Free and Open Landsat Data

This situation provided more fuel for those pushing for free data provision.
In June, 2007, in preparation for Landsat 8, advocates received support for
a pilot program making selected Landsat 7 imagery freely available online.
New Landsat 7 imagery would be added to this set of freely available data
as it was collected, and USGS planned to include older Landsat 7 data, as
well.[90] Six months later, in January, 2008, USGS released a new Landsat Data
Distribution Policy, stating that all Landsat data products accessed online
would be available at no cost. USGS would provide unrestricted access to
all users and promote open and unrestricted exchange and redistribution of
Landsat data products.[91] Online access would eventually include more than
two million images going back to 1972.[92]

Looking to emphasize both the importance of international data sharing as well as to promote its own activities as widely as possible, USGS
announced the new policy at the 2007 Group on Earth Observations Plenary, where it was hailed as a breakthrough.[93] The Secretary of the Interior, Dirk Kempthorne, discussed the new policy at a popular international

Geospatial Information Systems (GIS) user's conference, ensuring those with the tools and interest to make use of the data were aware of its availability.[94] The new policy was also highlighted in the US report on its support of the Global Climate Observing System (GCOS), providing further global awareness.[95] When President Obama announced the Open Government Directive (OGD) in 2009, calling on agencies to register open datasets on data.gov, USGS was quickly able to register tens of thousands of free and open images.[96]

After the policy was announced, data use exploded. Prior to the transition to a free data policy, the greatest annual distribution was in FY 2001, with 25,000 scenes sold at $600 each. In the year following the transition to free and open online data, USGS distributed more than a million scenes. In 2010, 2.5 million scenes were distributed. By 2012, USGS was distributing more than 250,000 images *a month*. There was a corresponding increase in the amount of research and number of applications developed, and, interestingly, a change in the nature of the data requested, as well. In particular, there was an increase in requests for images of the same area at many different times, allowing researchers to analyze global change in fine-grain temporal steps that would have been cost prohibitive in the past. Analysis over large areas, and the study of changes in forested environments, in particular, increased dramatically.[97] It became feasible to merge Landsat data with other data to create new, useful products, including carefully processed, long-term datasets needed for climate studies.[98]

A 2012 analysis by the National Geospatial Advisory Committee provided some insight into how this increased use translated to financial benefits. Analyzing just 10 major uses of Landsat data, including US government mapping, flood mitigation mapping, national agricultural commodities mapping (the USDA had returned to use of Landsat data after the announcement that data would be freely available), and others, the committee found that the total annual savings from those applications alone was between $178 million and $235 million.[99] A Booz Allen Hamilton report released the same year estimated the total value of the Landsat system at more than 1.7 billion annually.[100]

A 2013 National Research Council review of the Landsat program praised the open data policy, stating that the change had had "enormous benefits to science and to operational users" and resulted in the creation of value-added products that benefit society at large. The analysis found that "the economic and scientific benefits to the United States of Landsat imagery far exceed the investment in the system."[101] In 2014, the National Plan for Civil Earth Observations, prepared by the White House Office of Science and Technology, ranked the Landsat series third out of 145 high-impact observation systems (behind GPS and the weather radar system), and committed to implementing a 25-year program of sustained land imaging.[102]

Success Breeds Success

The increase in users and discovery of new uses created an appetite for more Landsat data, and the dramatic success of the policy, carefully documented and promoted by USGS, led to the additional budgetary support needed to fulfill these desires. Data from Landsat 1, 2, and 3, previously languishing on aging, incompatible hardware, was rescued from storage and processed. More than 100,000 new scenes were added to the archive through this process.[103] The Landsat Global Archive Consolidation (LGAC) program, originally conceived in 2006 as a highly desirable, but impractical, plan to consolidate data holdings from all international Landsat ground stations, suddenly found support after the passage of the new data policy in 2008. The impact of LGAC had the potential to be huge. Together, the 50+ international Landsat ground stations held more than twice as much data as the entire United States Landsat archive.[104]

Even with political support, implementation of LGAC was not easy. Many of the international ground stations had only operated for a limited period of time, and identifying contacts and determining the extent of holdings was a challenge in itself. Further, there was concern that international ground station operators would not be interested in providing data to USGS for inclusion in a free and open archive, particularly because they had paid an annual fee for the right to collect the data and many of the nations relied on data sales for income. That concern turned out to be unwarranted, as ground station operators saw USGS's offer to transfer data to modern storage devices and process it into standard products as beneficial for all involved. Eventually all of the international ground station operators, including those associated with stations that were no longer active, agreed to share data as part of the LGAC program. By 2015, nearly 2.2 million international images had been added to the Landsat archive, more than doubling its total holdings. Another 2.3 million are scheduled for inclusion as the LGAC effort continues.[105]

Landsat 8 and the Challenges of Big Data

Landsat 8 launched to much fanfare in February, 2013. Just days after it transitioned to operational status, Landsat 5, which had operated for an astonishing (and Guinness Book of World Records–holding) 29 years, transmitted its last image.[106] Landsat 8 carried new, advanced instruments and upgraded storage and transmission capabilities allowing it to rapidly accelerate the amount of data added to the archive. More data was collected by the Landsat 8 Operational Land imager (OLI) instrument in its first two years of operation than was collected from the primary imaging instruments of Landsat 4 and 5 over the course of 32 years.[107] By 2016, the Landsat archive was growing at more than a terabyte a day.[108]

The dramatic increase in data volume was a double-edged sword. As was the case with NOAA, the USGS servers could not handle distribution of the multiple petabytes of data held in its archive, the bulk of which were maintained offline.[109] In 2010, Google's Earth Engine, a cloud computing system designed for analysis of Earth science satellite data, began hosting historic Landsat data (from 1984 to 2012). Amazon Web Services (AWS), another cloud provider, in 2013 agreed to host global Landsat imagery from three distinct periods: the 1970s, 1990, 2000, and 2005. In March 2015, AWS announced it would make Landsat 8 data available through its simple storage service. All 2015 data was included as of July, 2016, with selected 2013 and 2014 scenes also available.[110] With a strong commitment to open data provision, concrete demonstrations of its value, and engagement with these new technological capabilities, free data distribution seems likely to continue.

Summary

The theory presented in chapter 3 argues that the mission and culture of the government agency generating the data are central to the development of a data sharing policy. However, as an operational program, Landsat did not fit with NASA's mission, and as a land-focused program, it didn't belong at NOAA—but as a satellite, there was no obvious alternative to these two options. So, over the 45-year history of the Landsat project, the system rarely had a stable home. At times managed by NASA, NOAA, USGS, DOD, and for almost a decade, by a private firm, it is no surprise that Landsat's data policy experienced a great deal of variation over time.

In its early days as an experimental craft developed and managed by NASA, Landsat data was shared freely or at the cost of reproduction and distribution. This aligned with the goals of national-level government actors, as they hoped to operate the satellite in a way that would allay national security concerns and maintain international support for an open skies policy. But once these concepts were well established, from their perspective, there was no incentive to continue free and open sharing. Unlike the case with weather satellites, there was no history of sharing land remote sensing data, no international organization expecting such a contribution, and only a very small existing user base advocating for continued data availability. So when national-level policy-makers in the administration and Congress expressed interested in privatization of the system, there were no major objections from the managing agencies, and warnings from analysts and experts about the largely public, nonappropriable value of Landsat applications went unheeded.

Given this situation, recognition that Landsat commercialization was not feasible could only come through trial and error. Even then, the failure could have been interpreted as proof that the satellite data was not worth the cost. The fact that the imminent failure of the only system in the world with a two-decade history of collecting global environmental data coincided with the rise in awareness of global climate change was a lucky coincidence that likely saved Landsat from cancelation. Still, without a natural home or a strong agency advocate, Landsat data policy continued to be set by Congress, which saw revenue generation, rather than science applications, as a key sign of program success. The program adopted a "marginal cost" pricing policy with relatively high fees that would allow it to recover costs for the data center.

It was not until Landsat found a permanent home at USGS, an environmental science agency, that officials began to consider a more open data sharing policy for Landsat. With past experience as proof that the system was not commercially feasible, and a lack of private-sector interest in commercialization or public-private partnerships, arguments against data sales were made easier. The promise of important scientific and value-added industry applications, the advancement of computing and Internet technology, and international trends toward open data sharing—particularly the example set by Brazil—helped to push USGS to abandon a few million dollars in annual data sales revenues and adopt a free and open data policy.

The predictions of a rapidly growing user base under this new policy were proven correct. Recognizing that funding for any program, but particularly Landsat, is never assured, USGS has continued to promote and study the policy and its impacts, even commissioning reports to place a dollar amount on the impacts of its data sharing efforts. With a large and growing user base, seemingly endless scientific studies, extensive use by US government agencies, and a growing value-added sector, any efforts to transition back to a commercial model for Landsat would be likely to face significant resistance.

9   US National Aeronautics and Space Administration

The National Aeronautics and Space Administration (NASA) was formed as a science and technology agency focused on engaging in peaceful activities in outer space. From its beginning, NASA favored broad sharing of its scientific data both domestically and with the international community. As NASA began to grow its Earth science activities in the 1980s, it continued this commitment to open data sharing and resisted pressure to engage in commercial activities that would have required data access restrictions. NASA actively advocated to remove even marginal cost fees on access to its data and developed a vast data distribution system that allowed all users to access data for free online. Throughout its history, NASA has been a leader in open data efforts on both the national and international stage.

Creation of NASA and Early Satellites

The United States' first civil satellite was developed in support of the International Geophysical Year (IGY), an effort initiated by the International Council of Scientific Unions that brought together 67 nations from around the world to coordinate Earth science research and data collection.[1] Explorer 1 launched in January 1958, carrying a few simple scientific instruments. Building on this foundation, the National Aeronautics and Space Administration (NASA) was formed in July 1958 to demonstrate a clear separation between civil and military space activities in the United States and highlight the peaceful, cooperative nature of US space activities. The National Aeronautics and Space Act of 1958, which created NASA, tasked it with examining problems that involved utilizing space activities for peaceful and scientific purposes. NASA was encouraged to engage in international cooperation, and the NASA administrator was directed to "provide the widest possible and appropriate dissemination of information concerning its activities and the results thereof."[2]

From the beginning, NASA's Earth observation program focused on practical benefits. Meteorological satellites were seen as having great potential, not only technically for monitoring and predicting the weather, but also politically, in promoting the values of an open society and gaining international good will through the free sharing of data.[3] For nearly 15 years, NASA's Earth observation program was focused primarily on the development of experimental weather satellites in cooperation with the Weather Bureau.[4] In 1972, NASA began development of the first civil land remote sensing satellites, again in cooperation with agencies that hoped to use them operationally. In the case of both weather satellites and the Landsat series, while the programs were under NASA's control, data was shared freely and international data access and use was actively encouraged.[5]

NASA as an Environmental Agency

The same broad trend of environmental awareness that led to the Landsat program in the early 1970s also helped to build support for other Earth science activities in NASA. The "blue marble" image taken by the final Apollo crew symbolized the transition: with the Apollo program coming to an end, NASA's space application programs, especially its Earth observation program, would have room to grow. In 1973, NASA Administrator James Fletcher testified to Congress that NASA could be called an environmental agency, explaining that virtually everything the agency did helped to understand and improve the environment. He suggested that it may be NASA's "essential task… to study and understand the Earth and its environment."[6]

Due to its close cooperation with NOAA in the development of weather satellites, as well as its own interest in Earth science studies, NASA was actively involved in the Global Weather Experiment (GWE). While the GWE was largely seen as a success with respect to weather, some argued that it had done little to advance its other objective—studying the physical basis of climate.[7] A 1975 report by the US Committee for the Global Atmospheric Research Program stated that there was ample evidence that climate does change, and that these changes can affect, and be affected by, human activities. "We have an urgent need for better information on global climate," they concluded.[8] Responding to these concerns, NASA examined the potential of its experimental satellites to contribute to key climate parameters and identified potential systems for data management. NASA concluded that processing, archiving, and distribution of climate data would require the most complex data management system ever developed by the agency.[9]

In 1978, NASA launched the first two dedicated Earth science research missions: the Heat Capacity Mapping Mission (HCMM) and Seasat. Together with Landsat, which collected visible and near infrared imagery, these three missions would take observations across the electromagnetic spectrum. HCMM measured infrared radiation emitted from the ground to examine thermal conditions at the Earth's surface. Seasat took measurements in the microwave portion of the spectrum useful for studying oceanographic phenomena.[10]

Among its five instruments, Seasat carried the first civil synthetic aperture radar (SAR). The instrument was capable of detecting three-dimensional phenomena, useful for monitoring the ocean waves and polar sea ice conditions. It could collect imagery during the day, at night, and even through clouds. The Seasat SAR instrument produced so much data that it could not be recorded on the spacecraft and was only collected and transmitted when in range of a ground station. Although the satellite failed after only three months in orbit due to a massive short circuit, it was widely praised as a highly successful and influential demonstration of the value of satellite monitoring for oceanography. Data from the mission was originally evaluated at a NASA Jet Propulsion Laboratory workshop in January 1979 and later shared with the full user community through NOAA's Environmental Satellite Data and Information Center.[11] HCMM data was available at the cost of dissemination from the NASA Goddard National Space Science Data Center (NSSDC), part of the World Data Center system.[12]

Mission to Planet Earth

The first World Climate Conference, held in 1979, reinforced the importance of understanding and addressing climate change and called for increased use of in situ and satellite-based observations for doing so.[13] NASA took the initiative in this area in 1982, proposing an international cooperative project to use space technology to address global environmental changes.[14]

NASA began to build a coalition to support the initiative within the US government, creating the Earth System Sciences Committee in 1983.[15] In 1985, the NASA Associate Administrator for Space Science and Applications published an editorial in *Science* calling for a new "Mission to Planet Earth." The new program would study the Earth as an interconnected system, employing the types of methods NASA used to study other planets in the solar system, investigating long-term physical, chemical, and biological changes on a global scale. This ambitious project would involve many organizations and countries.[16]

NASA's ability to contribute to understanding and addressing global challenges was demonstrated in 1985, when NASA scientists used observations from the Nimbus meteorological satellites to confirm the existence of an ozone hole over Antarctica. Images of the phenomenon generated by NASA became the visible symbol of the ozone problem for policy-makers and the general public.[17] When the NASA Earth System Sciences Committee came out with a report the following year outlining plans for the NASA-led Mission to Planet Earth (MTPE) program to study global change, policy-makers and the public could easily envision the agency as an important actor in this area.

The NASA report called attention to the agency's existing efforts, pointing out that it had recently launched new missions focused on measuring Earth's radiation budget and studying geodynamics, and was developing a new oceanography satellite for launch later that decade. The proposed program would continue this trend with a steady stream of specialized research missions through the year 2000. The centerpiece of the plan, the Earth Observing System (EOS), would take form in the 1990s, coordinated with the launch of the space station. NASA, in collaboration with NOAA and other agencies, would place a complex suite of research and operational Earth observation instruments on the station. The report emphasized as a key component the development of an advanced information system to allow wide distribution of data for scientific analysis.[18]

In the following years, a number of reports and reviews supported this plan, emphasizing the ability of NASA to contribute to understanding environmental issues. MTPE was listed first among four NASA leadership initiatives in a 1987 report by Sally Ride. "Championing this initiative," she said, "would establish the United States at the forefront of a world-recognized need to understand our changing planet."[19] A 1988 report by the National Research Council endorsed MTPE and reiterated the central role of data management and dissemination for ensuring optimum science return.[20] NASA's Mission to Planet Earth was designated as the largest component of the US Global Change Research Program (USGCRP) when the program was created in 1989.[21] A 1990 review of the future of the US space program led by a NASA advisory committee recommended MTPE as one of four key elements of NASA's civil space program and again noted that data management was of critical importance.[22] The mission received internal recognition, too. In 1993, MTPE became a separate NASA program office, on the same organizational level as Space Flight, Exploration, and Space Sciences.[23]

Amid the widespread support, some, particularly in the Office of Management and Budget (OMB) and Congress, were concerned about the program's price tag, projected at $30 billion over 30 years, with $17 billion of that needed by 2000. The size of MTPE's budget was second only to the space station.[24] There was also opposition from within the Earth science community and among some NASA employees. Despite plans for one third of MTPE funding to go to data management systems and another third toward research and analysis of the data, scientists repeatedly expressed concerns that there wasn't sufficient support to actually use all of the data that would be collected.[25] At a 1989 meeting of 500 scientists involved in EOS program planning, one researcher asked whether the program was a commitment to science or to satellite building, and received applause from his colleagues.[26] Some felt that a large number of smaller satellite systems would be more economical and provide better science return than a small number of very large systems.[27]

Congress agreed with this suggestion, directing NASA in 1991 to restructure EOS to fly instruments on multiple smaller platforms and reduce the overall cost from $17 billion to $11 billion. In 1992, the incoming NASA Administrator, Dan Goldin, announced his "faster, better, cheaper" mantra, and EOS underwent an internal rescoping effort that further cut the program budget, bringing it to $8 billion. EOS was decoupled from the space station program and downsized. It would now consist of three flagship missions, later named Terra, Aqua, and Aura, focusing on land, water, and atmospheric measurements, respectively. The program also included support for a series of smaller, Earth probe satellites focused on specific research missions.[28] A 1992 presidential directive reiterated that NASA's

Mission to Planet Earth was central to providing global observations that would determine the ultimate success of the USGCRP.[29] Even at a fraction of its original size, the program would be the most comprehensive system ever undertaken to monitor the Earth as an interconnected ecosystem, and the database to accompany it would be the largest in the world.[30] Nearly a quarter of the program budget, about \$2.6 billion, was earmarked for development of this data system.[31]

Dealing with Data

Since Apollo, NASA had been recognized as a leader in computing, and it was an early adopter of Internet technology, as well. NASA made data available through an online computer information system for the first time in 1987. Though only 2 percent of the total digital data archive was initially made available online, NASA saw a dramatic increase in data requests. Officials attributed this rise to the convenience of the new medium, which allowed immediate ordering and rapid turnaround, and did not require users to pay a fee for the storage medium or postage. A NASA publication explained that the new system essentially allowed the data center to remain open past normal working hours, allowing users to access data, or at least data catalogs, 24 hours a day for the first time ever.[32]

As NASA began to design the EOS Data and Information System (EOS-DIS) to accommodate data from the MTPE program, it built on this existing capability. EOSDIS would consist of a set of Distributed Active Archive data Centers (DAACs) around the country that each processed, archived, and disseminated data. The distributed design allowed NASA to take advantage of existing US capabilities; for example, one of the DAACs was the EROS data center in South Dakota that processed Landsat data. A central user interface would allow users to access all EOSDIS data online, without any need for the user to know the physical location of the dataset. The system would include new data collected by the EOS satellites, but could also accommodate archived data and even nonsatellite data. The goal of EOSDIS was to make a vast wealth of information easily accessible and affordable to a broad range of users within government agencies, academia, foreign countries, and industry.[33] A 1992 report by the Government Accountability Office (GAO) stated, "The intended scope of EOSDIS far exceeds that of any previous civil data management system."[34] Additionally, a 1994 review by the National Research Council warned, "If EOSDIS fails, so will EOS, and so may the US Global Change Research Program."[35]

NASA recognized that the ambitious new data management system needed to be supported by a clear policy on data access. From the beginning, NASA had been interested in providing broad distribution, and in March, 1991, the agency released its Earth science program data policy, which was designed to make data widely available as quickly as possible given the current state of technology and legal requirements.[36] The policy was to apply to all NASA centers and to any other organizations serving as DAACs for the EOS program. It stipulated that satellite data be made available "to all users and other entities for use in Federally funded research, development, and application programs and cooperative research... at a price not to exceed the marginal cost of reproduction and dissemination." Data could be provided at no cost "to a limited degree for purely educational or informational activities in the public interest," in the case of binding agreements with foreign or domestic organizations that included a quid pro quo, or when determined by a government agency to be in the national interest. NOAA would have free access to any data with potential operational applications for real-time processing, applications, and distribution to the operational community.[37] Building on its own policy, NASA took the lead in developing the data exchange principles for the US Global Change Research Program.[38] Commonly known as the Bromley Principles, these principles were publicly released later in 1991, promoting full and open sharing of global datasets at the lowest possible cost across the US government.[39]

The NASA policy also stated that data was to be made available as soon as practicable after acquisition, "without any period of exclusive access for any user group."[40] Until this time, it was common practice to provide principal investigators, the scientists who invest time and effort in designing the system, with a period of exclusive use of the data that would ensure they had the first opportunity to publish interesting results. The new policy was not applied to the first two Mission to Planet Earth satellites, so the 40 scientists affiliated with the Upper Atmosphere Research Satellite (UARS), launched in September, 1991, received exclusive access to data for up to two years.[41] In the case of the oceanographic TOPEX-Poseidon mission, developed in partnership with France and launched in August 1992, principal investigators had 12 months of exclusive access. NASA officials felt that eliminating this practice would allow faster and wider dissemination of the data and help to maximize data use and benefits.[42]

NASA's data sharing policy did include some restrictions on data use, stipulating that users not engage in commercial applications of NASA-provided data without authorization.[43] This restriction was required by the Land Remote Sensing Commercialization Act of 1984. Congress supported continued experimental remote sensing satellite development at NASA. However, to avoid competition with Landsat commercialization efforts, the Act required that data for commercial users only be sold en bloc through a competitive process to be marketed by a US commercial entity.[44]

A GAO report from June 1992 suggests that NASA did not put significant effort into developing the commercial aspect of its policy. The review found that NASA had not formally defined any plans for making data commercially available and had not considered commercial needs in the planning of its information system. NASA officials did not disagree with the finding, instead responding that the highest priority for the program was science and global climate change research, and that NASA did not plan to begin addressing commercial interests until later that decade.[45]

The Land Remote Sensing Policy Act of 1992, which transferred Landsat back to government control, repealed the requirements for NASA to restrict access to commercial users.[46] An updated version of NASA's policy, published in its 1993 EOS Reference Handbook, eliminated any mention of the need for special authorization and negotiated arrangements for commercial users, instead stating, "EOS data and products will be available to all users." This was in line with NASA's stated objective of maximizing data utility for scientific purposes and simplifying access to and analysis of EOS data. However, additional complications had been added due to the desire to coordinate with international partners who had other data policy priorities.[47]

International Compromises

NASA had always intended the Mission to Planet Earth to be an international program. Since 1986, it had been working with its space station partners—Europe, Japan, and Canada—on development of an International Earth Observation System (IEOS). When the original plans to place a coordinated set of instruments on the station fell through, the group turned to coordination of free-flying polar-orbiting environmental satellites. In addition to coordinating data collection plans, agencies agreed to develop a set of IEOS data exchange principles. However, as negotiations went on, it was clear that views on data dissemination varied widely among the partners.[48] NASA's 1993 policy reflected the state of negotiations at the time.

The new proposed policy stated that data would be made available to all users, but the conditions for access would differ depending on the intended use of the data. Four types of users were identified. Research users would be required to submit a brief proposal and sign a research agreement confirming that the data would not be sold and would only be redistributed to other researchers covered by a similar research agreement. Under these conditions, researchers would be granted access to the data required for their project at no more than the marginal cost of filling the request. Public-sector users, including operational agencies like NOAA as well as the larger organizations to which they belong, such as WMO, could gain real-time data access for noncommercial use.

NASA and its IEOS partners would also provide data to those who wished to use the data to conduct limited proof-of-concept studies demonstrating new techniques or testing the feasibility of operational applications. Similar to the researchers, these users would be required to submit an application and, if selected, would be required to sign an agreement confirming the data will only be used for the approved application. The terms for commercial users would vary depending on the sensor or dataset. While the first three categories of users would be able to access data at no more than the cost of fulfilling the request, commercial users could be charged a higher fee. Lacking further detail, the policy stated that procedures for commercial distribution would be in place prior to EOS launch.[49]

The new policy represented a significant increase in the conditions and restrictions related to data access and seemed to signal a decision to attempt some data sales to commercial users. However, this iteration of the policy largely reflected thinking in Europe and Japan, and NASA officials continued to be wary of any restrictions. Lisa Shaffer, the Deputy Director of NASA's Earth Science Modeling, Data, and Information Systems Program Office, wrote a paper that summarized the feelings of many at NASA. She stated that US cost recovery efforts with Landsat had "failed dramatically" to generate enough revenues to fully sustain an operational remote sensing system and argued that the broad availability of affordable, minimally processed data would provide opportunities for development of applications and services. There were other reasons to support open data sharing, as well. Shaffer raised the concern that charging any users, including commercial users, may be considered unfair, as they had already paid for the system with their tax dollars.[50]

She noted that in the United States, the increased importance of understanding global environmental challenges had "highlighted the importance of full and open access to all relevant data by the widest scientific and operational community on an international basis," adding that "no individual nation can afford to develop and maintain the necessary observation and analysis capability, nor can all the scientific progress be confined to one set of researchers." The ability to understand the changes in the global environment depended on open access to data. Rather than improving the cost-to-benefit ratio by trying to recover costs, she argued that governments should focus on increasing the benefits by providing the data in a way that maximizes the public return on investment. Harmonizing international data policy approaches would be easier, she suggested, if data was seen less as an economic resource to be protected, and more like a public good to be shared widely.[51]

By the mid-1990s, the IEOS partners were no closer to agreement on a set of data principles, and the IEOS coordination as a whole was dissolved in 1996.[52] In the 1995 EOS Reference Handbook, NASA had already begun to distinguish its own policies from those of its intransigent international partners. The EOS data policy directly referenced OMB Circular A-130, which had been released in 1993, echoing the document in stating that data would be available to all users without restriction at no more than the cost of dissemination, without regard to intended use. The NASA policy noted that while OMB A-130 "states that the government could charge for the cost of data dissemination," it also "gives individual agencies the right to charge less than the cost of dissemination, if the true cost inhibits the use of the data."[53] There was a growing push within NASA to exercise this right, providing data at no charge to maximize data use.[54]

In the late 1990s, implementation of Mission to Planet Earth began to ramp up. The Total Ozone Mapping Spectrometer (TOMS) and Tropical Rainfall Measuring Mission (TRMM) satellites launched in 1996 and 1997, respectively, and the launch dates of the three flagship missions were approaching. The Mission to Planet Earth office within NASA was renamed the Earth Science Enterprise in 1998, one of four strategic focuses of the agency. The name highlighted NASA's role in pioneering the emerging integrated Earth system science discipline, with an emphasis on global climate change.[55]

The wording and justification provided in the 1999 EOS Reference Handbook demonstrated NASA's growing commitment to maintaining an open data policy. In the introduction to the policy, NASA stated directly that it was now "the intent of the Enterprise to promote open access to its data by the general public, including the academic and industry communities" as a "fundamental feature" of the EOS program. Open access was expected to accelerate the progress of climate change research and result in the availability of more information for policymakers facing critical decisions related to the environment. NASA also believed sharing would promote the development of practical applications and commercial products using NASA data.[56]

NASA provided further justification for providing data at no cost, as allowed by OMB Circular A-130, stating that "cost can be one of the greatest barriers to broad access to data." The report called on the Associate Administrator to consider adopting nominal cost or no cost pricing. Responding to proponents of cost recovery, NASA acknowledged that this may "impair the ability of federal agencies to defray their operational costs and encourage users to request more data than they actually need," but asked that these challenges be weighed carefully against the benefits of discount pricing to the user community.[57]

One new addition to the policy was NASA's stance on purchases of commercial data to meet scientific objectives. The policy stated that NASA would make these purchases where cost effective, and would arrange, at a minimum, use, distribution, and duplication of the data for Earth Science Enterprise purposes by affiliated researchers.[58] This new addition to the policy was a result of NASA's first "data buy" experience, which had largely been viewed as a success.

Buying Commercial Data

The Sea-viewing Wide Field-of-view Sensor (SeaWiFS) instrument, launched in 1997 on Orbital Sciences' SeaStar/Orbview-2 satellite, was designed, built, and operated by the company to meet NASA's specifications. In return,

NASA signed an advance agreement to purchase data for a five-year period. Under the contract, NASA had access to the data for research use, while Orbital Sciences retained the rights for commercial sales. The same ocean color data useful for scientists studying oceanography and climate change, if processed and distributed quickly, could be used to identify the most productive locations for the commercial fishing industry.[59] To make this arrangement work, NASA had negotiated a special data distribution policy. Commercial and operational users were directed to purchase data directly from Orbital Sciences, which handled real-time data distribution. After a two-week delay designed to protect Orbital Science's commercial interests, data was provided to NASA. This data was then distributed to authorized users, selected by NASA, for research purposes only. Redistribution to non-authorized users was prohibited. After a five-year period, data purchased and archived by NASA was available for use without restrictions.[60] Researchers generally seemed to find this arrangement agreeable, and although it did not meet NASA's ideal of free and open data, the concessions were considered acceptable in return for the cost savings.

Congress tried to repeat this success with respect to a new SAR mission. NASA had expressed an interest in development of such a mission for years. Seeing the commercial SAR efforts in Europe, Canada, and Japan, the US Congress directed NASA to develop the mission as an industry-led effort with NASA as a data customer. NASA actively sought commercial partners, but US companies reported that the commercial market was not sufficiently mature to generate a financial return within a reasonable amount of time. Due to the inability to find an industry partner, NASA was not able to develop the mission. A 2001 National Research Council review argued that research SAR data, which it called "one of the most exciting remote-sensing technologies" for scientists, had been limited by these premature commercial efforts. It recommended that in the future, the agency should evaluate its priorities and determine whether data should be collected exclusively with public funding.[61]

Despite the failure of the SAR effort, Congress remained interested in commercial possibilities, and the Commercial Space Act of 1998 directed NASA to acquire Earth remote sensing data and services from a commercial provider to the extent possible while satisfying scientific and educational requirements.[62] This led to the creation the five-year experimental scientific data purchase (SDP) program, in which NASA purchased commercial satellite data to evaluate its potential for meeting Earth science user needs. The $50 million program supported imagery purchases from multiple systems, including Landsat and aerial platforms, but of greatest interest from a policy perspective was the purchase of IKONOS satellite data.[63] IKONOS was the first successfully launched commercial remote sensing satellite in the United States, owned by Space Imaging and lofted in 1999.

Through SDP-supported work, scientists found that the IKONOS data was useful for a variety of environmental research purposes. For example, the high-resolution data could be used as a kind of "virtual" ground measurement to help validate data from lower-resolution scientific satellites.[64] However, the experiment also brought to light challenges in public-private coordination on research, particularly issues related to data licensing and costs. The SDP data was purchased in bulk by NASA and provided for free to NASA-affiliated researchers. Redistribution and sharing of the images was allowed among these users, but data was not available outside of this group. The license also did not provide for the data to be archived and made available in the future, which many scientists saw as a significant drawback. Furthermore, while initial provision of IKONOS data was free through the SDP, most researchers recognized that when the program ended, they would not be able to afford to continue using the data.[65] Although two more US commercial companies launched their first remote sensing satellites during the SDP program period—Earthwatch's Quickbird in 2001 and Orbital Sciences' OrbView-3 in 2003—NASA's science data purchase program was not renewed after it ended in 2004.

Earth Science Prominence Rises and Falls, but Data Sharing Is Steady

As the SDP experiment was taking place in the background, the Earth Observing System was front and center. The first EOS flagship mission, Terra, launched in 1999, followed by Aqua in 2002 and Aura in 2004. Ten smaller missions launched over this same period. Researchers were eager to access this new data. The number of unique users requesting NASA data products on EOSDIS increased from about 400,000 in FY 1996 to more than 2 million in FY 2004. The volume of data delivered over that period increased by an order of magnitude, from just under 40 terabytes in FY 1996 to more than 700 in FY 2004.[66]

There was widespread recognition of the importance of NASA's Earth science efforts. President Bush announced the US Climate Change Science Program in 2002 to direct new investments in climate-related science and technology, and the 2003 strategic plan for the program acknowledged the central role of NASA's Earth observation and data management systems. It also expressed a commitment to supporting an open data policy.[67] In 2002, one of the first acts of the new NASA administrator, Sean O'Keefe, was to announce a new mission statement for NASA that gave top billing to NASA's Earth science activities. The new mission was:

To understand and protect our home planet
To explore the Universe and search for life
To inspire the next generation of explorers
... as only NASA can.[68]

O'Keefe highlighted NASA's new mission at the first international Earth Observation Summit, held in Washington, DC, in 2003. At the time NASA was operating 18 Earth science research satellites in orbit, carrying over 80 instruments. "Every day," he said, "we gather and distribute terabytes of Earth Science data and information to millions of researchers and others around the globe. It is easier today to access that information than ever, ever before." O'Keefe's speech emphasized the moral significance and weight of the Earth summit and its goal of creating a coordinated global Earth observation system. Quoting Albert Einstein, he said, "Concern for man himself and his fate must always form the chief interest of all technical endeavors... in order that the creations of our minds shall be a blessing and not a curse to mankind." NASA was motivated by scientific curiosity, but also by the desire to provide practical benefits. "We will use the knowledge derived from our Earth Observation Systems to help feed our people, to assure the availability of fresh water, and to protect vulnerable populations from natural and man-made disaster," he stated. "Our children and grand-children will be ultimately the beneficiaries of this work."[69]

The NASA Earth Science program's time in the spotlight, however, was short. In 2004, President Bush announced his Vision for Space Explo-ration. An ambitious plan to return humans to the Moon and then to Mars, it did not leave much room—or budget—to focus on Earth science.[70] NASA updated the agency mission again: "to pioneer the future in space exploration, scientific discovery, and aeronautics."[71] All mention of the "home planet" was removed.[72] The agency was restructured, as well. Earth and space sciences were placed within a new Science Mission Directorate (SMD), and Earth science activities were further merged with studies of the Sun to become the Earth-Sun System Division (ESSD).[73] Between 2000 and 2006, NASA's Earth science budget decreased by more than a third in real terms.[74]

In 2005, the National Research Council, which was developing a decadal survey report for US Earth observations, released an interim report warning that the US system of environmental satellites was "at risk of collapse." NASA's rapidly shrinking budget no longer supported already approved missions and programs with high scientific and societal relevance. "Mission after mission is canceled, descoped, or delayed because of budget cutbacks," the report found, largely as a result of the administration's new focus on space exploration.[75] The National Research Council's final report, released in 2007, stated that the concerns expressed in 2005 had greatly increased. With older systems reaching the end of their life-spans and a dearth of new satellites to be launched, the council estimated that the number of instruments would likely decrease by about 40 percent between 2006 and 2010.[76]

In this environment, international cooperation had become even more important, and the council recommended encouraging full and open data sharing whenever possible.[77] NASA's commitment to open data had continued throughout this period, domestically and internationally. It was a leader in the development of the GEOSS Data Sharing Principles in 2005, and actively supported increased data sharing within this forum. Its 2006 Earth Science Handbook, the first update since 1999, reiterated its commitment to "full and open sharing of all data with the research and applications communities, private industry, academia, and the general public." The introduction to the policy explained that "the greater the availability of data, the more quickly and effectively the user communities can utilize the information to address basic Earth science questions and provide the basis for developing innovative practical applications to benefit the general public." NASA's policy was designed to "maximize access to data and to keep user costs as low as possible."[78] Working in concert with this open policy, NASA's technical Earth science data sharing system was operating on a vast scale. In FY 2006, more than 2.5 million distinct users accessed the EOSDIS system, and 54.5 million data products—a petabyte of data—were delivered.[79]

With the entry of the Obama Administration in 2009, Earth science activities received increasing attention. A 2010 report by the Office of Science and Technology Policy acknowledged the concerns of the National Research Council and committed to substantial increases in NASA's Earth science budget.[80] An updated NASA mission statement in 2011 returned some high-level focus to these efforts as well, stating that NASA "drives advances in science, technology, and exploration to enhance knowledge, education, innovation, economic vitality, and stewardship of Earth."[81]

NASA gained further recognition as a leader in President Obama's Open Government Directive. The agency released a report describing NASA's history of openness and highlighting the many activities of NASA that meet and exceed the requirements of the directive, stating, "Data is at the heart of what we do," and asserting that openness is "part of NASA's DNA." The report noted that NASA already makes available exabytes (billions of gigabytes) of scientific data, which would now also be accessible on data.gov, along with new, additional data and information to be identified. In a section focusing on its Science Mission Directorate, NASA concluded, "We are a community of scientists and instill the principles of transparency, participation, and collaboration in everything we do to better understand out home planet, our sun, our solar system, and the universe beyond."[82] NASA expressed similar enthusiasm in responding to the administration's Open Data Policy in 2014, noting that NASA's Earth Science Program was among the first in the world to establish a free and open data policy.[83] By 2015, NASA was distributing more than 30 terabytes of Earth observation data to users each day.[84]

NASA Summary

From its founding, NASA has taken the view that sharing data as widely as possible provides the best return—for science, for the agency, and for the United States as a whole. It implemented this philosophy with regard to the weather and land remote sensing systems before they were transitioned to operational status, and data sharing was a high priority when NASA began to develop its own Earth science systems. Efforts by national-level actors to encourage NASA to engage with the private sector did not lead to major changes within the agency. Although its policy reflected legal obligations to restrict data access for commercial use, NASA never put significant effort into implementing commercial data sales. As soon as the legal requirement was lifted, NASA removed the restriction from its policy, and committed to making data fully available to all users.

NASA's ability to avoid the commercialization trends that affected NOAA and USGS was due not only to the agency's mission and culture, but also to the nature of the data itself. Unlike operational data collected by the other two agencies, data from NASA's experimental scientific satellites was unlikely to have any immediate commercial value. The fact that NASA satellite programs are often one-off systems, designed to address a particular research question, also makes them less well suited for commercial users, which generally need assurance of continuous data collection to develop a product and facilitate a market. NASA's few experiments in engaging the commercial sector, with the SeaWiFS project and the Science Data Buy (SDB), while generally considered successful, did not reveal any significant opportunities for cost savings or new capabilities.

Instead, NASA has continued to strengthen its commitment to free and open data, promoting its policy within the US government and internationally. When the Obama administration announced the Open Government Directive and the Open Data Policy, NASA was quick to highlight its existing activities and to make further commitments in the area, leveraging national priorities to pursue policies it had already deemed desirable.

10 US Defense, Intelligence, and Commercial Satellites

In the United States, commercial remote sensing satellite development has been closely tied to activities in the defense and intelligence sector, and these developments have in turn affected civil Earth observation satellite development and data sharing policies. As would be expected, sharing of reconnaissance and commercial satellite data has been much more limited than sharing of civil Earth observation data. However, there have been numerous examples in which archived reconnaissance data has been declassified and made broadly available, and there have been sustained efforts to broaden access to commercial satellite data purchased by US intelligence agencies, making this data accessible to all government agencies and their affiliated researchers. Further, many commercial entities have engaged in data sharing of their own volition, in projects designed to increase awareness and usefulness of their products or to address acute humanitarian needs.

Reconnaissance Satellites and Data Sharing

Explorer 1, developed as part of the International Geophysical Year (IGY), was the first satellite ever launched by the United States, but it was not the top priority satellite program. In 1955, the Joint Chiefs of Staff recommended to President Eisenhower that a large surveillance satellite be developed. They were supportive of developing a small scientific satellite in support of the IGY, only "so long as the small scientific satellite program does not impede development of the large surveillance satellite."[1] The first reconnaissance satellites used photographic technology, capturing images on film that was then returned to Earth to be processed and analyzed. The military and intelligence value of these satellites was demonstrated quickly.

The first successful flight and recovery of a reconnaissance satellite system by the United States, in 1960, resulted in more imagery of the Soviet
Union than had been gathered in two years of flights by the high-altitude
U-2 reconnaissance aircraft. By 1972, The United States had launched
more than 100 reconnaissance satellites, and its efforts expanded as more
advanced technologies were developed.[2]

Not surprisingly, access to reconnaissance data has generally been highly
restricted due to national security concerns. National security officials don't
want rivals to know what information they have collected or even what
information intelligence satellite systems are capable of collecting, both
of which could potentially be deduced from the resulting satellite imagery or metadata. Despite these concerns, there are numerous examples of
reconnaissance data being declassified and publicly released in support of
environmental studies.

The end of the Cold War in 1989 coincided with growing public concern about global warming, leading then-Senator Al Gore and Senator Sam
Nunn to push for the release of archived reconnaissance satellite data for
use in studies of climate change and ecology. Private groups, including the
Council on Foreign relations, also lobbied for the idea.[3] NASA's Mission to
Planet Earth was in development, but launches of the flagship systems were
still almost a decade away, and only a handful of research-oriented Earth
observation satellites were operating at the time. There was some hope that
reconnaissance satellite data could help to fill the gap. In 1992, President
Bush signed a directive to facilitate access to reconnaissance data for scientists.[4] This resulted in the creation of the Measurements of Earth Data
for Environmental Analysis (MEDEA) group, which allowed a collection of
scientists to gain security clearances to examine and request satellite reconnaissance imagery for environmental studies.

A 1997 report by the Rand Corporation examined the possibility of
developing a more stable arrangement for use of intelligence data for environmental purposes. In addition to the challenges of balancing national
security and environmental concerns, the report noted that efforts were
stymied by a lack of funds, personnel, and facilities. A further impediment
was the opposition of the nascent US commercial remote sensing industry, which was just beginning to launch its own high-resolution satellites
and saw the release of intelligence data as potentially harmful to its interests. The report recommended that the administration provide additional

funding for civil environmental applications of intelligence data, increase efforts to declassify environmental datasets, and engage in a dialog with industry.[5] Instead, for reasons not made public, MEDEA stopped operating altogether in 2000. The group was re-formed in 2008 to focus on using the data to help answer public policy questions about the potential national security impacts of climate change. The program continues to provide ad hoc access to a select group of scientists, rather than broader, more systemized access.[6]

The work within MEDEA did lead to some broader releases of data. In 1995, President Clinton signed an Executive Order declassifying imagery obtained by reconnaissance satellites operating between 1960 and 1972—more than 800,000 images. This provided scientists with land remote sensing data stretching back more than a decade before the launch of the first Landsat satellite. The order also required that the Director of Central Intelligence periodically review imagery from other systems with the objective of making as much imagery as possible available to the public, consistent with the interests of national defense and foreign policy.[7] Based on this, additional imagery, collected by reconnaissance satellites that operated between 1971 and 1984, was declassified in 2011. As of 2016, declassified images that have already been processed can be accessed online for free through the USGS EROS Data Center, the same organization that provides Landsat data. Requests for unprocessed images, which require the film negatives to be located, scanned, and digitized, are subject to a fee of $30 per frame.[8]

Defense Meteorological Weather Satellite Program

In addition to its extensive reconnaissance satellite program, in 1962, the DoD launched the first Defense Meteorological Satellite Program (DMSP) satellite. The DMSP polar-orbiting weather satellites were complementary to those operated by NOAA, carrying different instruments and operating in a different orbit to provide data of particular interest to military decision-makers. The data from these satellites was made available to NOAA and to scientists on a limited basis beginning in 1972.[9] By 1994, all DMSP data was unclassified and archived at the National Geospatial Data Center in Colorado. NOAA had operational access to the data via DMSP ground stations. A limited set of DMSP data, including temperature and moisture measurements, was made freely available to international users through the WMO's World Weather Watch system, although delays and the possibility of restrictions made the data unsuitable for operational use.[10] In 2001, real-time high-resolution DMSP data was made available to many US allies.[11] As of 2016, minimally processed data from the DMSP instruments was made freely available to all users on NOAA's CLASS database and the National Center for Environmental Information.[12]

National Security and the Commercial Sector

National security officials have always closely followed developments in the commercial remote sensing sector to identify both risks and opportunities. In 1978, President Carter developed a classified space policy that made it US policy to encourage commercial exploitation of space capabilities, but required that Earth remote sensing systems be authorized and supervised by the government. The policy limited civilian systems to a resolution no better than 10 meters without special authorization. The policy envisioned systems that were designed in such a way that during national emergencies, the US government could take control or at least deny use of the system to enemies.[13]

While it was US policy to encourage commercial remote sensing, a 1984 report by the Office of Technology Assessment (OTA) raised a number of national security issues. It noted that US commercial systems could reveal information about sensitive US facilities. Limiting sensor resolution or screening data could help to avoid this, but these requirements would conflict with the commercial need to provide high-quality data as quickly as possible. In line with the 1978 policy, military and intelligence agencies would need to address "the steps to be taken to preempt and operate commercial systems in times of national emergency." While raising these issues, OTA recognized they may be moot. Any limitations developed in the United States to address these concerns would not apply to high-resolution commercial systems under development in France, Japan, and elsewhere.[14]

In early 1984, as the United States was preparing for the transition of management of the Landsat system to a private entity, the Office of Technology Assessment (OTA) noted that together, the military and intelligence agencies were the largest customers of Landsat data. Landsat data was used to complement the higher-resolution, more narrowly focused data acquired from reconnaissance satellites. The military utility of commercial imagery was further demonstrated during the Persian Gulf War. DOD purchased more than five million dollars' worth of Landsat data and also bought commercial data from the French SPOT satellite, which had launched in 1986. It used the data to provide broad area coverage of locations of military interest.[15] The value of Landsat to the military was seen as so significant that when the system was brought back under federal management in 1992, the DOD was appointed a co-owner of the program.[16] Even though commercialization of the Landsat system itself failed, the program paved the way for the success of other commercial remote sensing systems.[17] Together with the rise of the international remote sensing market, as well as technological developments, lessons learned from the program led three US companies—WorldView, Lockheed Martin, and Orbital Sciences—to request commercial remote sensing licenses by 1993.[18]

In 1994, President Clinton released a presidential directive that reiterated the US government's commitment to support and enhance US industrial competitiveness in the field and stated that license requests for systems with capabilities available, or planned for availability, on the world marketplace would be considered favorably, signaling that companies would not be held to a 10-meter resolution limit. It required that the US government be provided access to all satellite tasking from the previous year, that the company use a downlink format that allowed government access and use as required by national security, and that the company limit data collection and/or distribution during periods when national security might otherwise be compromised, a requirement generally referred to as "shutter control."[19]

Although their names changed, and all three companies experienced initial launch failures, the same companies that received the first licenses in the early 1990s successfully launched the first commercial high-resolution land imaging systems in the United States. Lockheed Martin, together with its partner, Space Imaging, launched IKONOS in September, 1999, EarthWatch (previously WorldView) launched QuickBird in October, 2001, and Orbital Sciences' Orbimage launched OrbView-3 in 2003 (OrbView1 and 2 were ocean-focused systems developed in cooperation with NASA).

Commercial Data Purchases: ClearView, NextView, and EnhancedView

In April, 2003, President George W. Bush signed the US Commercial Remote Sensing Space Policy (CRSSP), which replaced the 1994 directive. The policy stated that the US government would develop a long-term sustainable relationship with US industry, relying to the maximum extent on US commercial remote sensing capabilities for filling imagery and geospatial needs for military, intelligence, foreign policy, homeland security, and civil users. The policy reiterated the commitment to allow the US industry to remain competitive on the international market, and retained the options for shutter control and government preemption of systems in emergency situations.[20]

The military and intelligence community had actually preempted this policy by a few months with the January, 2003, announcement of the ClearView program. Under ClearView, the National Geospatial Intelligence Agency (NGA) awarded initial contracts of $72 million to DigitalGlobe (formerly EarthWatch) and $120 million to Space Imaging to provide high-resolution satellite imagery over a three-year period. Additional data buys could occur under the contract, with a maximum value of $500 million per company over five years.[21] Orbimage was awarded a ClearView contract in 2004 following the successful launch of its high-resolution satellite in June, 2003.[22]

Interestingly, one of the major incentives for the military and intelligence communities to purchase high-resolution commercial imagery is the ability to share the data. Commercial data is unclassified and can be provided to allies, coalition forces, and troops on the ground. The ClearView contract included a broad license that allowed NGA to share imagery with other US federal agencies. Temporary sharing was possible with state and local government, foreign governments, international organizations, and nongovernmental organizations if working with the US government on a joint project.[23]

Shortly after awarding the ClearView contracts, NGA announced plans for the NextView program.[24] NextView contracts would give NGA more control over satellite tasking and image processing. In addition to guaranteed data buys, the contract also subsidized the development of new satellites, providing about half the cost of developing each of the $500 million next-generation systems. In 2004, NextView contracts were awarded to DigitalGlobe and Orbimage.[25] Space Imaging, which did not receive an award, was acquired by Orbimage, and the merged company was renamed GeoEye.[26] Supported by NextView, DigitalGlobe launched WorldView-1 in 2007, and GeoEye launched GeoEye-1 in 2008. WorldView-2, financed without NGA support, was launched in 2009.[27]

In August, 2010, NGA announced the EnhancedView program, awarding contracts worth $3.5 billion to DigitalGlobe and $3.8 billion to GeoEye. Both contracts included data purchases over a 10-year period, to be renewed each year, as well as funding to offset capital investments. NGA explained that the contracts were carefully structured to ensure both US commercial remote sensing firms were in sufficient financial health to survive and that the contracts would not "unduly tip the balance" within the industry.[28] However, just two years later NGA announced that due to budget shortfalls, only DigitalGlobe's contract would be renewed. NGA also stated that it would not provide the full funding originally planned to support the construction and launching of the GeoEye-2 satellite.[29] The next day, the companies announced they would be merging.[30]

After the merger, DigitalGlobe was the only US commercial provider of high-resolution remote sensing imagery. In 2014, the company launched its WorldView-3 satellite, which has a spatial resolution of 30 centimeters, the best commercially available resolution in the world, and WorldView-4 (previously GeoEye-2) launched in November, 2016. Public filings reveal the extent to which DigitalGlobe is reliant on US and foreign government customers. In 2015, business with the US government accounted for 63.7 percent of the company's revenue. International defense and intelligence customers accounted for 17.3 percent. The company reported that its sales in these sectors had "never been stronger." Commercial sales, which includes international civil government sales, accounted for 19 percent of revenues and was not seeing similar growth.[31]

"New Space" Commercial Remote Sensing

In recent years, there has been a new wave of interest in commercial remote sensing based on the decreasing costs and increasing capabilities of small satellites. Founded in 2010, Planet Labs, generally referred to simply as "Planet," plans to operate a constellation of more than 100 satellites, each about the size of a shoe box, that together are capable of imaging any spot on the Earth in 3- to 5-meter resolution every day.[32] The company has successfully launched more than 100 satellites already, but only about half of these remain operational. Frequent launches and rapid refresh of technology are part of Planet's business model. In 2015, Planet acquired the Canadian remote sensing company, BlackBridge, a deal that included the company's five RapidEye remote sensing satellites as well as its imagery distribution network and customers.[33] Terra Bella, purchased by Google in 2014, had launched seven satellites by the end of 2016, each about the size of an oven, capable of collecting both imagery and video.[34] Consolidation has hit this new market already: In February, 2017, Planet announced that it had reached an agreement with Google to acquire Terra Bella. As part of the agreement, Google will enter into a multiyear contract to purchase Earth-imaging data from Planet.

As they developed their initial constellations, these companies stated that they were targeting commercial customers, but government customers may be interested as well. In 2015, NGA released a new commercial geo-intelligence strategy that seeks to explore the capabilities of new small satellite companies. The NGA Director specifically mentioned both Planet Labs and Terra Bella as companies of interest. It plans to fund a series of experiments to explore possible uses of the data.[35]

Commercial Remote Sensing and Data Sharing

Commercial satellite operators generally do not share their imagery for free, but there are some exceptions. For example, although it markets "First-Look," an online subscription service for crisis response and emergency management, DigitalGlobe has provided free access to imagery for disasters on multiple occasions.[36] The DigitalGlobe Foundation, a nonprofit organization originally created by GeoEye in 2007, provides imagery grants to students and faculty around the world. Users must submit an application and redistribution of the imagery is restricted.[37] In addition to supporting environmental research, these projects increase awareness of DigitalGlobe products and potential applications.

In 2015, PlanetLabs debuted Open California, a two-year archive of satellite imagery of California freely available under a creative commons license, with the goal of encouraging research and experimentation with their data.[38] The company also maintains the RapidEye Science Archive, which provides free access to RapidEye data for German researchers, a legacy of the satellites' original development under a public-private partnership with the German space agency.[39]

Defense, Intelligence, and Commercial Summary

With respect to military data sharing, the barriers are clear: sharing data poses national security risks, and agencies, understandably, have a strong culture of secrecy. In the case of commercial entities, while a limited amount of data sharing may serve to increase awareness of their products, widespread or systematic sharing would undermine their business models. Even in cases where systematic sharing may be possible, designing and implementing these types of arrangements is simply not a priority. It is interesting, however, that even given these disincentives, sharing does occur in some instances.

In the case of military and intelligence users, this may come from prodding on the part of national-level policy-makers and potential users, as was seen with declassification efforts. However, since this is not an activity the agency is likely to pursue on its own, declassification will remain an ad hoc activity, occurring at irregular intervals based on the interest, influence, and level of effort of outside groups and agency leadership. Some sharing occurs because it is necessary for national security and makes it easier for military and intelligence agencies to achieve their missions. For example, the negotiation of licenses that facilitate easy sharing with allies and coalition partners was the result of active effort on the part of the NGA. This data sharing was necessary and beneficial for the agency. Availability for other federal government users is primarily a by-product, in part driven by cost concerns by national-level policy-makers. A similar situation is seen on the part of commercial providers. Data is shared in an ad hoc manner, to support high-profile natural disasters, for example, or to raise awareness about products, but these activities are limited in scope and other opportunities for sharing are left unexplored.

11 European Organization for the Exploitation of Meteorological Satellites

The European Organization for the Exploitation of Meteorological Satellites (EUMETSAT) was formed in 1986 to take responsibility for operation of European weather satellites. Prior to the organization's creation, these satellites had been operated by the European Space Agency and data was shared freely in accordance with meteorological and research community norms. However, EUMETSAT's creation corresponded with increasing efforts among national meteorological agencies to recover costs through commercial sales of data and products. To enable these activities, EUMETSAT quickly moved to restrict access to its own data, adopting both legal and technical restrictions. EUMETSAT worked to gain acceptance for these more restrictive policies in international fora, particularly the WMO.

Over the course of the 1990s, however, it became clear that data sales were not generating significant revenues for national meteorological agencies, and they were slowing the growth of private-sector meteorological activity in Europe. European leaders reversed course and began to push for the adoption of more open data policies. EUMETSAT followed this trend, gradually increasing the amount of data it made freely available, eventually reaching a point at which nearly all of the satellite data the agency collected was freely available to all users.

International Contributions

Impetus for the first joint European Earth observation satellite came from the British Secretary-General of the World Meteorological Organization (WMO), D. A. Davies. In a 1967 speech to the WMO, he argued that a European meteorological program would bring immediate practical advantages and provide data of scientific interest. A geostationary meteorological satellite would not only benefit Europe, but provide useful data for Africa, he noted, constituting a form of indirect technical assistance to developing countries. The satellite would also make a meaningful contribution to the WMO's World Weather Watch plan, which European nations had unanimously supported.[1] This was Europe's chance to make a global impact in its first foray into Earth observations.

At the time of Davies' speech, the WMO was just beginning to plan the Global Weather Experiment (GWE), with the goal of testing global observing systems in support of the World Weather Watch Program.[2] A major goal for the GWE was to establish monitoring of the entire globe using a constellation of five geostationary satellites. This provided Europe with the potential to play a key role in global efforts, contributing a geostationary satellite and working to provide this global coverage in coordination with the United States and Japan. This opportunity to play a role on the international stage helped to solidify support for the European meteorological satellite program.[3]

Meteosat-1 was chosen as the European contribution, building on a program originally started as a national effort in France. The satellite was launched in 1977.[4] In accordance with the GWE plans, datasets from the satellite were held at the World Data Centers in the United States and the Soviet Union and provided at the cost of reproduction and dissemination to any user who requested them.[5] The European Space Agency, which had been formed in 1975 when the European Space Research Organization (ESRO) merged with the European Launcher Development Organization (ELDO), held meetings in 1979 and 1980, attended by more than 80 scientists from a dozen countries, to share information with scientific users.[6] ESA launched Meteosat-2 in 1981.

From the beginning, the intention had been to develop the first meteorological satellites at ESA and then transition them to operational status, but there were challenges to doing so. First, it was not clear what organization should be given operational responsibility, or if a new entity would need to be created. Also, despite the promising results of early meteorological satellite systems and the numerical weather forecasting techniques they enabled, the technology was still relatively new and very expensive compared to traditional meteorological practices. Many in the meteorology community saw these new technologies and techniques as experimental, and did not want to use limited resources from their operational National

Meteorological Services (NMSs) to pay for them. The result of this was tension between ESA and meteorologists and a slow transition to operational status for European meteorological satellites.[7]

The US policy of providing meteorology data free of charge affected the European policy in multiple ways. On one hand, the availability of free data from the United States decreased the incentive for European states to invest in their own systems. On the other hand, some European meteorologists expressed concern that if other nations did not begin to develop systems and reciprocate, the United States may not continue this practice.[8] It was clear that the United States was considering alternate arrangements already. In 1983, as the transition of meteorological satellites to operational status was being discussed in Europe, the Reagan administration announced a proposal to privatize both Landsat and the US weather satellites. Privatization of the weather satellites was rejected by the US Congress, but the consideration of the option was sufficient to gain attention in Europe.[9]

EUMETSAT Founded

The convention creating the European Organization for the Exploitation of Meteorological Satellites (EUMETSAT) was signed in 1983 and entered into force in 1986. The convention began by recognizing the fundamental importance of the task at hand—stating that meteorological data contributes to forecasts that ensure the safety of the population. Addressing previous skepticism, the convention argued that meteorological satellites had "proved their aptitude and unique potential," and noted the WMO's plans to feature satellites in development of the World Weather Watch. It stated that the experimental Meteosat program had proven Europe's capability to "assume its share of responsibility" in the WMO's Global Observation System. It also reminded members that no national or international organization planned to collect meteorological satellite data in Europe's primary zones of interest, and that the magnitude of resources needed to support these activities was beyond the means of any single European country. EUMETSAT was the only solution for meeting European meteorological needs.[10]

Accordingly, the convention stated, "the primary objective of EUMETSAT is to establish, maintain and exploit European systems of operational meteorological satellites, taking into account as far as possible the recommendations of the World Meteorological Organization." The system would also support meteorological research and benefit European industry by "taking maximum advantage of the technologies developed in Europe" (although in contrast to ESA, it would emphasize low cost, not distribution of work among its member nations, in its contracting). EUMETSAT was governed by a council of representatives from the National Meteorological Services, which determined its activities and set its budget. Plans approved by the council would be implemented by a director aided by a small staff—fewer than 10 people at the outset.[11] Investments of the original 16 participating nations were determined based on internal negotiations, with Germany, France, and the United Kingdom together providing more than half of the initial contributions. In 1991, EUMETSAT transitioned to a payment system based on each country's Gross National Product.[12]

Move to Cost Recovery

By the time EUMETSAT was established, meteorological satellites, and the World Weather Watch as a whole, had proven its value for improving weather forecasts, but almost immediately the high costs of maintaining operational systems put the program, and assured data continuity, in jeopardy. The international meteorological community struggled with the disproportionately high cost of weather satellites compared to other meteorological data collection methods. Some argued that the burden on WMO members that were also satellite operators was too high. There were considerations in both the United States and Europe of reducing the capabilities of currently planned systems or undertaking international programs to share costs. In facilitating these discussions, the WMO reiterated that free and unrestricted access to meteorological data remained a basic principle.[13]

WMO's pronouncement was primarily in response to a third option under consideration for dealing with budget challenges: increasing commercialization of government activities. In the 1980s and early 1990s, many European countries began to apply this concept to their National Meteorological Services, requiring these agencies to sell data and products to recover some of their operational costs. Some agencies created commercial branches specializing in creation and marketing of these products and services. These organizations competed with private companies, which also developed meteorological products and services, generally based on data and products provided for free from the NMSs themselves. Often, if access to data was restricted from one NMS, it could simply be obtained from another. Agency officials argued that this constituted unfair competition, and that the private companies should pay for access to data to contribute toward the cost of the observational systems. This would place them on a more equal footing with NMSs, which had to price their own goods consistent with efforts to recoup the costs of developing observational systems.[14] In addition to being a practical necessity, agency officials argued this was a more appropriate way to treat data paid for by taxpayers: the data are owned by all citizens, not just those who use them; therefore the government should attempt to reduce the costs for the general population by charging users.[15]

In keeping with international meteorological norms, data from the first European weather satellites had been provided as a direct broadcast from the satellites, available to anyone with the appropriate ground station equipment.[16] However, consistent with the changing interests of its members, EUMETSAT moved to restrict access beginning in 1988, developing its first resolution on data dissemination and charging. The policy acknowledged that EUMETSAT's meteorological satellites made an important contribution to the WMO World Weather Watch and that the free exchange of data between NMSs is a worldwide accepted principle within WMO. It also recognized that there was interest in satellite data and products outside of NMSs and, in particular, a growing interest in commercial use of satellite data which created the need to avoid unfair competition with respect to the distribution of and charging for EUMETSAT data. Under the policy, EUMETSAT member state NMSs would receive free access to EUMETSAT data, products, and services, and they would have full control over the distribution of these products and services within their own national territory. It would be up to each individual NMS to develop its own data policy for sharing, selling, or restricting EUMETSAT data within its own country.[17]

NMSs in nonmember states and international organizations, including the WMO and the European Center for Medium Range Weather Forecasting (ECMWF), which had been established in 1975 to develop numerical weather forecasts for Europe, would have free access to a limited set of data defined by the EUMETSAT council. The data would be provided for internal use only and could not be redistributed without EUMETSAT approval. If these organizations wished to receive additional data, they would need to pay a fee or develop a bilateral agreement with EUMETSAT. Data access by other users, particularly commercial users, would be subject to a fee that would take into account factors such as the intended use and the market price of the data. Scientific, educational, and humanitarian users could apply for a fee waiver.[18]

The resolution did not address archived data, which is of particular interest to researchers. Although EUMETSAT had full ownership and utilization rights to Meteosat data, ESA still operated the data archive, which had established fees for access to data ranging from 13 deutsche marks (about $15 in 2016) for a photograph to 450 deutsche marks (about $600 in 2016) for a computer compatible tape with "real-time window data." The EUMETSAT resolution declared that the prices would be increased by 10 percent annually until the charge reached the marginal cost of reproduction and mailing.[19]

EUMETSAT held its first workshop on legal protection of meteorological satellite data in 1989 to determine how to implement the necessary copyright over its satellite images.[20] International copyright conventions refer to national legislation, but due to the nature of satellite technology, it was not clear whether Meteosat images were subject to copyright under the national laws of any of EUMETSAT's members. EUMETSAT addressed this in two ways. First, it amended its constitution to assert worldwide exclusive ownership of all data generated by EUMETSAT's satellites or instruments. Second, given the remaining uncertainty of this legal protection, it also resolved to develop technical protection, passing a resolution in 1990 stating that it would pursue encryption of its satellite data.[21] The European Commission eventually clarified the legal question in March 1996, issuing the directive on the legal protection of databases. The directive acknowledged that databases require the investment of considerable human, technical, and financial resources, and asserted that copyright was an appropriate form of exclusive right for authors who have created databases.[22]

In 1990, EUMETSAT elaborated on its data policy, creating the concept of a "general license" that would give a nonmember NMS full control over redistribution of the data in its own state: essentially the same rights over the data as a EUMETSAT member state. The license could be acquired by agreeing to provide equivalent data to EUMETSAT—an option available to meteorological satellite operators—or by paying an annual fee equivalent to 90 percent of the relevant EUMETSAT membership fee, determined based on GNP.[23] EUMETSAT subsequently developed bilateral data sharing agreements with the National Meteorological Services in all meteorological-satellite-operating nations. Agreements with the United States, Russia, China, and India were signed between 1995 and 2000. Agreements with South Korea and Japan occurred later, in 2006 and 2007, respectively.[24]

The 1990 resolution also stated that NMSs that were interested in only a limited subset of data, and all other users outside of EUMETSAT member states, including scientific and education institutions, commercial entities, and personal users, could obtain a "limited license." Limited licenses would provide access to data for internal purposes, but would restrict any redistribution to third parties. The type of data covered, the length of the license, and the cost of the data would be negotiated on a case-by-case basis taking into account the intended use of the data, the market price of the data, and other considerations.[25]

A EUMETSAT official explained in 1992 that the restrictions contained in its data policy were necessary because "certain National Meteorological Services are under tremendous pressure to ensure that services are provided on a fully commercial basis and that income is used to offset costs." Cost recovery wouldn't be possible if these agencies had to compete against private entities that use the data but do not contribute to the cost of data collection. In addition to allowing EUMETSAT to recover some costs and enable commercial activities by its members, the new policy also provided an incentive for European countries to become and remain members of EUMETSAT. Ensuring that the benefits of membership were worth the costs was seen as critical for the sustainability of the organization.[26] There are some indications that the incentive structure worked: Austria joined as the 17th member of EUMETSAT in 1993.[27]

Despite these actions, EUMETSAT insisted that it remained committed to international sharing of meteorological data. In fact, EUMETSAT officials stated that data sharing needed to be not only continued, but increased, and explained that, perhaps paradoxically, restrictions in its data policy were a necessary element of achieving that goal. The basic argument was that it was the NMSs that paid for the development and maintenance of observational systems, and the commercial sale of data and products was necessary to provide funding for NMSs. If commercial activities and cost recovery were more successful, the resources of NMSs would increase, and they could increase the amount of data they collected and exchanged, even if that data was subject to some fees or restrictions. Conversely, failure to raise revenues could put continued operation of observational systems in jeopardy, potentially reducing the amount of data available to share. It was this logic that led European officials to make the seemingly contradictory statement that "in order that data exchange may continue to grow, it is essential to safeguard and increase the relevant databases."[28]

EUMETSAT and Climate Change

Part of the reason that EUMETSAT argued that data collection and sharing needed to increase was its awareness of the growing importance of global climate change. WMO had sponsored the Second World Climate Conference in 1990, and EUMETSAT recognized the critical role of meteorological satellites in collecting global environmental data vital to understanding this issue. The organization went so far as to develop amendments to its founding convention in 1991 stating this and adding contribution "to the operational monitoring of the climate and the detection of global climatic changes" as an extension of its primary objective.[29] European Union law recognized the importance of environmental issues, as well. A 1990 directive on the freedom of access to information on the environment reinforced the need to improve public access to information held by environmental authorities. It called for information relating to the environment to be provided at a reasonable cost within two months of a request.[30]

That same year, EUMETSAT agreed to the international Committee on Earth Observation Satellites (CEOS) data sharing principles for global change and environmental research, which promoted sharing among CEOS members and identified maximum data use as a fundamental objective.[31] Researchers relied primarily on archived data, which had been maintained by ESA since the Meteosat program began in 1978 and which EUMETSAT continued to provide to ESA within five months after collection.[32] EUMETSAT delayed and eventually abstained from voting on CEOS data sharing principles related to operational environmental data use in 1994, noting that its own policy was still evolving.[33]

In 1992, EUMETSAT began its Meteorological Data Dissemination (MDD) program to improve accessibility of Meteosat data in Africa. Because geostationary satellites must be placed above the equator, the data collected by Meteosat was arguably more relevant to Africa than it was to Europe— more than 70 percent of the land area covered by the satellite was of the African continent. EUMETSAT developed data relay stations, held meetings, and organized training sessions to strengthen the capacity of African nations to use satellite data. Bilateral agreements were developed in 1996 to provide free EUMETSAT data to the African Centre of Meteorological Applications for Development (ACMAD) and the Centre Regional de Formation et d'Application en Agrométéorologie et Hydrologie Opérationnelle (AGRHYMET).[34]

Resolution 40 Compromise

Despite arguments made by EUMETSAT and others about the ability for restrictions to enhance data sharing, the continuing trend toward commercialization caused concern in the WMO, which continued to argue that free and unrestricted data sharing was essential to its activities. In 1991, it reported, "Commercial meteorological activities have the potential to undermine the free exchange of meteorological data and products between national Meteorological Services." There was concern that without free and open data sharing, the World Weather Watch, the heart of the WMO system, would be in jeopardy.[35] European countries insisted there was middle ground: a way to preserve the free and unrestricted international sharing of data underlying the WMO systems and to allow for national efforts of cost recovery and commercialization. They advocated for a two-tiered system, which would preserve free and unrestricted exchange of the data that is most important for global weather prediction, consistent with existing practice. Other data, they argued, could be restricted without harming the international system.[36]

This compromise was formalized in WMO Resolution 40 in 1995. The resolution reiterated "the continuing fundamental importance" of the exchange of meteorological data and products, and recognized "the increasing requirement for the global exchange of all types of environmental data." It mentioned the importance of meteorological data and international data sharing in providing services that support safety, security, and economic benefits, and called attention to the dependence of the research and educational communities on access to meteorological data. The resolution also recognized the trend toward commercialization of many meteorological activities and "the requirement by some Members that their NMSs initiate or increase their commercial activities."

The resolution adopted a policy committing WMO members "to enhancing the free and unrestricted international exchange of meteorological data and related data products," and laid out a tiered policy. The policy required free and unrestricted sharing of essential data, to include, at minimum, data and products needed to support "protection of life and property and the well-being of all nations." Sharing of additional products was also encouraged, but the resolution noted that WMO members may restrict the redistribution of these "nonessential" data and products to enable commercial sales or cost recovery.[37]

In anticipation of this change, in 1994, EUMETSAT updated its data policy. The set of data, products, and services freely available to WMO and nonmember NMSs was expanded and no longer subject to restrictions on redistribution. A further set of data and products was made available without charge for official use by NMSs of nonmember states, subject to licenses preventing commercial use or redistribution to third parties. EUMETSAT's highest-quality data, near-real-time Meteosat data updated every hour or half hour, would also be made freely available in the event of disasters or other emergencies. Educational and scientific users who would not use the data commercially or operationally were provided access to all real-time EUMETSAT data free of charge.[38]

EUMETSAT implemented a new fee structure for nonmember NMS's interested in gaining access to EUMETSAT's highest-quality near-real-time data for their official use, with the cost of a license calculated based on GNP. Countries with a GNP above $3,000 would pay 50 percent of the equivalent member state contribution for access to hourly data or 60 percent for half-hourly data. Countries with a GNP below $2,000 could access this data without charge. Countries with a GNP above $2,000 but below $3,000 would pay a fee based on GNP, but reduced according to a linear scale developed by EUMETSAT.

A separate schedule of fees was developed for commercial and other users, with prices based on the data to be accessed and its intended use, differentiating between service providers offering value-added services and broadcasters showing EUMETSAT data or imagery on television. EUMETSAT noted that service providers that also use data for broadcasting would be subject to both fees. Any nonmember nations wishing to redistribute data within their territory, regardless of GNP, would be assessed an additional fee established for service providers and/or broadcasters, as appropriate.[39]

ECOMET and the Consolidated Data Policy

In 1995, the same year that WMO passed Resolution 40, a group of European nations formed the European Cooperation in Meteorology (ECOMET) economic interest group. Through ECOMET, European countries would make meteorological data available across Europe under the same conditions, developing a joint catalog of data and products and their prices. While nations were still free to set their own fees, members agreed on an initial goal of recovering 3 percent of combined infrastructure costs—there was no expectation that government sales of data and products would fully, or even significantly, cover the costs of observing and analysis systems. ECOMET explicitly recognized the division of essential and additional data defined by WMO, and committed to ensuring free exchange of the essential data.[40]

Interestingly, even this economically -focused group stated that the first of five objectives was "to preserve the free exchange of datasets and products between members of the grouping within the framework of WMO regulations." ECOMET would also help to maintain and improve infrastructure, expand the availability of meteorological information, and increase the use and improve the distribution of data, products, and services. Last on the list was ECOMET's objective "to create the conditions for members of the grouping to develop their economic activities."[41] Once again, the emphasis was on the potential for commercialization and cost recovery to increase data availability, rather than reduce it.

In 1998, EUMETSAT developed a new, consolidated data policy, explicitly referencing WMO Resolution 40 and taking into account a number of amendments made to its 1991 policy in the intervening years that had been designed to ensure fair competition. Access to EUMETSAT data remained free for official use by the NMSs of its member states, and those states would continue to have control over access to real-time data by third parties within their own territory. However, the cost and conditions of access were now uniform, defined by the EUMTSAT Council. Member states were also required to treat their own commercial activities in an equivalent way to commercial service providers, ensuring that data access fees were paid by both and competition was fair.[42]

The updated policy also called out some efforts to increase data availability. ECMWF would be provided all data without charge, defined as no more than the cost of reproduction and delivery, although redistribution of the data, including to its member states, was prohibited.[43] Archived data would be provided to all categories of user at no cost through EUMETSAT's new Meteorological Archive and Retrieval Facility (MARF). However, to avoid an unmanageable load and consequential degraded service, the amount of data that could be requested in a single or successive orders was limited. Once the limit was reached, users would be required to wait two weeks before making another request.[44] In 1999, EUMETSAT applied similar conditions to access to real-time and archived data from its Meteosat Second Generation (MSG) satellites, the first of which, Meteosat-8, launched in 2002.[45]

National and Regional Support for Open Data

After years of promoting commercialization of government functions, and successfully gaining accommodations for this policy in international organizations, the position of European government leaders was beginning to change. In 1999, the European Union published a "Green Paper," a report aimed at prompting debate and discussion, on the use of public-sector information in Europe. The paper argued that public-sector information played a fundamental role in the function of the market, and that data should be made clearer and more accessible to potential users. It argued that European companies were at a serious competitive disadvantage to their American counterparts, which were benefiting from a highly developed, efficient public information system—inaccessibility of weather data was given as an example.[46]

The emerging international and European private meteorological sector agreed with this assessment and provided further elaboration of the problems from their perspective. A paper written by officials from a private weather services company in Finland and printed in the Bulletin of the American Meteorological Society in 2000 argued that "ECOMET functions primarily as a price-fixing cartel for the National Meteorological Service," noting that the prices for satellite data were particularly excessive. They leveled the same complaint at the NMSs that had earlier been aimed at them: commercial branches of NMSs were engaging in unfair competition. They noted that the public and commercial activities of NMSs are often not strictly separated, with commercial branches benefiting from colocated facilities and privileged access to data. They argued that it was unfair for the NMSs to profit from publicly funded data while commercial entities, which pay taxes, are forced to pay again. They contrasted this with the policy adopted in the United States in which meteorological data collected by the government was seen as a public good and made freely available to all users. The article listed a number of industry groups—the Coalition for the Open Exchange of Global Data, the Association of Environmental Data Users in Europe, and the Association of Independent Weather Services—that had mobilized to lobby against the European policy.[47]

In 2000, a report prepared for the European Commission on commercial exploitation of Europe's public-sector information directly compared the results of open policies in the United States to cost recovery efforts in Europe. The analysis found that the European value-added sector was significantly smaller than the corresponding sector in the United States. Not only were cost recovery efforts against the interests of private industry, they also seemed to be resulting in a net loss for European governments: by their calculations, the tax revenue from a larger commercial value-added sector would be greater than the loss of revenue from eliminating user charges. Once again, weather data was called out as an example, with overpricing and unfair competition from the government identified as particular problems.[48] Additional reports in the United States and the academic sector reinforced this finding.[49]

Taking note of these findings, the EU released a directive on public access to environmental data in January 2003, replacing its earlier 1990 directive with a much more strongly worded statement in favor of broad data sharing. The goal of the new directive was to ensure that environmental information was made available as soon as possible for free or at a reasonable cost in order to achieve "the widest possible systematic availability and dissemination to the public." The directive was designed to encourage transparency, increase public awareness in environmental matters, and lead to a better environment. Generally, reasonable cost was defined as a fee not exceeding the actual costs of producing the material in question. However, the directive stated that a market-based charge would be considered reasonable in cases where commercial sales are necessary in order to guarantee the continuation of collecting and publishing such information.[50]

Later that year, the European Union issued a directive further reinforcing these ideas, encouraging public-sector bodies to make data available in a timely manner at or below the marginal cost of reproduction and dissemination. The European Union Directive on the Reuse of Public Sector Information, generally known as the PSI Directive, allowed as an upper limit pricing that includes "a reasonable return on investment" in the price of the data, above the cost of collecting, producing, and disseminating the data, but stated that excessive prices should be precluded. The directive argued that public-sector information was an important resource in the move to an information and knowledge society and stated that making public-sector documents available to the public was part of the fundamental right to knowledge and a basic principle of democracy.[51]

EUMETSAT Expands Activities and Sharing

Many NMSs also provide national hydrological services. Noting this, EUMETSAT agreed in 2001 to establish an optional program to allow its members to participate in the Jason-2 Ocean Surface Topography Mission (OSTM) along with NASA, NOAA, and the French Centre National d'Etudes Spatiales (CNES). Aligning with US practice, EUMETSAT adopted a free and open policy for Jason-2 data, categorizing all data from the satellite as "essential" in accordance with WMO Resolution 40.[52] In 2008, EUMETSAT decided to participate in Jason-3, which made data available under the same conditions.[53]

In 2006, EUMETSAT launched its first polar-orbiting satellite, Metop-A. Plans for the EUMETSAT Polar System (EPS) stretched back more than 20 years to the 1980s, when the United States contemplated reducing the number of polar-orbiting satellites that it maintained. As part of this plan, the United States would operate polar-orbiting satellites that cross the equator in the afternoon, which were most relevant to US weather developments. Europe was the natural candidate for taking over operations in the morning orbit, as the data collected by those satellites would be of greater interest for Europe. After years of discussions and changing plans, including a phase in which meteorological instruments would have flown on the space station in coordination with NASA's Mission to Planet Earth, in 1998, NOAA and EUMETSAT signed an agreement to develop the Initial Joint Polar System. Under the arrangement, NOAA and EUMETSAT would coordinate launches to provide continuous global coverage of the afternoon and morning orbits, and each satellite would carry some instruments developed by each agency.[54]

The agreement also described the data sharing policy that would govern the joint system. From the beginning, the data policy leaned toward the US system. Data from all instruments on the satellites operated by NOAA and from NOAA instruments operating on the Metop satellites would be subject to NOAA's data policy. Only data from the Metop instruments on the Metop satellites would be subject to control and potential restriction by EUMETSAT under its data policy.[55] The agreement was extended in 2003 in the lead-up to the first Earth Observation Summit—the high-level meeting that eventually resulted in the creation of GEOSS. The NOAA Administrator stated that the agreement set a strong precedent for the free and open exchange of data.[56]

When EUMETSAT released its official data policy for the EPS system in 2006, a significant amount of data was made openly available. EUMETSAT classified all real-time direct readout data from the satellite as "essential" under WMO, meaning that data of each region collected by local ground stations as the satellites passed overhead would be available to all users without charge or restrictions. A collection of global and regional products was also made available under the same conditions. Additional minimally processed data from the satellite and all archived data and products could be accessed without charge or conditions on use, but required signing a license agreement, and redistribution of this data was restricted.[57]

Growing European Support for Open Data Sharing

The European Union continued to push for increased data sharing. In May 2007, it issued a directive establishing the Infrastructure for Spatial Information in the European Community (INSPIRE). Building on the PSI Directive and the directive on public access to environmental information, the directive stressed the importance of interoperability, discoverability, and accessibility of data. It stated that access to metadata and the ability to view geospatial data should be made available free of charge. However, it stated that the reuse of this data for commercial purposes could be restricted, and charges could be levied for data viewing if necessary to secure their maintenance, particularly in cases involving very large volumes of frequently updated data. The ability to download or transform data as well as data-related services could also be subject to a reasonable charge.[58] The directive required that all member states adopt measures for sharing geospatial datasets for public tasks that have an impact on the environment. Member states could license and charge for these datasets, but should avoid any restrictions likely to create practical obstacles to their use, and any charge should be kept to the minimum required to ensure the necessary quality and continued supply of data. Exceptions to sharing could be made in cases where public safety or national security may be compromised.[59] Overall, in tone and spirit, the directive advocated for more open data provision, but in practice, it allowed for potentially significant restrictions.

The Network of European NMSs, EUMETNET, which had many of the same members as EUMETSAT, in 2009 produced the Oslo Declaration. This document noted the INSPIRE and PSI Directives and their impact on expectations for NMSs. It also stated that the Internet and other technologies had "changed radically the expectations from the general public and the access to and possible sharing and use of data, products, and services." EUMET-NET members committed to progressive expansion of both the datasets that they provided on a free and unrestricted basis as "essential" data consistent with WMO Resolution 40 as well as the datasets that are licensed for reuse by the private sector. They aimed to lower prices and adopt user-friendly online access systems.[60] EUMETSAT officially adopted the Oslo directive in 2009.[61]

EUMETSAT highlighted its existing efforts in this area, rebranding its satellite archive as the EUMETSAT Data Center in 2009 to better reflect the service it provided to the user community, including not only storage, but also processing and dissemination of data for all users.[62] EUMETSAT had operated a satellite archive since 1995, as an important aspect of its mission to support the collection of long-term data records for climate change. Data in the archive stretched back to 1981. From 2008 to 2009, the amount of data retrieved by users jumped from 400 terabytes to more than a petabyte.[63]

Copernicus and Increasing Open Data

Also in 2009, EUMETSAT signed the EUMETSAT/ESA Framework Agreement on Global Monitoring for Environment and Security (GMES), later renamed Copernicus. Copernicus was originally conceived in 1998 as a joint European Commission (EC) and ESA program to provide data needed by European decision-makers to address environmental policy issues.[64] Given the potential for overlap with its own activities and the high-profile nature of the program, EUMETSAT had a significant interest in becoming involved.[65] Under the June, 2009, framework agreement, EUMETSAT would contribute data from its missions to the Copernicus program. It would also operate three "Sentinel" satellites focused on oceanography, atmospheric composition, climate, and global land monitoring. Participation in Copernicus required that EUMETSAT provide the relevant data in accordance with the Copernicus full and open data policy.[66]

Membership of EUMETSAT expanded over time, particularly following expansion of the European Union as a whole—although the memberships of the two organizations still differed. In early 2016, EUMETSAT had 30 member states.[67] Its most advanced geostationary satellite, Meteosat-11, launched in July, 2015. It also operated two polar-orbiting satellites, Metop-A and -B, and was participating in the Jason 2 and 3 oceanographic satellite programs.[68]

As of early 2017, nearly all EUMETSAT data was freely available to all users. Some of the freely available data is provided with no restrictions and without the need for a license, consistent with the definition of "essential" data in WMO Resolution 40. This includes a subset of data from the Meteosat geostationary satellites, direct readout data from all instruments on the Metop satellites, and all data from the Jason-2 and -3 oceanographic satellites. Additional data is freely available to all users, but subject to user registration and covered by a license that restricts redistribution of minimally processed data. These conditions are designed to allow EUMETSAT to monitor use of its data and maintain close communication with its users. Much of the data from the Meteosat and Metop satellites falls into this category.[69]

As of early 2017, the only data for which EUMETSAT still charged a fee was the near-real-time half-hourly and quarter-hourly data from the Meteosat geostationary satellites. Commercial users, including commercial activities of member and nonmember NMSs, and nonmember countries at or above the "Upper Middle Income Value" level as defined by the World Bank are required to pay. Countries that have negotiated bilateral agreements with EUMETSAT, which includes all of the global meteorological satellite operators, nations below the "Upper Middle Income Value" threshold, and research and educational users can all access this data for free, subject to a license and restrictions on redistribution. Data is provided free to all relevant users in the case of a tropical cyclone or natural disaster.[70] Given the limited application of fees, revenue from EUMETSAT data and product sales was quite low, approximately €2 million in 2015, and accounted for less than 1 percent of the agency's budget for the year.[71]

EUMETSAT Summary

As in the United States, European meteorological satellite programs were developed in coordination with WMO and the international meteorological community, embracing the tradition of free and open data sharing. However, Europe's start also took place at a time of tight government budgets and increasing government commercialization efforts. EUMETSAT had the added challenge of attracting and retaining national participants to fund its programs. These economic challenges generally dominated the historical normative arguments for continuing free and unrestricted data exchange.

Although European agency officials advocated for acceptance of cost recovery policies on the international level within WMO, the wording used suggests that they saw this as a policy of necessity, imposed by national-level decision-makers. Advocacy for cost recovery was often couched in terms of NMSs being "pressured" or "required" to recover costs or engage in commercialization. NMSs argued that it was something they had to do in order to maintain and increase their existing data collection efforts, knowing that without continued collection, there would be no data to share, freely or otherwise. Further, European agency officials repeatedly expressed support for free and open international data exchange. EUMETSAT took a leadership role in promoting the responsibility of meteorological agencies to provide critical data for understanding climate change and recognized that this would require an increase, not a decrease, in international data sharing.

As time went on, it became apparent that cost recovery efforts would not be a silver bullet, allowing agencies to self-fund the cost of expensive infrastructure. Industry groups complained about unfair competition and negative effects of these policies on private-sector growth. Arguments about the economic benefits of adopting open policies, rather than cost recovery, were made stronger by the use of the United States as an example, particularly in meteorology. The spread of computers and the Internet further strengthened these arguments. These trends began to lessen, and eventually reverse, the pressure from national- and EU-level decision-makers to attempt to recover costs. This allowed national agencies, and by extension EUMETSAT, to gradually reduce the amount of data subject to fees, resulting in the present situation in which the majority of data is freely available.

Unlike the United States, much of EUMETSAT's data is subject to licensing restrictions that prohibit redistribution of the data. This reflects differences in national and regional law between the two entities, with the EU enabling, and originally encouraging, copyright of government databases, while the United States prohibited it. It also reflects ongoing economic pressures on the part of EUMETSAT. Requiring all users to request data directly from EUMETSAT ensures that the agency maintains good statistics on data dissemination. In the absence of revenues, these statistics are the primary tool EUMETSAT has to demonstrate the importance of its data sharing policies and systems to its members, and by extension, their national-level policy-makers.

Like NASA, the European Space Agency (ESA) is a science and technology agency that primarily develops research-oriented satellites. Throughout its history, ESA has seen the value of sharing this data with the research community. However, the development of ESA's first nonmeteorological Earth observing satellite corresponded with a global trend of commercial remote sensing activity, and ESA's early data policies reflected its own interest in participating in this trend. ESA developed policies that restricted access to data to allow commercial sales. Even with regard to research access, ESA placed restrictions on access, such as the need to submit full research proposals as a requirement for data access, which further limited data use. As commercial efforts proved to be less lucrative than hoped and evidence suggested existing restrictions were hindering research use, ESA gradually decreased the amount of data sold and increased the accessibility of its data for researchers and others. A breakthrough in ESA's data policy came in 2009, when it first adopted an open data policy in support of the European Copernicus program. In the following years, ESA applied this policy to nearly all of its satellite data.

Creation of ESA and the First Satellites

The European Space Agency (ESA) was established in 1975 through the merging of two existing organizations—the European Space Research Organization (ESRO) and the European Launcher Development Organization (ELDO). The new organization was designed to strengthen European cooperation in the development of space research and technology and improve worldwide competitiveness of European industry. The ESA convention also noted that the human, technical, and financial resources required for activities in the space field were beyond the means of any single European country.[1] In the view of its initial 11 member states, cooperation was Europe's key to space.

ESA launched its first satellite, Meteosat-1, just two years after it was formed. Meteosat-1 was followed by Meteosat-2 and -3, launched in 1981 and 1988, respectively, after which time the meteorological program was transferred to EUMETSAT to be maintained on an operational basis. ESA had promoted use of these early satellites, organizing users' conferences in 1979 and 1980, and provided data from the Meteosat series at or below the cost of reproduction and dissemination for all users.[2] This was in line with ESA's convention, which stated that member states and the agency should ensure that any scientific results of its activities would be published or otherwise made widely available after prior use by the scientists responsible for the experiments.[3]

Earth Resource Satellite 1: Serving Multiple Communities

In the same year that Meteosat-1 was launched, the ESA Council recognized the need for Europe to develop its own remote sensing satellite program, outside of meteorology.[4] The first step in doing so was establishing the Earthnet program, which would acquire, process, archive, and distribute data from foreign satellites collected by European ground stations. Through this program, Europe gained expertise in working with remote sensing imagery. Earthnet also laid the framework for the ground segment, data dissemination systems, and data policies for future European satellite systems. Landsat data collected by Earthnet was made available at prices in line with those charged in the United States, which, according to ESA, required subsidization. Data from experimental satellites, including Seasat and HCMM, were available for free to principal investigators and at the cost of reproduction to other scientific users whose investigations were "considered meritorious" by ESA and NASA."[5]

Earth observation was designated as an optional activity within ESA. This meant that, unlike mandatory programs, states could opt out of the programs—only a majority needed to approve for a project to move forward—and financial contributions could be based on gross national income or on another arrangement agreed to by the participating states.[6] Almost all member states, with the exception of Ireland, participated in ESA's first nonmeteorological Earth observation satellite, the Earth Resources Satellite 1 (ERS-1). Canada and Norway also joined the program as cooperating states.[7]

Launched in 1991, just a year after the IPCC released its first assessment report, ERS-1 was seen as part of an effort to address important global environmental challenges. The satellite carried a synthetic aperture radar (SAR) instrument, the first on a civilian spacecraft since NASA's short-lived Seasat program flew more than a decade earlier, in 1978. The instrument made it possible to take global measurements regardless of cloud and sunlight conditions. It was also capable of collecting data on sea state, sea surface temperature, sea surface winds, ocean circulation, sea and ice levels, and more. ESA noted that ERS-1 data could contribute directly to the efforts of the IPCC, in helping to develop a more accurate representation of the interaction between the ocean and atmosphere in climate models, one of the major challenges identified in the first IPCC report.[8]

The SAR instrument provided new opportunities, but also posed challenges. Due to power constraints, the SAR instrument could only be operated for an average of 10 minutes per orbit, so data collection only took place at the request of users and had to be carefully planned. The Earthnet ERS Central Facility, responsible for most interactions with ERS data users, maintained an ERS Global Activity Plan based on user requests, which could also be consulted online to determine if the desired observation was already being taken or if a new observation needed to be requested. The large volume of data the instrument produced precluded onboard recording and storage, so the SAR instrument only provided data via direct readout.[9] ESA signed memorandums of agreement with the United States, Japan, India, Canada, Australia, Brazil, Indonesia, China, and others to allow reception, archiving, processing, and distribution of ERS-1 data, ensuring the possibility of near-global collection of SAR data.[10]

Although important, environmental research was not ERS-1's only mission. Planning for the ERS system had occurred during a wave of commercial remote sensing activity in the 1980s, beginning with the commercialization of the US Landsat system in 1984, followed by the launch of France's commercial SPOT satellite in 1986, and the commercial sales of Soviet satellite imagery beginning in 1987. Japan, India, and Canada all had satellite systems under development with expected commercial applications. In 1988, the Eurimage company was created to market and distribute selected

Earthnet products.[11] Given these trends, it is no surprise that operational and commercial applications were part of Europe's plan for ERS from the beginning.[12] As part of this plan, ERS-1 was considered a preoperational system: it would be expected to demonstrate the operational and commercial capabilities of the program, including the ability to recover some of the costs of the program by commercializing selected ERS products.[13]

From the beginning, ESA's ground segment plans, including the requirements for foreign ground stations, were designed to meet the needs of both research and commercial users. ESA ground stations had the ability to do rapid preprocessing of data that resulted in a series of "fast delivery products" that could be provided to users within three hours of observation, an attribute of particular interest for commercial users. The development of long-term archives essential for environmental research was also a priority. ESA committed to archiving ERS data for at least 10 years after the satellite stopped functioning. If a foreign ground station was not capable of long-term archiving, it was required to send a copy of the data to ESA.[14] ESA claimed copyright over all ERS data, and access was provided through licensing agreements.[15] The ground station agreements ensured that data collected at foreign stations would also be handled in a way consistent with ESA's data policies, respecting ESA copyright and implementing uniform pricing.[16]

In return for their contributions to the program, ESA member states could access ERS-1 data at the cost of reproduction and dissemination and were to be exempt from any charges designed to provide a return on investment. Selected data was also provided free of charge to meteorological organizations of participating states, EUMETSAT, and ECMWF, with the understanding that they were to be used only for forecasting purposes, and could not be redistributed to third parties without ESA's consent.[17]

Because it was introducing new instruments with new capabilities, ESA officials recognized that it would be necessary to do some work to develop a user base and demonstrate the satellite's potential. The organization released an Announcement of Opportunity for proposals to exploit ERS data and received hundreds of responses. However, while the scientific response was strong, there were fewer proposals focused on developing practical applications that might have operational or commercial uses. A second Announcement of Opportunity was released, focused specifically on applications.[18] Investigators associated with the more than 200 projects chosen through these announcements were also able to request and access data free of charge.[19]

ESA began its commercial efforts in earnest after just a year of satellite operations, signing a deal with the Eurimage, RADARSAT International, and Spot Image Consortium (ERSC) in 1992. ESA provided the consortium with worldwide commercial distribution rights for ERS-1 satellite data. Each of the companies in the consortium had experience in marketing satellite data—for Europe, Canada, and France, respectively—and ESA hoped to take advantage of this experience for the promotion, marketing, and distribution of its own data.[20]

ESA fixed the prices to be charged for end users, and the percentage of revenue to be allocated to the distributers. Prices ranged from 250 euros ($300 in 2016) for a medium-resolution SAR scene to more than 2,000 euros ($2,400 in 2016) for a more advanced, highly processed SAR product.[21] These prices were designed to evolve over time, increasing as acceptance of the product's value for operational or commercial applications became apparent.[22] At the outset of the arrangement, ESA noted that researchers not selected as part of the Announcements of Opportunity process could purchase data from these distributors at commercial prices. In 1994, ESA amended this, introducing special pricing for research users at about 30–50 percent of commercial levels.[23]

International Data Sharing and Data Continuity

In 1992, the same year it signed the agreement with ERSC, ESA founded the Earth Sciences Division to provide scientific support for current and planned missions and ensure that the scientific community became fully involved in the selection of future missions. ESA also endorsed the CEOS data exchange principles for global change data, which stated that maximizing the use of satellite data was a fundamental objective, and that a common goal was the provision of data to global change researchers at a price primarily reflecting the cost of fulfilling the user request. Importantly for ESA, the principles also acknowledged that the acquisition and processing of satellite data involves major investments, that the resulting data has value, and that the constraints of mission operations and available resources may require different mechanisms for data sharing in different programs.[24]

ESA had just begun its commercial activities with ERS-1 a year earlier, and it was determined to continue the effort. In support of this goal, ESA recognized that data continuity was crucially important for attracting operational and commercial customers. It wasn't practical for these users to invest in expensive processing equipment or to develop complex techniques for development of value-added products if data might only be collected for a couple of years.[25] In response to this issue, ESA launched ERS-2 in April 1995 carrying the same instruments as ERS-1, with the addition of the Global Ozone Monitoring Experiment (GOME) sensor. ERSC continued to manage marketing and dissemination of data to commercial users, and the number of ERS SAR scenes sold increased steadily, from 325 in 1992 to 3,074 in 2000.[26] National public-sector agencies and intergovernmental organizations—particularly those responsible for agriculture, environmental monitoring, and meteorological services—were the primary operational users of ERS data.[27]

ESA also continued to emphasize scientific exploitation of the data—noting that data continuity was important in studies of climate change, as well. The agency selected 340 research teams in response to an Announcement of Opportunity to receive initial data from ERS-2 in tandem with ERS-1.[28] By 2001, ESA had released several Announcements of Opportunity, selecting a total of more than 1000 projects from scientists around the world, and resulting in more than 8000 papers or articles published.[29]

Challenges to Implementation

While ESA pointed to the number of research papers published and SAR scenes sold as evidence of success in serving multiple user groups, some argued that restrictions and fees were having a negative impact on research use. A 2001 report by the US National Research Council noted that data from the joint NASA-CNES Topex/Poseidon ocean-altimetry instrument had been used much more extensively than data from similar instruments on ERS-1 and -2. The report attributed this in part to the fact that Topex/Poseidon data was openly available to all users while access to ERS data was much more restricted. Approval for access to Topex/Poseidon was typically granted within 24 hours of filling out the online form. To gain access to ERS data, scientists were required to submit a formal research proposal, which, in the case of non-European researchers, could include only a request for data and no additional funding. The approval process for these proposals generally took about a month. If approved, researchers were required to sign a formal agreement that included a list of the names of all people who would access the data, and redistribution was limited to this group. The NRC found that these restrictions and requirements discouraged many scientists from attempting to use the data.[30]

European attitudes toward public data, which had once strongly favored commercial efforts by government agencies, were beginning to change. In 1999, the European Union published a green paper titled, "Public Sector Information: A Key Resource for Europe" that raised questions about the ethics and efficiency of cost recovery policies. The paper questioned whether public organizations have the right to charge for the provision of information, given that it was produced at taxpayer expense. However, it also recognized that if a small section of the public wishes to use the information, it may not be ethical for their activities to be subsidized by the rest of the population. Regardless of the ethical issues, the paper suggested that restrictive data policies in Europe had put European companies at an economic disadvantage compared to their American counterparts, where data was made freely available.[31] A study commissioned by the European Union the following year suggested that if Europe made its data freely available, it would generate more government revenue through taxes on the value-added industry than it had in direct cost recovery efforts.[32]

Balanced Development: Differentiating by Use

ESA maintained its commitment to data continuity, and ERS-1 and -2 were followed by the launch of Envisat in 2002. At the time of its launch, Envisat was the largest civilian Earth observation satellite ever built—about the size of a school bus. It included more advanced versions of the instruments flown on ERS-1 and -2 as well as new instruments and was capable of taking measurements relevant to oceans, land, ice, and atmosphere.[33] The Envisat ground segment and data distribution system was upgraded, as well. Electronic data was available via direct broadcast for users with the appropriate receiving equipment, and products smaller than one megabyte could be accessed via the Internet. Electronic data could also be provided on CD-ROM or other physical media.[34]

To accompany the new technology, ESA developed a new data policy that it hoped would address weaknesses of the ERS data policy. The objectives were to "maximize the beneficial use of Envisat data" and "stimulate a balanced development of science, public utility, and commercial applications." With the new data policy, ESA stated that it would increase support for scientific activities and provide continued support to operational applications. It also aimed to support the development of a globally competitive Earth observation industry in its participating states and foster a transition toward a sustainable market, led by user demand.[35]

The policy reiterated ESA's ownership of the data, including any derived products "to the extent that the contribution of Envisat is substantial and recognizable."[36] ESA's legal right to the data had been strengthened by the 1996 European Commission Directive on the legal protection of databases, and subsequent national laws, which stated that collections of numbers, facts, and data were eligible for copyright protection.[37]

The biggest change to the Envisat data policy was the introduction of a new framework that differentiated conditions of access by use, rather than by user, and established two use categories. Category 1 use included research and applications development, including research in preparation for future operational use, and ESA internal use. The ESA Earth Observation Program Board would be responsible for identifying and approving category 1 use on the basis of a peer review processes. All other uses, including operational and commercial use, fell into category 2.[38] The Envisat data policy was developed in 1998 in anticipation of the launch, and the policy was applied to ERS-1 and -2 data beginning in 2000.[39]

ESA distributed data for category 1 use itself. By 1998, ESA had already released an Announcement of Opportunity and selected 674 proposals for use of early Envisat data.[40] In the past, the only recourse for researchers whose projects were not selected as part of an Announcement of Opportunity was to purchase data from the ERSC consortium at discounted market prices. Under the new policy, ESA would accept unsolicited proposals outside of the Announcement of Opportunity process. Like the Announcement of Opportunity submissions, these proposals would be subject to peer review. If accepted, they would be eligible for access to data at category 1 prices, set by ESA "at or near" the cost of reproduction, which included the costs of media and shipping.[41] Some low-volume products could be downloaded at no cost from password-protected servers.

The category 1 price for the higher-volume ASAR scenes was typically between 100 and 300 euros ($120–$360 in 2016) per image.[42] With the additional approval of the Earth Observation Program Board, these fees could also be waived. Similar to the previous policy, users would be required to sign an agreement limiting use of the data to the purpose agreed to in the proposal. They were prohibited from redistributing data to third parties and were required to report on and publish research findings.[43]

National stations run by ESA member states and foreign stations operated under international agreements with ESA could also be authorized to provide data to category 1 users.[44] Authorization required that these entities follow similar procedures for selecting proposals, including peer review and evaluation, and place the same conditions as ESA with respect to restrictions on redistribution and reporting.[45] Scientific users who wished to avoid the proposal process and reporting requirements had the option of working with the distributing entities, either purchasing data at commercial prices or negotiating with these entities for more favorable data access conditions.[46]

Not all of the changes increased data access. Under the ERS data policy, meteorological services associated with the WMO had been able to access ERS data and products free of charge through the WMO Global Telecommunications System network. However, ESA determined that the arrangement did not provide sufficient transparency concerning the ultimate use of the data—some believed that meteorological agencies had been redistributing data contrary to ESA's policies—and this practice was discontinued. Under the new policy, meteorological services would follow the same data access procedures as other approved category 1 users, downloading data from password-protected servers.[47]

An interesting exception to the two-category system involved data from the three Envisat instruments focused on atmospheric observations. ESA recognized that foreign satellites carried similar instruments and made the data available at marginal cost. Because this would preclude the possibility of commercial sales of this data, ESA determined that data from these instruments would be made available for all uses—research, operational, or commercial—under category 1 conditions. Approved ESA category 1 users would be able to download the data from these instruments for free from password-protected servers.[48]

Data for category 2 use would be accessible through commercial distributing and specialized operators directly appointed by ESA.[49] With the goal of ensuring a competitive market, ESA had ended its arrangement with ERSC and instead expanded its arrangement to two commercial providers. The two new consortia chosen by ESA to distribute the data were SARCOM, led by Spot Image, and EMMA, led by Eurimage.[50] Unlike the previous policy, the new Envisat policy allowed distributors to set their own prices, although ESA retained the right to fix a ceiling level on the market price. These entities would in turn pay ESA an amount comparable to the price for category 1 use. This allowed for significant profits on the part of the distributing entities. EMMA, for example, sold a three-day "emergency," newly acquired Envisat ASAR image, for 2,600 Euros ($3,000 in 2016), about 10 times the category 1 cost.[51] This arrangement was designed to help promote the creation of a sustainable European commercial remote sensing sector. The Envisat data policy explicitly stated, "the creation of a revenue stream to ESA is of less importance."[52]

Only in special circumstances, and with the prior agreement of ESA, would commercial distributors be allowed to set prices below the category 1 price. Foreign stations would be subject to an access fee for the right to receive ESA data, and all distributors would be subject to a fee for SAR, ASAR, and MERIS instrument programming requests. (The power requirements and large volumes of data generated by the Advanced SAR instrument, as well as the Medium Resolution Imaging Spectrometer [MERIS], required that the instruments be tasked according to user requests.[53]) With an aim to ensure fair competition in the market, ESA required that commercial distributors provide equal opportunity of access to ESA data and products to all users, including other market operators, service providers, and value-added operators.[54]

ESA and its commercial partners expected that data and products from the radar and optical imaging instruments would be of most interest to operational and commercial users. SARCOM and EMMA anticipated applications in areas such as forestry, ship routing, coastal management and fisheries, and risk management to form the primary commercial markets.[55] Other national and foreign stations licensed to sell data for category 2 use typically focused on niche markets. For example, the Canada Centre of Remote Sensing operated the only station capable of collecting and distributing ERS SAR Fast Delivery Products over an area of interest to the

Canadian Ice Service, so it was licensed to focus primarily on servicing this market.[56]

In its policy, ESA committed to archiving the data for at least the active lifetime of Envisat, and retained the right to obtain a copy of all Envisat products archived at national and foreign stations. National and foreign stations that didn't have the capability to maintain an archive were required to offer the acquired datasets to ESA for archiving in its facilities. For the most part, access to archived data was provided under the same conditions as those applied to near real-time Envisat data as part of the category 1 and 2 framework.[57] Special conditions applied to access to and archiving of data from three of the instruments—the Scanning Imaging Absorption Spectrometer for Atmospheric Cartography (SCIAMACHY), the Advanced Along-track Scanning Radiometer (AATSR), and the Doppler Orbitography and Radio-positioning integrated by satellite (DORIS)—which were each developed and provided by a single nation. Raw data from these instruments was provided to the nation that developed the instrument at no charge, and the instrument developer would have the option of owning the archives of the instrument at the end of the mission if ESA was not in a position to ensure maintenance of the archive itself.[58]

Natural Disasters

In addition to the updates to its official data sharing policy for ERS-1, ERS-2, and Envisat, ESA also undertook a major effort to increase international satellite data sharing in the event of natural disasters. Together with the French space agency, ESA initiated the International Charter on Space and Major Disasters in 1999, with the recognition that satellite images could be used to see where damage had occurred and help to more effectively and efficiently target recovery efforts. The Canadian Space Agency, the Indian Space Research Organization, EUMETSAT, NOAA, the National Space Development Agency of Japan (NASDA), China, Russia, Argentina, and Brazil all joined the Charter in subsequent years, contributing data from their own space systems. Under the terms of the agreement, any authorized user could call a single number to activate the charter, leading to mobilization of space assets from all members. Participants would take images of the affected area and provide them to relevant authorities and aid organizations.[59]

The Living Planet Program

In the mid-1990s, as development of the Envisat mission was underway, ESA began to look ahead. The agency argued Europe should not become reliant on others to obtain data that underlay its positions on international environmental issues, and that a robust European Earth observation program would give Europe the ability to speak independently on issues of global concern and provide data that is needed to validate conformance with international environmental conventions, such as the Kyoto Protocol. It promoted a coordinated European approach to Earth observation from space, an idea that the European Parliament officially supported in 1998.[60]

To meet these goals, ESA proposed the "Living Planet" program in 1999. The new program would focus on pursuing scientific knowledge, enhancing the quality of life, providing Europe with an independent Earth observing capability, and promoting European industry and value-added services. ESA would distinguish between science missions, referred to as Earth Explorer missions, and prototype operational missions, referred to as Earth Watch missions. Many of the Earth Explorer missions would relate directly to the work of the IPCC. Earth Watch missions would be designed to eventually become freestanding services outside the agency.[61]

ESA argued that while promotion of European industry was one goal, the implementation of a strategic European Earth observation system could not be left to the private sector. It also complained that the US policy of "anchor tenancy," in which the government provided a guaranteed market to private remote sensing companies, placed the European industry in an impaired situation and hampered its competitiveness. The Living Planet Program would help to address this, developing an industrial capability to take on the global market.[62]

Rather than continuing to develop large, complex satellites like Envisat, Living Planet missions would be smaller and more focused, and would launch more frequently, a strategy that aimed to decrease the cost of the Earth observation program by about 25 percent annually.[63] The Earth Explorer missions were subdivided into core missions and opportunity missions. Core missions were larger, reflecting widespread consultation with the science community. Opportunity missions were smaller, lower cost, and aimed at addressing immediate questions. Rather than requiring approval and funding for each individual satellite as an optional program, the Living Planet program would be approved as a whole, under a mechanism called the Earth Observation Envelope Program (EOEP). EOEP was funded as an optional program in five-year increments, with support for the full life cycle of the missions, including planning, research, development, launch, and data exploitation.[64] This new arrangement would save ESA members time and help to protect the continuity of the program.

Earth Explorer's Science Focus

In 2002, ESA released the Earth Explorer Data Policy, which stated that because these were science missions, data from all Earth Explorer missions would be provided according to the category 1 use procedures described in the Envisat Data Policy. Proposals would be subject to peer review, and, if accepted, data would be provided at, or close to, the cost of reproduction. Subject to further review, the fee could be waived altogether by the ESA Earth Observation Program Board. Data that could be downloaded from the ESA servers, for which the marginal cost was essentially zero, would be free to approved users. Restrictions on use and redistribution and requirements for reporting remained the same. Although it wasn't expected, the ESA Earth Observation Program Board added that it would consider amending the policy if an organization or private entity demonstrated interest in a commercial distribution scheme. The policy allowed special arrangements to be made for cases in which member nations provided additional contributions, for example, building and providing an instrument for inclusion on the satellite at no charge to ESA.[65]

Copernicus and European Data Policy Trends

Parallel to the Earth Explorer missions, the Earth Watch program was also taking form. Part of this program consisted of ESA's contribution to Europe's third-generation Meteosat weather satellite, as well as its first polar-orbiting weather satellite series, Metop, which the agency was developing under an agreement with EUMETSAT. A second, larger component was to be developed in partnership with the European Union, as part of the European Global Monitoring for Environment and Security (GMES) program, renamed Copernicus in 2012. Endorsed by the European Parliament in 1998, Copernicus would apply the operational mindset of meteorology to a much broader set of environmental domains, combining in situ and space-based data to develop and provide information services for a wide variety of activities and organizations. Copernicus would include services for agricultural monitoring, fisheries, emergency response, humanitarian aid, border surveillance, and more. The program included three main components: in situ measurements, space-based observations, and operational services. The space component of Copernicus was cofunded by ESA and the European Commission, with funding for the operational phase expected to be primarily the responsibility of the European Union. In 2007, the European Union released a European Space Policy that identified Copernicus as one of two flagship European programs.[66]

The space component of the program included five families of satellites, called the "Sentinel" series. Each Sentinel satellite would focus on a different instrument and mission. Sentinel-1 would carry a radar instrument, continuing ESA's long history of measurements of this type. The instrument would be capable of collecting data with a resolution of five meters, allowing it to see in significantly more detail than the Envisat radar instrument, which was capable of 30-meter resolution. The Sentinel 2 satellites would carry a multispectral imager particularly useful for land imaging. With a 10-meter resolution and 13 spectral bands, it was more advanced than the US Landsat 8 system, which had a resolution of 15 meters and collected data in 11 spectral bands. Sentinel-3 would carry instruments focused on land and oceanographic measurements of particular importance to climate change, including sea and land surface temperature, and ocean and land color. In 2012, ESA and the EU decided to also include the follow-on Jason Continuity of Service (Jason-CS) oceanography mission as part of Copernicus, continuing the successful cooperation between the United States, France, and Europe in this area. Sentinel-4 and -5 would focus on atmospheric monitoring, from geostationary and polar orbit, respectively. Under an agreement with EUMETSAT, these missions would be composed of instruments flying on European weather satellites already planned to operate in these orbits.[67]

As plans for the Copernicus program were developed throughout the first decade of the 2000s, support for increased access to data was growing, particularly on the European level. In January, 2003, the European Commission updated its 1990 directive on public access to environmental information. It called on member states to make available and disseminate environmental information to the widest extent possible, in particular using information and communication technologies.[68] Later that year, it issued a directive on the reuse of public-sector information, the "PSI Directive," which encouraged public-sector bodies to make their documents and information available for reuse in support of the right to knowledge, a basic principle of democracy.[69] European governments and the European Commission were highly supportive of the Earth Observation Summit, first held in 2003, which resulted in the creation of the Group on Earth Observations (GEO) and later the Global Earth Observation System of Systems (GEOSS). European leaders stated that Copernicus would be the European contribution to GEOSS.[70]

In 2006, the EU released a directive establishing an Infrastructure for Spatial Information in the European Community (INSPIRE). INSPIRE aimed to harmonize procedures for discovering, accessing, and sharing spatial information across a large number of domains and levels of public authority to assist environmental policy-making. The directive noted a direct connection between this effort and the Copernicus initiative, and encouraged member states to take advantage of the data and services resulting from the program.[71]

Updates to ESA Data Policies

There were continued indications that ESA data policies were restricting research and operational use of satellite data. The NASA moderate-resolution imaging spectrometer (MODIS) instrument, included on the Terra and Aqua satellites launched in 1999 and 2002, respectively, was very similar to the ESA MERIS instrument included on Envisat, but data from the NASA instruments was much more widely used in the ocean color community.[72] The number of publications using MERIS data was significantly lower, even taking into account the number of years in orbit, and the MERIS publications were overwhelmingly written by European researchers.[73] Scientists complained that MERIS data was difficult to obtain, restricting the ability to carry out comparisons and data merging efforts.[74]

In 2007, ESA streamlined the process for access to data systematically provided online, which included recently collected and archived data from the majority of the instruments. Rather than submitting a full scientific proposal subject to peer review—often a lengthy process—users could submit a project title and short description on the ESA website, which would go through a short review process. If approved, the user would be required to agree to ESA's data access terms, and then the requested information could be downloaded for free. Full research proposals were still required for category 1 data that was subject to acquisition or dissemination constraints, such as new ASAR data that needed to be scheduled for collection, or full-resolution MERIS data that was too large for systematic online storage and dissemination.[75]

By September 2008, acceptance notification for fast registration applications was immediate, and only SAR and ASAR data still fell within Category 2 use. Regardless of category, all users were required to sign terms of use stating that the data would only be used for the purpose described in the proposal, that the results would be widely published, and that reports on progress would be provided to ESA.[76]

The new policy procedures were put in place just as the Earth Explorer program was beginning to ramp up. CryoSat was expected to be the first Earth Explorer Mission, continuing ESA SAR observations begun with ERS-1, but it was lost in a launch failure in October 2005. A replacement, CryoSat-2, was launched in 2010. The Gravity field and steady-state Ocean Circulation Explorer (GOCE) and the Soil Moisture and Ocean Salinity (SMOS) missions were both launched in 2009, and Swarm, a mission to map Earth's geomagnetic field, launched in 2013. Beginning in 2010, Earth Observations became ESA's highest funded activity, with spending greater than that provided for launch vehicles.[77]

ESA Embraces Open Data

ESA had taken incremental steps to improve access to its scientific data, but spurred by trends on the European level, it took a dramatic step with regard to the Copernicus data policy. In October 2009, ESA adopted the "Joint Principles for a GMES Sentinel Data Policy," endorsing free, full, and open sharing of data from the Sentinel satellites. The policy was designed to maximize the beneficial use of Sentinel data for the widest range of applications. It also aimed to strengthen Earth observation markets in Europe, in particular the downstream sector. All Sentinel data would be provided to all users free of charge, with no distinction made between public, commercial, and scientific use, nor between European and non-European users.[78] Users could access the data online subject to user registration and acceptance of terms and conditions that required the source of the data to be appropriately cited, but did not restrict redistribution of the data.[79] The policy noted that in the event that security restrictions apply to specific Sentinel datasets, special operational procedures would be activated.[80] The 2009 policy was a historic break from previous policies, fully embracing the open data model.

A few months later, in May 2010, ESA released a new ESA Data Policy revising its existing practices to align more closely with the newly announced Sentinel Data Policy. The new policy would apply to past, present, and future ESA missions, including ERS-1, ERS-2, Envisat, and the Earth Explorer missions. Under the new policy, ESA made distinctions primarily based on the data, rather than on the use or user, differentiating between a free dataset and a restrained dataset. Even the restrained dataset would be available free of charge online for research users—the restraints referred to the process required to gain access to the data, which allowed for some continued efforts at commercial sales.

The majority of ESA data was contained in the free dataset. Online access would be immediate, subject to registration and acceptance of the ESA terms and conditions. No proposal describing the use of the data was required, but the data could not be redistributed to a third party without requesting and receiving permission from ESA. The restrained dataset included SAR data from the ERS and Envisat missions and requests for datasets of very large volume, which could not be handled online. SAR data remained subject to the previous category 1 and 2 framework. Category 1 users would need to submit a proposal, which would be peer reviewed. If selected, the data would be made available at no cost. Commercial users could purchase data from distributing entities.[81] SAR data remained available for purchase through the ESA distributing entities, EMMA and SARCOM.[82]

Commercial Misgivings, Then Support, for Open Data

ESA's move to an open data policy for Copernicus was driven largely by trends and guidance on the European level, and ESA fully expected the European Commission to approve the new Sentinel Data Policy in its September 2010 regulation on Copernicus initial operations. This did not occur. The regulation referenced INSPIRE and GEOSS and called for full and open access to Copernicus data subject to international agreements and security restrictions. However, the Commission reserved the right to introduce restrictions on data access and put off the final decision on the Copernicus data policy to the future.[83]

This surprising hesitation stemmed primarily from concerns voiced by commercial remote sensing companies in Europe. A number of companies, including the UK-based Disaster Monitoring Constellation International Imaging (DMCii) company, Spanish Deimos imaging, and Italian e-Geos saw Copernicus as a threat, because Sentinel data was closely comparable to the satellite data that they were currently selling. Data from the Sentinel satellites would actually be of higher quality than the 22-meter resolution, three-spectral-band satellite data marketed by Deimos, and free provision of Sentinel data would almost certainly put the small company out of business. The largest commercial satellite data provider in Europe, Airbus Defence and Space, recognized the Copernicus program would not include the type of high-resolution imagery that it offered from its SPOT and Pleiades satellites. However, the company voiced concern that some users would settle for less-than-optimal imagery if it were free, rather than purchasing commercial data, which could hurt Airbus's sales. A European Commission official concerned about the effect of the Copernicus program on these companies noted that there was still room for compromise, stating his belief that "free and open doesn't always mean without charge."[84]

In response, ESA commissioned a report on the issue from the European Association of Remote Sensing Companies (EARSC), a membership-based organization that included both commercial satellite data providers and value-added providers. The resulting report, released in December 2012, was titled, "About GMES and Data: Geese and Golden Eggs," and it expressed strong support for a free and open policy, equating attempts to generate public revenue from data sales with killing the goose that laid the golden eggs. It argued that public-sector data was valuable, but government bodies that sell their data "discourage innovation and creativity, inhibit the creation of new businesses, and generally earn less for the government than if they were to give it away for free." It also pointed to the trend toward free and open policies among European governments for all types of government-collected data, as well as the support for open data demonstrated by the European Commission itself, which had passed a directive establishing the European open data portal in December 2011.[85]

The report began by arguing that the special economic characteristics of public-sector information—its nonrival and nonexcludable nature—fit the definition of a public good and meant that data collection was best financed through general taxation. It explained further that the wide availability of the Internet meant that the marginal cost of providing data was essentially zero and that the cost structure for producing information—expensive initial investments, but very low marginal costs—favored mass production and led to natural monopolies. The report looked at existing studies of the value of public-sector information, including the PIRA report from 2000, which argued that free data provision would be more lucrative for European governments than cost recovery. Additional studies conducted by other firms in the interim had come to similar results. The report cited a number of case studies in other sectors in which cuts to data access fees had led to large increases in demand for information. Finally, it used the example of Landsat, the closest analog for Sentinel data, to show that the move to free data provision could result in a dramatic increase in data access and use.[86]

The report did note the risk to commercial data providers, stating that "there is no doubt that free data coming from the Sentinels will have an impact on their data models." However, they argued that this threat would be counterbalanced by the expected growth in the total market for Earth observation services, stimulating greater awareness and use of commercial data. It also noted that data from high-resolution providers may be purchased to meet GMES needs. The report suggested that additional methods for mitigating the damage to private business models should be pursued, although it didn't offer suggestions as to what these methods might entail.[87]

One restriction the report did support was limiting free access to European users only, rather than making data globally available. It argued that European taxpayers' money had been used to produce GMES data and the economic and social benefits within Europe would be watered down if non-European entities were allowed access on the same terms. The organization recognized that these restrictions would require economically inefficient barriers and red tape, but argued that these were justified, as the

European Commission had a special responsibility to support European industry.[88]

EARSC recognized that European leaders may not like this recommendation. It acknowledged that other countries, particularly the United States and Brazil, were making data freely available on a global level, that GEOSS was further encouraging this trend, and that free global sharing of Sentinel data would promote an image of Europe as a global leader. It also recognized that the Sentinel data is global in nature and could help promote markets for EO services in other parts of the world, just as it would in Europe. As a compromise, the organization suggested that if the European Commission and ESA decided to make Sentinel data freely available to global users, this should be done through the framework of international agreements that ensure reciprocal access to international datasets. These agreements should include a limitation on the performance of future satellites (necessary to protect existing European commercial data providers) and technical controls that could be used to cut off data access if the free data policy was abused by other nations.[89]

In July, 2013, EARSC followed up with a position paper reiterating its support for a free and open data policy, but also raising the importance of clarity on access policies for commercial users. For example, the organization wanted to know whether private entities would have the right to request tasking and programming of Sentinel satellites for the creation of commercial products, what data and products would be produced, and how they would be made available. EARSC again argued for a "preferential system of access" for the European services industry that would ensure data access is possible without bottlenecks or delays resulting from non-EU commercial or public users.[90]

Copernicus Data Is Officially Open

In July, 2013, the European Union made its decision, stating in a new regulation that "access to Sentinel data should be free, full, and open in line with the Joint Principles for a Sentinel Data Policy" adopted by ESA. For the most part, it did not accept EARSC's recommendation to privilege European companies, and instead stated that data from dedicated Copernicus systems, such as the Sentinel satellites, may be used worldwide without limitations, and would not be subject to restrictions on redistribution, with or without modifications. It added that as the European contribution to GEOSS, Copernicus would be fully compatible with GEOSS data principles.[91] Registration would be required to download the data to make it possible to collect user statistics. However, registration would be free, and users would be accepted automatically, ensuring the requirement would not deter people from accessing data and information.[92]

In one concession to European remote sensing companies, the policy stated that in the case that technical limitations resulted in an inability to serve all users, requests from individuals or organizations associated with nations or groups that contributed to the Copernicus system would have priority. To enable this, users affiliated with these groups would be required to go through a more in-depth form of registration that would provide their identity, area of activity, and country of establishment.[93]

Access could also be restricted if release of the data posed a security issue. In particular, satellite systems collecting data with a resolution of 2.5 meters or better, or with other particularly advanced features, such as high spectral resolution or a very large number of spectral channels, would automatically be reviewed against criteria to determine sensitivity of the data. In carrying out the review, the EU would balance security benefits of restricting data against the environmental, social, and economic benefits of open dissemination of the data. It would also take into account the availability of data from other sources in considering the likely efficacy of such restrictions. A strict registration process, with unequivocal identification of the user, would be put in place to provide access to data restricted for security reasons.[94]

Sentinel launches began in April 2014, with the launch of Sentinel 1A. This was followed by Sentinel 2B, 3A, and 1B in June, 2015, February, 2016, and April, 2016, respectively. By April, 2016, nearly five petabytes of information had been acquired from the Sentinel 1A satellite alone, and about four million downloads had been undertaken by more than 30,000 users registered for access to Sentinel data.[95] One of these users, Google, was downloading and systematically archiving all of the Sentinel 1A data. The company provides the SAR imagery on its Google Earth Engine, along with Landsat and other data, enabling cloud-computing-based analysis.[96]

In 2015, the European Commission and the United States signed an agreement establishing a dedicated channel for US access to data from the Sentinel satellites. The agreement did not specify any imagery that would be received from the United States in return. "Europe has benefited enormously from free access to Landsat data for many years. During that period, there was very little going in the other direction. This agreement is a recognition of that," the director of ESA's Copernicus Space Office explained. "NASA's data is free anyway," he added.[97]

ESA Summary

The European Space Agency was formed primarily as a research-oriented organization. It helped to develop the first European meteorological satellites as part of a coordinated international effort, both internal to Europe and externally. Data from these early systems was freely shared internationally as part of this effort, in line with meteorological tradition, and in order to help ESA establish a user base for its satellite data.

As ESA began planning its own research programs, the importance of climate change was rising, and it was also a time of intense international interest in commercial remote sensing systems. Accordingly, ESA planned its first systems, ERS-1 and -2, to serve multiple needs: scientific research, operational services, and commercial applications. To achieve this, and to demonstrate a clear benefit to its member states, ESA adopted a data policy that differentiated access procedures and cost based on the use of the data and restricted redistribution. The goal of the commercial effort was primarily focused on promoting the growth of a European commercial remote sensing sector and avoiding interference with national-level commercial remote sensing efforts. This was reflected more explicitly in the Envisat data policy, in which ESA licensed multiple commercial firms to market and sell data, charging these firms only marginal cost for data access.

As it became clear that these efforts were not generating significant revenue or promptly leading to a commercial market, and as scientists complained about the challenges in accessing data, the amount of data that ESA attempted to sell decreased. None of the data from the science-focused missions in the Living Planet Program were licensed for sales. Enabled by improved distribution capabilities and decreased costs resulting from the spread of the Internet, ESA made its data more widely available, often at no cost if accessed online.

This trend was reinforced by support on the national and EU level for open data provision, rather than data sales. In 2010, ESA adopted a policy that was almost entirely open, making the majority of its data freely available online for all users, although redistribution of the data was subject to ESA approval. High-volume requests that could not be provided online and SAR data remained restricted. This was in part due to technical constraints in distribution technology, but also reflected the ongoing efforts by commercial distributing entities to sell SAR data. Support from the growing European value-added sector contributed to the 2013 EU endorsement of ESA's decision to provide all standard Copernicus data free of charge under an open data policy, ensuring that data from this vast new program would be made widely available.

13   Japan Meteorological Agency

Japan saw the development of meteorological satellites as a way to use space activities to provide practical benefits to its citizens while also making a visible global contribution that would bestow prestige on the nation. The Japan Meteorological Agency (JMA) enthusiastically accepted these advanced tools, and worked diligently to maintain funding for an operational series of satellites, working with a variety of partners within the Japanese government. Throughout the history of the program, JMA has been dedicated to sharing data with other National Meteorological Services, following WMO guidelines. It has also adopted relatively open policies with respect to research and commercial access to its data, and places no restrictions on use or redistribution. However, its technologies and processes for disseminating satellite weather data to nongovernment users have not progressed significantly over time. As of 2017, the low level of development of these data access systems stands in stark contrast to its advanced meteorological satellite technology.

Early Years and International Impact

As an island nation in the Pacific Ocean with a diverse geography, Japan faces a wide range of weather conditions, from typhoons and torrential rains to heavy snow, and Japan has a long history of government-supported meteorological observation to address these challenges. The Tokyo Meteorological Observatory, later renamed the Central Meteorological Observatory (CMO), was established within the Department of Interior in 1875, and Japan issued its first national weather forecast in 1884.[1] By 1906, Japan had more than 100 meteorological stations, including stations in China and Korea, that provided information to the central office via telegraph.[2] In

1952, as the operational capabilities of meteorology were on the rise, and the space age was growing near, Japan passed the Meteorological Service Act to guide the development of meteorological services in the country. CMO was renamed the Japan Meteorological Agency (JMA) in 1956.[3]

Japan's Council of Space Activities was created in 1960, but other than sounding rocket experiments at the university level, the nation was not very involved in space activities in the first decade of the space age. In 1969, when Japan created the National Space Development Agency (NASDA), an overriding goal was to establish the nation as a leader in science and technology, on par with other advanced nations. Japan moved quickly to catch up with the state of space technology and make its new status clear to other nations.[4] The WMO's Global Weather Experiment (GWE) offered a good opportunity to achieve the latter goal. Japan had been a member of the WMO since 1953, and providing a geostationary weather satellite for this program would place the nation on par with the United States and Europe, which were providing the other components of the global constellation.[5]

To achieve the goal of catching up technologically and actually developing the satellite, Japan worked closely with the United States. In 1969, shortly before the creation of NASDA, the United States agreed to permit US industry to provide space technology and equipment to Japan.[6] Japan's first Earth observation satellite, the Geostationary Meteorological System (GMS), also referred to as Himawari (Japanese for sunflower), was built by the US Hughes Space and Communications Company, the same company that developed the US Geostationary Operational Environmental Satellites (GOES). Development was completed in coordination with Japan's Nippon Electric Company (NEC) and NASDA. GMS was launched in 1977.[7] After an initial checkout period, the satellite was transferred to JMA for operation.[8] Cloud imagery collected by the satellite was disseminated directly to ground stations in the region via fax.[9] China, South Korea, Australia, and others were able to use GMS data to support their own meteorological services.[10] In support of the GWE, data was also provided to the Global Data Centers in the United States and Soviet Union, where it could be accessed at the cost of reproduction and dissemination by global researchers.[11]

Even with the success of the first satellite, JMA had to justify the substantial costs to gain support for continued development and operation of the series. In doing so, JMA emphasized plans to contribute to the WMO's

Global Atmospheric Research and World Weather Watch programs. The agency aimed to ensure compatibility with US and European geostationary systems, particularly to meet the requirements of worldwide exchange of data. JMA reached out to NOAA in the United States to request information on its long-term planning to help ensure this compatibility. JMA also carried out an analysis of the economic benefit of its meteorological satellites, showing that the benefits far outweighed the costs.[12] JMA did ultimately receive approval to continue the program, with GMS-2 through GMS-5 built with only very minor modifications to the original design and launched between 1981 and 1995.[13] Data remained available through direct transmission and the WMO Global Telecommunication System. Archived data could be accessed by writing to the Japan Weather Association.[14]

Commercial Interactions and Continued Support

In 1993, Japan amended the Meteorological Services Act to establish the Certified Weather Forecaster System. It enabled commercial-sector providers of meteorological services to gain authorization from the JMA Director General to implement their own forecasting services and retain certified weather forecasters.[15] JMA also adopted a policy of providing meteorological data and products to the private business sector, mass media, and others on an open basis. The Japan Meteorological Business Support Center (JMBSC) was created in 1994 as a "general incorporated foundation"—essentially a government-registered organization operating on a nonprofit basis—to provide services relevant to the promotion of private meteorological business. JMBSC provided real-time and non-real-time JMA data and products to users at marginal cost.[16] The policy led to a steady increase in the number of certified weather forecast service companies.[17] In fact, private-sector meteorological companies in Europe pointed to the relatively large private weather industry in Japan, along with the United States, as evidence that Europe should abandon its cost recovery efforts and adopt open policies for its meteorological data.[18]

JMA had requested in 1987 that NASDA examine options for the second-generation Geostationary Meteorological Satellite series.[19] However, despite the success of the first generation, JMA had difficulties securing funding to support the follow-on series. This was resolved by partnering with another organization within the Ministry of Land, Infrastructure, and Transportation (MLIT), the Japan Civil Aviation Bureau, which was interested in developing a space-based GPS augmentation system that would improve the precision and reliability of GPS in the area, allowing GPS navigation to be used for aviation purposes. It was determined that this instrument could be flown on the meteorological satellite. The result was the jointly developed Multifunctional Transport Satellite (MTSAT), which was expected to include two satellites. Unfortunately, the first MTSAT was lost in a launch failure in 1999.[20]

GMS-5, the last of the first-generation meteorological satellites, operated well beyond its five-year design life, until May, 2003, but this was still two years prior to the launch of Japan's MTSAT replacement, MTSAT-1R. Japan requested that NOAA move its GOES-9 geostationary satellite to help fill this gap. The United States agreed to do so, and GOES-9 was placed over Asia from May, 2003, to July, 2005, providing coverage until MTSAT-1R was launched and ready for operational use. MTSAT-2 was launched in 2006.[21] Data from the satellites was made available via direct broadcast and through the WMO Global Telecommunications System. Near-real-time data could also be accessed online, but the service was restricted to National Meteorological Services (NMSs) due to the limitations of JMA's server capability and network bandwidth.[22] Private entities, researchers, and others could access data through JMBSC.

Following the MTSAT launch failure and the long delay in launching MTSAT-1R and -2, the Japan Civil Aviation Bureau declined to partner with JMA on future satellite projects, and once again, the agency was faced with challenges in finding adequate funding to maintain operations.[23] However, in 2008, Japan passed the "Basic Space Law" that aimed to implement a more user-driven space policy. The Japanese Strategic Headquarters for Space Policy laid out a basic plan to align with the new law. The plan recognized the value of meteorological satellites in providing information relevant to everyday life. It also acknowledged the role that meteorological satellites had played on the international stage, including providing data to 30 countries in the Asia-Pacific region over the past 30 years.[24] Development of Earth observation and meteorological satellites, and in particular the Himawari 8 and 9 satellites, was one of five priority efforts to be pursued under the new plan.[25]

In March, 2011, Japan was hit by what is referred to in Japan as the "Great East Japan Earthquake and Tsunami," which caused significant damage, including the disaster at the Fukushima Nuclear Power Plant. JMA, which had underestimated the size of the wave that caused the disaster, undertook significant analysis after the fact to determine what lessons could be learned.[26] JMA's geostationary meteorological satellites also took on more significance after the disaster.[27]

At the 2011 WMO Congress, JMA increased its commitments to data sharing activities—at least on an official, government-to-government level. JMA was designated as a Global Information System Center and its Meteorological Satellite Center was designated as a Data Collection or Production Center.[28] A Global Information System Center is responsible for collecting data and information from national and other data centers, aggregating those data and products, and exchanging them with other Global Information System Centers for global dissemination. These centers maintain a catalog of all data and products for global exchange in accordance with WMO standards and provide access for public and private networks. Data Collection or Production Centers provide specific roles in enabling the generation of data and products for international distribution.[29] Data cataloged and cached by the Global Information System Center in Tokyo is available for use by registered users of National Meteorological Services.[30]

Next-Generation Satellites, Last-Generation Data Sharing

The first of Japan's third-generation meteorological satellites, Himawari-8, was launched in October 2014, and the second, Himawari-9, two years later in November, 2016. The satellites have significantly increased capabilities compared to past systems, providing a greater volume of data more frequently and with higher spatial resolution. Unlike past satellites, Himawari-8 and -9 do not carry equipment for direct dissemination of data. Instead, all imagery is collected by JMA and made available for NMSs online via its HimawariCloud service. Several Japanese universities and research organizations have also been authorized to access data via the Himawari-Cloud service and are redistributing the data for research purposes on a best-effort basis. Users that do not have sufficient Internet capability can instead access the data with a receiving terminal from HimawariCast, JMA's telecommunications broadcast. JMA also makes PNG and JPEG versions of Himawari imagery available on its website for public and meteorological use.[31] Private entities, researchers, and others can access near-real-time and archived imagery by contacting JMBSC and making a request.[32] JMBSC will provide data to these users via FTP, but access is subject to fees.[33]

JMA Summary

Japan got a late start in space activities compared with the United States and Europe. When it did decide to join the sector, development of meteorological satellites offered a number of advantages. This was a field in which Japan had a long history and in which satellite data would provide a clear benefit to society. The ability to contract with US firms meant that Japan could leap ahead to the most advanced technology. Aligning with the Global Weather Experiment and the development of the World Weather Watch allowed Japan to make an important and highly visible global contribution—on par with Europe and the United States—with its very first satellite.

However, the Japan Meteorological Agency (JMA) stands out largely for its consistency over time. Although it had challenges maintaining financial support for the satellite system itself, over nearly four decades, JMA has never significantly altered its data sharing practices. Its focus has always been fairly narrow, providing data for its own uses and to other national meteorological agencies in near real time, within the WMO framework. Data was made available for use by the private sector, researchers, and others at marginal cost, and as of 1994, provision of data to these users was outsourced to the Japan Meteorological Business Support Center. However, consistency is not always a good thing. The human-in-the loop data request process and the marginal cost pricing that were appropriate to prevailing technology in 1994 were unnecessarily limiting in 2017.

The Himawari-8 and -9 satellites are arguably the most advanced geostationary satellites ever launched. JMA has remained committed to sharing the data freely with NMSs in accordance with WMO systems, developing an advanced new cloud computing system for this purpose. JMA does not actively limit access to the data, and there are no restrictions on redistribution, but as of early 2017, it still did not have a dedicated portal or other method to provide broad and easy access to its data. JMBSC continues to make data available to researchers, private entities, and others, but the technology for sharing data with these users has not progressed in line with satellite systems and data sharing for official purposes.

14 Japan Aerospace Exploration Agency

The Japan Aerospace Exploration Agency began its Earth observation program with the development of meteorological satellites, the data from which was widely shared in line with international norms. When Japan began its nonmeteorological Earth observation satellite programs, however, it incorporated a commercial element into its program, and adopted a tiered policy that attempted to balance scientific research with cost recovery and promotion of the commercial remote sensing sector. As the prominence of environmental issues increased in importance globally and within Japan, there were recommendations from multiple entities for increasing the amount of data that was shared. However, numerous reorganizations within the government and seemingly endless shifts in partnerships among JAXA and other Japanese government agencies with varying goals stymied efforts to adopt a new policy. Finally, in 2013, JAXA adopted a new data sharing policy that had the practical effect of making the vast majority of its data freely available. However, although the balance had shifted, the policy still allowed for both open sharing in support of research and restrictions that would allow continued commercial efforts.

Formation of the National Space Development Agency

In preparation for large-scale space activities, Japan underwent a flurry of legal and policy activity in 1969. The Space Activities Commission (SAC) was formed to develop a coordinated space policy and strategy for the nation.[1] The Japanese Diet passed a resolution limiting the use of space to "nonmilitary" applications, a much stricter interpretation of the "peaceful purposes" clause of the Space Treaty than used by other nations. The United States agreed to allow the transfer of space technology to Japan, making it possible for Japan to leap ahead in technological development.[2] Most importantly, Japan passed a law creating the National Space Development Agency, generally known as the NASDA Law. The purpose of the new agency—the predecessor to the Japan Aerospace Exploration Agency (JAXA)—was to develop and launch artificial satellites and rockets, exclusively for peaceful purposes, "thereby contributing to the promotion of space development and utilization." The new agency was placed under the Science and Technology Agency (STA).[3]

With this organizational and legal infrastructure in place, Japan began to develop its first satellites. In the field of Earth observations, Japan first focused on developing a geostationary meteorological satellite in collaboration with the Japan Meteorological Agency (JMA). Working with companies in the United States, and providing a contribution to the WMO Global Weather Experiment and World Weather Watch, this effort provided practical and strategic benefits to the nation. NASDA participated in the development of the Geostationary Meteorological Satellites before transitioning them to JMA for operational use.[4] Data from the satellites was shared freely as part of the Global Weather Experiment and was in line with regular WMO practices.

Advancing Technology and Tiered Data Sharing

Like ESA, NASDA began its nonmeteorological Earth observation program by accessing data from other nations. Through an MOU with NASA, NASDA established a Landsat ground station and began receiving Landsat data in 1979. NASDA's policy was to make Landsat images freely available to users who would contribute to the development of Earth observation satellite systems, such as government institutions and universities. In fiscal year (FY) 1979, NASDA provided free data for 157 research projects carried out by 37 government organizations, 32 universities, and 5 others. Users not contributing to Earth observation development could purchase the data at "reasonable prices" from the Remote Sensing Technology Center (RESTEC). RESTEC was also in charge of promoting remote sensing, carrying out research and training activities, and disseminating publications on the topic.[5]

NASDA had been studying concepts for its first domestically developed Earth observation satellite, outside of the meteorological program, since

1976. STA and RESTEC gathered together experts to determine what system would contribute to addressing both national and global problems, and the group recommended pursuing a marine observation satellite with plans for a complementary land observation satellite as a follow on.[6] The Marine Observation Satellite 1 (MOS-1) was launched in 1987. In addition to its scientific goals, MOS-1 was also classified as a technology experiment. NASDA aimed to gain experience in the fundamental technologies for Earth observation satellites and in techniques for launching satellites and placing them in desired orbits.[7]

NASDA distributed MOS-1 data in a number of ways. Data was provided directly to principal investigators associated with the program. Memorandums of Understanding (MOUs) signed with the European Space Agency (ESA), Thailand, Australia, and Canada also allowed for direct reception and use of the data by those countries.[8] The agreements noted that intellectual property rights over the data would remain the property of NASDA.[9] Foreign and other general users could also access data from RESTEC at the cost of reproduction and dissemination.[10] MOS-1b was launched in 1990, carrying the same sensors as MOS-1 and continuing the MOS-1 mission.

Japan was a member of CEOS, and in 1991 and 1992, approved the organization's data sharing principles for global change research. The principles stated that a common goal was provision of data to global change researchers from all missions at a price reflecting only the cost of fulfilling the user request, but recognized the differing policy goals of maximizing use of the data and of shifting funding responsibility to users or other sources.[11]

The latter point became relevant when Japan launched its Japan Earth Resources Satellite (JERS) in February 1992. JERS carried an optical imager and a synthetic aperture radar (SAR). Together with Europe, which had launched its own Earth Resource Satellite, ERS-1, in 1991, Japan was one of the only sources of SAR data in the world, and both entities engaged in commercial sales. The commitment to commercial activities was reflected in the organizational structure for the project. JERS was a joint program between STA and Japan's Ministry of International Trade and Industry (MITI). NASDA, under STA, was responsible for the satellite framework and the launch, but the design and development of the remote sensing instruments was carried out by research groups within MITI.[12] Thailand and the United States both signed MOUs allowing direct access to JERS data using national ground stations.[13]

Japan's First Official Satellite Data Policy

Following the launch of JERS, NASDA released its first data sharing policy in July, 1992. The policy implemented a tiered system, similar to that adopted by ESA for its ERS satellite.[14] The policy differentiated among three groups. Principal investigators and other NASDA partners were able to access data for free, with the understanding that any results of their work were jointly owned by NASDA. Other researchers could access data at the cost of reproduction. RESTEC sold data at market prices to commercial and general users.[15]

Under the policy, RESTEC was free to set its own prices, and NASDA charged a royalty fee. Redistribution of data was not allowed without a special arrangement with NASDA. In adopting a tiered policy, NASDA aimed to support development of the commercial remote sensing sector in Japan.[16] There was also a feeling that such a policy was the most appropriate for a system built with government funds. The agency felt it had a responsibility to demonstrate that the data it distributed was resulting in benefits for taxpayers, whether these were in the form of scientific research reported by principal investigators or revenues generated through commercial sales.[17]

NASDA's next Earth observation project, the Advanced Earth Observing Satellite (ADEOS), launched in August 1996. ADEOS was a very large, NASDA-led project initiated to engage multiple domestic and international partners. The design of the satellite was modular, allowing the agency to accept instruments from multiple sources. After considering proposals from the United States, France, Italy, and Australia, NASDA chose to incorporate NASA's ozone and ocean surface wind instruments and France's imaging instrument. ADEOS also included an instrument developed by MITI to monitor carbon dioxide and methane and two ozone-monitoring instruments proposed by the Japanese Environmental Agency. Two NASDA-developed instruments, the Ocean Color and Temperature Scanner (OCTS) and the Advanced Visible and Infrared Radiometer (AVNIR), were also included. Unfortunately, a problem with the power system led the satellite to fail prematurely in July, 1997, after less than a year of operation.[18]

NASDA also participated in two missions in cooperation with NASA during this period. The first was the Tropical Rainfall Monitoring Mission (TRMM), launched in 1997, which aimed to improve understanding of weather and climate dynamics. JAXA developed the precipitation radar instrument and provided the launch.[19] In addition to collaborating on TRMM, MITI, in coordination with NASDA, contributed the Advanced Spaceborne Thermal Emission and Reflection Radiometer (ASTER) high-resolution imaging instrument for inclusion on NASA's Terra satellite. Access to ASTER data was restricted in accordance with MITI data policy.[20]

Both of these projects originally fell within the International Earth Observation System (IEOS) effort, which sought to coordinate Earth observations efforts among the early space station participants: the United States, Europe, Canada, and Japan. A 1996 MOU between the United States and Japan regarding the ASTER cooperation captured the current state of the IEOS Data Exchange Principles being negotiated at the time. The principles stated that data from all IEOS systems would be available to the participating agencies and their designated users at the lowest possible cost for noncommercial use, including research, applications, and operational use for public benefit. Redistribution of the data to third parties would not be permitted.

Mirroring ESA's policy at the time, research users were to be chosen through an Announcement of Opportunity process and provided only the data required for their proposed project. The principles stated that each agency would make its own arrangements with regard to commercial and other uses outside of the three specifically addressed in the policy. The agreement noted that the data exchange principles would be revised as necessary to remain consistent with updates being discussed among the parties.[21] In reality, the principles were never finalized, and all agencies involved instead developed their own independent data sharing policies.[22]

Administrative Changes

NASDA experienced significant administrative changes in the first decade of the new century. As part of a government-wide decision to reorganize the state ministries and agencies, in 2001 the Science and Technology Agency (STA), which oversaw NASDA, was merged with the Ministry of Education creating the Ministry of Education, Culture, Sports, Science, and Technology (MEXT). The Space Activities Council, which had previously led space policy development, was also incorporated into MEXT. High-level space policy became the province of the newly formed Council on Science and Technology Policy, chaired by the Prime Minister.[23]

In December, 2002, a new law was passed creating the Japan Aerospace Exploration Agency (JAXA). JAXA created one central body for space activity by merging NASDA with the Institute of Space and Aeronautical Science, which had previously operated under MEXT as the lead on space science projects, and the National Aerospace Laboratory, which focused on aviation and rocketry research and development. The new agency was given three objectives: development of academic research at universities or other institutes, enhancement of the level of space-related and aeronautical science and technology, and the promotion of space development and utilization.[24]

Throughout these changes, Japan continued to launch new Earth observing satellites. A Japanese instrument designed to monitor the water cycle was launched on NASA's Aqua satellite in April, 2002. In December, 2002, NASDA launched ADEOS-2, once again carrying instruments from the United States and France in addition to its own domestically developed sensors. Unfortunately, like its predecessor, the satellite failed after less than a year in orbit. During this time, the data sharing policy for access to Earth observation satellite data remained largely the same as before. Data was provided at the cost of reproduction for Earth science research through agreements between JAXA and these scientists. RESTEC distributed data to other users, including commercial users, at a market price, paying a royalty to JAXA for the data.[25]

Balancing Climate and Commercial

In the first decade of the new century, a string of natural disasters, as well as growing awareness of climate change issues, led to a rise in the importance of environmental monitoring in Japan. The Deputy Minister of MEXT was present at the First Earth Observation Summit in Washington, DC, in 2003. Japan hosted the Second Earth Observation Summit in Tokyo, and the Prime Minister gave the opening remarks, stating, "science and technology is the key to achieving both economic growth and environmental protection." He concluded with a comment on the importance of the effort being undertaken: "It is my sincere hope that this meeting will lay the foundation of the Earth observation system, which will support the future of human-kind."[26] The meeting resulted in the agreement on the framework for the creation of the Global Earth Observation System of Systems (GEOSS).

The Japan Council for Science and Technology Policy followed up with a national Earth Observation Promotion Strategy in December, 2004. Future Earth observations in Japan would be driven by user needs, providing data that is not only useful for researchers, but also necessary for the government to make decisions and develop policies on environmental issues. The plan included data that is useful for industry and data that affects the everyday lives of the general public. The strategy stated that data should be "swiftly publicized both inside and outside Japan in order to achieve the effective use of the results and to facilitate a contribution to international society."[27]

Building on the strategy, the Space Activities Commission, now within MEXT, published an Earth Observation Satellite Development Plan in July, 2005. It emphasized the international importance of Japan's efforts, particularly to GEOSS. Contributing to GEOSS would allow Japan to "fulfill its duties to international society." Once again, the importance of data sharing was recognized. Mirroring its existing policy, the document noted that users could be divided into two categories, those who use the data for public purposes and research and others who plan to use the data for commercial use. The policy noted that data for the first group should be made available at the cost of copying the data, resulting in "almost no charge" when accessing data online. However, for commercial users, the commission argued free data provision would not be appropriate, as Japan must avoid undercutting data sold by private companies.[28]

In January, 2006, JAXA launched the Advanced Land Observing Satellite (ALOS). The satellite was developed in collaboration with the Ministry of Economy, Trade, and Industry (METI), the successor to MITI in the 2001 government reorganization. It carried both high-resolution imaging instruments and a SAR instrument, collecting the type of data that was sold commercially by companies and governments in the United States, Europe, and Canada. Accordingly, JAXA adopted a data policy for ALOS that was "most appropriate to the characteristics and purposes of using high-resolution optical sensors... and a radar sensor."

JAXA set up the "ALOS Data Node Concept," which involved selecting one distributor in each region of the world that would have the responsibility of processing, archiving, and distributing ALOS data in its area. JAXA would be responsible for distribution in Asia, with Thailand's space agency, the Geo-Informatics and Space Technology Development Agency (GISTDA), identified as a subnode for the area. ESA would manage European and African data access, NOAA would be responsible for the Americas, and Geoscience Australia would distribute data in Oceania. Under the arrangement, these organizations, as well as selected agencies within Japan, such as the Geographical Survey Institute, the Ministry of Environment, the Ministry of Agriculture, and the Japan Coast Guard, would receive a limited volume of data for free. ALOS node organizations could access additional data at marginal cost. These organizations could then distribute data to commercial users.[29] ALOS data would be provided for free in the event of a natural or man-made disaster, in accordance with the International Charter on Space and Natural Disasters, which JAXA joined in 2005.[30]

JAXA's commitment to environmental monitoring was displayed with the launch of the Greenhouse Gas Observing Satellite (GOSAT) in January, 2009. GOSAT had been developed in cooperation with the Japan Ministry of the Environment and the National Institute of Environmental Studies. In contrast to ALOS, the atmospheric-monitoring sensors had little commercial application and were primarily useful for monitoring and studying climate change. Despite this, the GOSAT data policy, which technically fell under the purview of the National Institute for Environmental Studies, not JAXA, maintained a distinction between research and commercial users, stating that research users could access the data for free online while commercial users would be charged a fee. Data would be subject to copyright and redistribution to a third party was prohibited. The development and sale of value-added products would be allowed, given that the commercial fees for data access had been paid.[31,32]

Reorganization Continues

Discussions regarding the proper organization and structure of space activities in Japan had not ended with the creation of JAXA in 2003. Following years of debate, another major adjustment came in 2008 with the passage of the "Basic Space Law." The change was largely driven by the perceived need to change the practice, in place since 1969, that limited Japan to non-military space activities. After a North Korean missile overflew Japan in

1998, Japan had developed two Information Gathering Satellites, launched in 2003. However, doing so required administrative gymnastics to classify the satellites as nonmilitary, but still justify limiting satellite development to domestic companies and restricting access to the resulting data.[33] The new law updated the interpretation to allow military use of space for non-aggressive purposes—a definition more in line with that used by the rest of the international community.[34]

The law also aimed to transition Japan's space activities from the traditional focus on research and technology development to a more strategic, user-focused policy.[35] To help achieve this, the law established a Strategic Headquarters for Space Policy at the Cabinet level. In 2009, this new organization released a Basic Plan for Space Policy, based on the Basic Law. The Basic Plan called for the creation of a satellite data utilization system to facilitate easy access to remote sensing data. The technical development was to be accompanied by the development of a standard data policy that would balance the goal of promoting data use and the needs of the commercial market.[36]

When JAXA's Global Climate Observing Mission–Water 1 (GCOM-W1) was launched in May 2012, data from this satellite was available on a tiered basis, similar to the policy adopted for GOSAT. Despite the recommendation of the Basic Plan, a standard data policy had not yet been developed. The delay in acting on this recommendation was largely caused by ongoing political turmoil within Japan in the intervening years, with control of the Diet, and visions for priorities in space activities, changing multiple times in three years.

Just a month after the GCOM-W1 launch, a new law was passed, again restructuring Japanese government space activities. The Space Activities Commission, created in 1969 and placed within MEXT in 2003, was abolished. A newly created Space Policy Commission within the Cabinet Office, reporting directly to the prime minister, was made responsible for coordinating both policy and budget development for all space activities in Japan. To help ensure that JAXA programs were aligned more directly with user needs, administration of JAXA would be shared between MEXT, METI, and the Ministry of Internal Affairs and Communication. Whereas previously MEXT had exclusive control over program planning and budgeting for JAXA, now all three agencies, and the Cabinet Office, would be able to promote their own preferences with regard to JAXA, with the Cabinet Office making the final decisions with regard to programming and budget.[37]

A Move Toward Open Data

In July 2012, Japan released its Open Government Data Strategy. While e-government and open government initiatives had been in development in Japan since 2004, the need for data access and reuse had been dramatically illustrated during the Great East Japan Earthquake and Tsunami, and the accompanying Fukushima nuclear disaster in 2011, spurring additional interest and effort in the area. The policy aimed to increase transparency, promote public-private collaboration and economic stimulus, and improve government efficiency.[38] METI was a leader in this area, developing its own open data portal in 2012. The national open data portal was opened in December, 2013.[39]

Prior to the 2012 law, there were some indications that JAXA was moving toward a free and open data policy, following the example of agencies in the United States, Europe, and elsewhere. Speaking about JAXA's contribution to GEOSS in February, 2011, Osamu Ochiai of JAXA's Space Applications and Promotion Center, stated, "I think JAXA's satellite data hasn't yet reached a stage where it is widely utilized. By making our Earth observation data freely available and encouraging its use, I think we can make it the international standard not just for GEO but for broader use as well." He noted that climate change, disasters, and water were key areas where Japan hoped to contribute, but that the agency needed to lower the barriers to entry to get people to use the new data.[40] By contrast, METI expressed an interest in moving toward dual-use Earth observation systems and promoting commercialization of Japan's space industry.[41]

A new Basic Plan for Space Activities, released in January 2013, reflected both of these views. It noted that remote sensing efforts in other countries, including SPOT Image in France and Worldview and GeoEye in the United States, were developed in partnership with the government, either through anchor tenancies or other arrangements. It also noted that NASA and ESA operated satellites focused on global environmental observation for academic and public purposes. It found that insufficient efforts were being made to adopt policies, and in particular data policies, to promote these types of public private partnerships in Japan and to promote utilization of the data in the value-added industry, and it reiterated the need for a standard data policy.[42]

This time the recommendation was implemented quickly: in August, 2013, JAXA released a new data policy. The new policy represented a compromise that would allow JAXA to openly share data from its scientific satellites, without completely giving up its commercial efforts, and had the effect of making the vast majority of its data freely available. Rather than differentiating based on user, the new policy differentiated based on the technical attributes of the data, particularly spatial resolution. Data with a low or medium resolution would be distributed under a "full and open access" policy. JAXA released the beta version of its Globe-Portal (G-Portal), for free online access to its Earth observation data that same year, and it has been steadily adding data to this portal since that time, beginning with the types of data most frequently requested. The data can be used for any purpose, including commercial activities, with proper acknowledgment of JAXA as the data source, but the amount of data that can be downloaded is limited due to network constraints.[43]

Under the new policy, access to high-resolution imagery and data from SAR instruments on ALOS, which had the greatest potential for commercial sales, would be restricted. JAXA stated that a limited amount of this data would be provided free of charge for scientific research or disaster management and additional data would be available at marginal cost for government users who develop a cooperative agreement. Commercial users would need to purchase data at a market price set by private distributors chosen by JAXA.[44] Distribution of data from some satellites and instruments, including GOSAT and ASTER, still falls under the policies of other Japanese agencies. Despite its scientific nature, GOSAT remains subject to a tiered policy that includes fees for commercial users.[45] Data from the ASTER instrument, which launched on NASA's Terra satellite in 1999, and had previously been restricted, was made freely available in April, 2016.[46] Continuing the process that began with the passing of the Basic Space Law, a law is under development that would normalize satellite data sharing across agencies.[47]

Summary

When NASDA was formed, it was focused primarily on technology development and establishing Japan as a leader in the space sector. Its Earth observation satellite program began with the development of meteorological

satellites built in cooperation with US companies, giving Japan international visibility with its first satellite. As part of the Global Weather
Experiment, data from the satellite was made freely available. After participating in the meteorological satellite program, NASDA quickly transitioned
to development of cutting edge technology, with its ocean-monitoring
satellite, MOS-1, followed by the JERS, which carried one of the first SAR
instruments. Data from these satellites was made available according to a
tiered policy, aimed at ensuring a return on taxpayer investment in the
form of scientific research and revenue from sales. There was also a desire to
promote the growth of the Japanese commercial remote sensing activities.
With its initial 1992 data policy, NASDA officially applied this tiered policy
across its satellites.

The creation of JAXA in 2003 consolidated space activities, but did not
fundamentally change the character of the space agency or its management. It retained its technological focus and remained within a ministry
focused on science and technology. Not surprisingly, the data sharing policy remained largely the same through this transition, as well. The rising
importance of climate change and other global environmental challenges
and Japanese participation in GEO contributed to a growing interest within
JAXA in sharing data more openly.

The passage of the 2008 Basic Space Law and the subsequent 2012
reshuffling of Japanese government space activities had a greater effect on
JAXA than the 2003 reorganization. National-level policy-makers and a
wider array of ministries, including METI, attempted to impose a broad set
of goals for space activities in JAXA. This resulted in another compromise
data policy, embracing free and open data sharing for scientific data, while
still protecting the narrower set of data that had commercial potential.

15 Brazil, Russia, China, India, South Africa

This chapter presents summaries of remote sensing activities and data sharing policies for Brazil, Russia, India, China, and South Africa—the BRICS nations—to complement the full case studies on agencies in the United States, Europe, and Japan. These nations differ in size, history, and structure, providing a broader sense of satellite data sharing around the world. The appendix includes short summaries of satellite remote sensing activities and data sharing efforts for all other nations that have operated a remote sensing satellite from 1957 to the beginning of 2016.

Brazil

Brazil has been involved in just four remote sensing satellites, all in collaboration with China, but it has had a major impact on international satellite data sharing discussions. Brazil was one of the first nations to sign a cooperative agreement with NASA to access Landsat data via a dedicated ground station, in 1973.[1] The US decision to commercialize the system in the 1980s, as well as delays in follow-on Landsat system development, caused concerns in Brazil about potential interruptions in service. Alternative imagery, such as that available from France's SPOT satellite, was prohibitively expensive. Facing these concerns, Brazil decided to develop its own satellite, and signed an intergovernmental agreement with China to begin the China-Brazil Earth Resources Satellite (CBERS) program in 1988.[2] The first CBERS satellite was launched in 1999, followed by CBERS-2 in 2003, CBERS-2B in 2007, and CBERS-4 in 2014.

The Brazilian space agency—the Instituto Nacional de Pesquisas Espaciais (INPE)—and the Chinese Center for Resources Satellite Data and Applications (CRESDA) were responsible for developing applications for

CBERS data. Original plans called for sales of CBERS imagery on a commercial basis; however, demand was quite low. Brazil sold fewer than 1,000 CBERS-1 images a year, and changes in price were not effective in increasing sales. Given the low return on investment, in 2004, INPE adopted a free and open access policy, providing full-resolution images online for Brazilian users. Data distribution jumped to 10,000 images per month.[3] It was the first nation to implement this type of open data policy for land remote sensing satellite data, and the change had a major impact on agency officials and politicians in other nations, particularly the United States.

While domestic data access was free, Brazil and China still hoped to generate revenue from international sales. A policy released in 2004 allowed for the licensing of international ground stations capable of receiving CBERS data. These ground stations would pay an annual fee as well as a per-minute fee for downlinks of raw data within the footprint of that station. The raw data could then be processed into image products and sold to users within their respective national market. Ground stations operated by INPE and CRESDA would have unlimited access to data collected within their footprint, and could each determine their own policies for distribution of that data. Any distribution to third parties would be done solely on the basis of an international price list, to be jointly developed by INPE and CRESDA.[4]

Despite these initial plans, commercial international CBERS ground stations did not proliferate, and instead China and Brazil increased the amount of data made freely available internationally. In 2008, China and Brazil agreed to provide free CBERS images to all Latin American and African countries, and agreements were signed for the operation of ground stations that would improve data collection over Africa. Shortly after, INPE began to provide data from its Landsat ground station for free as well. By the end of 2009, INPE had distributed more than a million free satellite images.[5] In 2010, Brazil and China decided to extend free access to CBERS data to users in all countries, with plans for additional international ground stations that would cover the majority of the Earth's land mass outside of the polar regions.[6] As of June 2016, CBERS-1 products are no longer being provided by INPE, but all CBERS-2, -2B, and -4 data can be accessed for free via the INPE online portal after registration.[7]

Russia

Russia, along with the United States, has one of the longest-running space programs. Remote sensing activities are coordinated among the Russian Space Agency (Roscosmos) and the Planeta Research Center within the Russian Federal Service for Hydrometeorology and Environmental Monitoring (Roshydromet). Neither organization released official data sharing policies over the course of their programs, and while meteorological data was generally made available in line with WMO principles, other Earth observation data was not readily available to the international community. However, Russia also does not systematically restrict access to its data and has shown a willingness to provide data to a variety of types of users upon request. In recent years, both the meteorological and space agencies have begun developing data portals that aspire to make data more readily accessible to the international community, but none of these was fully operational as of the beginning of 2017.

Meteorological Satellites and Data Sharing

The early Soviet space program focused much of its effort on dramatic space "firsts" and the development of capabilities needed for human space flight, and didn't launch its first unclassified remote sensing satellite until 1964. However, the Soviet Union was engaged in international remote sensing data sharing discussions from the beginning of the space age. In 1958, before weather satellite development had begun in the Soviet Union, the head of the Soviet Central Institute of Forecasts was appointed to the newly formed Panel of Experts on Meteorological Satellites along with a colleague from the United States and others. In 1961, the panel released a report on the value of satellites for meteorology that laid the foundations for the World Weather Watch (WWW) program. As part of the plan, one of three World Meteorological Centers was set up in Moscow as a hub for archiving and sharing weather data.[8]

Moscow also developed a bilateral agreement with the United States during this period, agreeing in 1962 to coordinate its meteorological satellite launches with the United States and engage in data sharing. A direct link for sharing meteorological data, the "cold line" was established in 1964.[9] In 1966, the Soviet Union announced that it had launched its first experimental meteorological satellite, and satellite data sharing over the link began. In

1969, the first operational Soviet weather satellite, Meteor-1, was launched. Twenty-five Meteor-1 satellites, each carrying a TV camera and an infrared imager, were launched between 1969 and 1977. Beginning in 1971, with the launch of Meteor-1 through Meteor-10, the Soviet Union used Automatic Picture Transmission (APT) to allow Soviet and Western ground stations to downlink pictures from the satellite in real time.[10]

The second generation of Russian meteorological satellites, Meteor-2, included 21 satellites flown between 1975 and 1993. The Meteor-2 series incorporated a number of improvements that increased the volume and quality of the data collected and increased the satellite life-span. This was followed by the Meteor-3 series, which incorporated additional improvements and included six satellites launched between 1985 and 1994. The last two of these satellites carried foreign payloads: the fifth Meteor-3 was equipped with NASA's Total Ozone Mapping Spectrometer (TOMS) and the sixth Meteor-3 carried a French Scanner for Radiation Budget (ScaRaB) and a German navigation instrument.[11]

Russia launched its first geostationary weather satellite in the mid-1990s, but planning for the satellite began more than 20 years prior. In 1972, Russia joined WMO's Coordination Group for Meteorological Satellites (CGMS) and announced plans to provide a geostationary satellite for the Global Weather Experiment. Many nations were hoping that Russia would help to fill the gap in geostationary data over the Indian Ocean. However, with a large portion of its land mass at high latitudes, geostationary satellites were not as useful for Russia as for some other nations, and hence not a high priority internally. Programmatic issues and complex military requirements for the system also contributed to delays. In the end, Russia's first Geostationary Operational Meteorological Satellite (GOMS), also known as Elektro, did not launch until 1994, and problems with the orientation and imaging systems meant that it was not able to provide high-resolution images. Russia had not yet launched another geostationary satellite when GOMS failed in 1998, and through coordination with the other members of CGMS, Europe agreed to move its Meteosat-5 satellite to cover the region.[12]

The collapse of the Soviet Union led to a lack of funds that essentially halted work on the meteorological satellite program. Russia experienced a seven-year gap between low-Earth-orbit system launches, during which time Russian scientists relied heavily on data from foreign satellites.

In 2001, Russia launched a new generation of low-Earth-orbit weather satellites. The Meteor-3M satellite incorporated a number of important upgrades: it was the first to be placed in a sun-synchronous orbit, the standard for global weather satellites because this special orbit allowed the satellite to view areas of the Earth under the same lighting conditions day after day, and it had a much longer life-span than previous meteorological satellites. The satellite carried not only meteorological instruments, but also remote sensing instruments, including NASA's Stratospheric Aerosol and Gas Experiment III (SAGE III) sensor, which provided data useful to both weather and climate. Unfortunately, there were technical issues with the satellite, and by 2003, the head of the Russian program said that the meteorological instruments had failed completely.[13]

Russia next moved on to the Meteor-M series, with the first two satellites launched in 2009 and 2014. These satellites carried a suite of meteorological instruments. Russia noted that these instruments would also produce data useful for climate studies. The satellites were unique among global weather satellites because they also each included a synthetic aperture radar.[14] However, no images from the radar were ever released, and unofficial reports suggested that the instrument had failed. Russia also restarted its geostationary program at this time, launching Elektro-L1 in 2011 and Elektro-L2 in 2015.

Throughout this time, Roshydromet has not placed any official policy restrictions on access to archived or real-time meteorological data, and it has generally made data available in accordance with WMO guidelines. A 2014 publication by Roshydromet officials notes that the center maintains an online catalog, but in practice this catalog is not maintained. Instead, the agency has generally opted to provide real-time data via dedicated lines.[15] For example, in 2015, the organization developed a bilateral agreement with EUMETSAT to exchange near-real-time satellite data.[16] Requests for archived data are also accepted, but data is not readily available online.

Earth Observing Satellite Development
In addition to its meteorological satellites, Russia has launched a series of remote sensing satellites focused on land and ocean observations. Satellite land remote sensing was of particular interest to Russia, which has large, sparsely populated areas within its territory. Russia also highlighted the

importance of operational environmental satellite programs for improving understanding of global environmental change. These efforts began during the Soviet era.

The Meteor-Priroda series, which included six satellites launched between 1974 and 1981, built on the success of the Meteor program, extending observations to focus on monitoring of Earth resources. The Resurs-O1 series of satellites, with four launches between 1985 and 1998, continued these observations. The Okean-01 series, with its first launch in 1986, measured the conditions and dynamics of the ocean. In total, eight Okean satellites were launched between 1986 and 1999. The Okean satellites included APT technology, so domestic and foreign users with receiving stations could access real-time data directly from the satellites.[17]

After a hiatus following the breakup of the Soviet Union, the Resurs series was continued, with three launches occurring between 2006 and 2014. Ukraine took over the Okean program, renaming it the Sich program, and launched three satellites between 1995 and 2011. Russia also began a small satellite program, launching Monitor-E in 2005 and Kanopus-V, developed under contract with SSTL, in 2012.

Russia did not adopt official data sharing policies for these satellites and typically did not make data available outside of the Soviet Union on a regular basis. However, the Russian space agency has recently shown an interest in developing data dissemination technologies. In 2010, Roscosmos launched a geoportal to provide access to information for domestic and foreign users.[18] It is possible to view satellite images via the portal, but to order remote sensing data, users are directed to contact the Russian Research Center for Earth Operative Monitoring (NTs OMZ). As of early 2017, there was not yet an automated process in place for ordering images from the data portal, and high-resolution imagery is available only to official registered users.[19]

Military, Intelligence, and Commercial Satellite Development

As in the United States, military and commercial remote sensing activities in Russia have historically been closely linked. The Soviet Union has been launching reconnaissance satellites since the beginning of the space age, with hundreds launched over six decades. In 1987, officials in the Soviet Union announced that they would sell photographic satellite images collected by some of these satellites to countries outside the Soviet bloc. With a resolution of 5 meters, image quality was high, but because the film had to be returned from orbit and processed, delivery could take days to months.[20] In 1992, Russia announced it would allow two Russian firms to sell data with 2-meter resolution, better than any other offering on the global market. Once again, the photographic images were collected by satellites originally designed and used by the Russian intelligence community. As such, some requests for imagery were delayed or denied based on national security. Archived images, from before the satellite was used commercially, were particularly sensitive.[21] By 1998, Russia was selling imagery and synthetic aperture radar data from a wide array of civil and military satellites and had declassified imagery from some past military systems.[22] In 2000, it announced that 1-meter resolution data would be available for sale.[23] As of early 2017, high-resolution imagery from the Resurs-DK and Monitor-E satellites could be ordered from the Russian SOVZOND company.[24]

Russia has also been making efforts to increase private space activities, opening the Space Technology and Telecommunications Cluster at Moscow's Skolkovo innovation park in 2010. In 2012, Dauria Aerospace, formed in the cluster, was founded as the first private remote sensing satellite company in Russia. With offices in Russia, Germany, and the United States, the company hoped to take advantage of a global talent pool in developing a constellation of eight small Earth observation satellites.[25] In 2014, Dauria joined the PanGeo Alliance, planning to coordinate imagery sales with other remote sensing companies around the world.[26] However, by 2015, the collapse of the ruble and other economic difficulties in Russia led the company to abandon plans to build its own satellites and focus instead on hardware exports.[27]

India

The Indian space program has developed both meteorological, land remote sensing, and Earth science satellites. Unlike most other meteorological agencies, the Indian Meteorological Department significantly restricted access to data from India's geostationary meteorological satellites for decades, mostly due to security concerns. A main focus of its nonmeteorological program was on land observation satellites, with a strong focus on commercial applications. In accordance with this focus, India adopted a restrictive data sharing policy that focused on enabling data sales. It also took a conservative stance with regard to the potential security issues related to satellite data dissemination, putting in place detailed procedures to address these concerns. In recent years both the meteorological and space agencies have begun to increase the amount of data they make available online, and both operate online databases or portals to facilitate free access to a significant amount of data.

Early Satellites and Meteorology

India's space agency, the Indian Space Research Organization (ISRO), was established in 1969. From the beginning, the agency was focused on using space technology to provide practical benefits for the country. By 1971, India was using data from foreign satellites in its Satellite Meteorology Center, and in 1979, India set up a Landsat data receiving station. India's first indigenous remote sensing satellite, Bhaskara 1, was launched the same year. Both Bhaskara 1 and Bhaskara 2, which launched in 1981, collected imaging data that contributed to natural resource monitoring, but both were primarily useful in providing hands-on experience working with satellite technology.[28]

In 1982, ISRO launched a geostationary meteorological and communications satellite, the Indian National Satellite 1A (INSAT-1A), developed in partnership with the Indian Meteorological Department. The first satellite failed in just a few months, but was quickly replaced by INSAT-1B in 1983. The ability to view developments over the Indian ocean greatly increased India's weather forecasting and early warning capabilities. By 1988, no tropical storm in the area went undetected, and India was able to provide 24–48 hours of warning for cyclones.[29]

Beginning in 1986, the United States requested that India provide near-real-time access to the half-hourly data from the INSAT satellites, noting that this data is essential for global climate modeling and weather forecasting. The data would provide information about conditions over the Indian ocean not available from other satellites. Indian military authorities objected, due to concerns that the data could be used to monitor ship and aircraft movements. Other government officials felt detailed information on the Indian monsoon could have commercial implications and thus were reticent to make the data widely available. India agreed to share low-resolution, one-day-old data on magnetic tapes.[30]

INSAT-1C was launched in 1988 and INSAT-1D in 1990. India continued to share only a limited amount of INSAT data with the international community.[31] India upgraded its capabilities with the five-satellite INSAT-2 series, which launched between 1992 and 1999. During this time, India agreed to provide increased access to near-real-time INSAT data for some US scientists. Through an agreement with NASA, US scientists could submit a project proposal, and, if approved by India, near-real-time data would be supplied for the duration of the project. Use of the data for other purposes and redistribution of the data to third parties was prohibited. In return, NASA provided data from its Tropical Rainfall Measurement Mission (TRMM) satellite as well as its archive of climate and weather data going back to 1978.[32]

In 1998, India and the United States signed an agreement that allowed NASA and NOAA access to near-real-time INSAT data through a dedicated link between NASA in Washington, DC, and the Indian Meteorological Department in New Delhi. India received a direct link to NASA and NOAA satellite databases in return. The agreement filled what had come to be known as "the Indian gap" among US scientists. Indian officials stated the impetus for change came from the scientific community, which felt there was a need for a joint cooperative effort. Decreasing military tensions in the region and increased availability of data from foreign geostationary satellites were also important.[33]

In 2002, India launched METSAT, later renamed Kalpana-1, as its first dedicated meteorological satellite. Unlike previous Indian weather satellites, it did not perform a communications satellite mission. INSAT-3A launched in 2003, and INSAT-3D was launched in 2013. (INSAT-3B and -C did not carry meteorological instruments.) As of 2016, data from Kalpana-1, INSAT-3A, and INSAT-3D was made available through the Indian Meteorological and Oceanographic Satellite Data Archival Center (MOSDAC). All users could access satellite image files without registration, but only a selected group of registered users can access digital satellite data.[34]

Civil and Commercial Remote Sensing

In parallel with developments in the meteorological satellite program, ISRO was developing the first India Remote Sensing (IRS) satellites. Building directly on the experience with the experimental Bhaskara satellites, IRS-1A was launched in 1988 and IRS-1B in 1991. The data from these satellites was comparable to that provided by Landsat, but data from IRS satellites wasn't made available outside India until the launch of IRS-1C in 1995. At this time, India signed an agreement with EOSAT to market IRS data internationally. Carrying multiple sensor types, and being capable of collecting images with a resolution of 5.8 meters, IRS-1C was the highest- resolution civilian remote sensing satellite on the global market—outperforming both Landsat and SPOT. Users within India were able to access data from the National Remote Sensing Agency (NRSA) in Hyderabad. The US-based EOSAT company had rights to market IRS-1C data to the international user community. IRS-1D launched in 1997, continuing with the same policy.[35]

Concurrent with IRS satellite development, India also launched a series of smaller, less expensive, experimental remote sensing satellites, called the IRS-P series. IRS-P2 launched in 1994 and IRS-P3 in 1996. These two satellites were primarily focused on natural resource monitoring. IRS-P4, also known as Oceansat, was launched in 1999. A follow-up mission, Oceansat-2, was launched in 2009. The Oceansat satellites carried instruments for monitoring ocean color and atmospheric conditions.[36]

In 2001, India released a national Remote Sensing Data Policy. The policy stated that due to national security concerns, data with a resolution of better than 5.8 meters would be subject to government screening before distribution to ensure that sensitive areas were excluded. Government agencies, or private organizations recommended by a government agency, could access imagery with a resolution of 1 meter or better through the same screening process used for 5.8-meter-resolution data. Other users—including private users without a recommendation and foreign users—wishing to access data with 1-meter resolution or better would require clearance from an interagency High Resolution Image Clearance Committee.[37]

India continued to launch remote sensing satellites at a rapid pace. ISRO launched IRS-P6, also called ResourceSat-1, in 2003, improving on the capabilities of IRS-1C and -1D and ensuring continuity for the system.[38] ResourceSat-2 followed in 2013. IRS-P5, also known as CartoSat-1, was launched in 2005, featuring cameras with 2.5-meter resolution.[39] Additional CartoSat satellites were launched in 2007, 2008, and 2010. In 2009, India launched its first radar reconnaissance satellite, the Radar Imaging Satellite 2 (RISAT-2), built by Israel Aerospace Industries. An indigenously built radar system, RISAT-1, launched in 2013. The high-resolution imagery and all-weather radar products are useful for civil, commercial, and intelligence purposes.

ISRO also developed experimental and scientific satellites during this period, including the Technology Experiment Satellites (TES) in 2001 and the Indian Mini Satellite (IMS-1) in 2008, both with basic imaging capabilities. Megha-Tropiques, launched in 2011, was a joint project with France to study the water cycle and energy exchanges in the tropics and their impact on climate and weather. It was followed in 2013 by another Indo-French satellite, SARAL, focused on ocean circulation and sea surface elevation.

Data Sharing Policy Developments

In 2009, ISRO created a geo-portal called Bhuvan. Originally populated with basic maps, it grew to include a large number of datasets. Selected data from IMS-1, Oceansat-2, ResourceSat-1, and CartoSat-1 was available as of early 2017. Other data, particularly from selected instruments on CartoSat-1 and -2, ResourceSat-1 and -2, RISAT-1, and Oceansat-2 missions, was sold by the National Remote Sensing Center.[40]

In 2011, India released a new Remote Sensing Data Policy. Like the 2001 policy it replaced, the new policy emphasized the need to balance national security concerns with the many benefits of remote sensing data use. It adjusted the resolution limits to allow sharing and sales of data with a resolution of up to 1 meter without requiring special clearance. Under the policy, government users, including government educational and academic institutions, could also access data with better than 1-meter resolution without the need for special permission, as could private-sector agencies supporting development activities, if recommended by at least one government agency. Other users were required to get permission from the interagency High Resolution Image Clearance Committee. Alternatively, these users could access data with a resolution better than 1 meter by negotiating a sale and nondisclosure agreement between the user and NRSC.[41]

In 2012, the Indian Department of Science and Technology released the National Data Sharing and Accessibility Policy. The policy encouraged data sharing to maximize use, avoid duplication, and improve decision-making. It requires all departments to review their datasets to determine which can be shared openly and which require some restrictions, and required that metadata be shared on data.gov.in. It also expressed support for budget provisions necessary to support efforts to make data accessible.[42]

The 2016 Indian National Geospatial Policy expanded on this, emphasizing the increasing importance of geospatial information in particular. It stated that all geospatial data, of any resolution, disseminated through agencies and service providers, should be treated as unclassified and made accessible. According to the policy, this access may be open or restricted, and subject to registration or authorization as appropriate. The policy mandates that all government institutions make data available to all other government institutions at no cost. Pricing for access for other users is at the discretion of the data owner.[43]

China

China began launching meteorological satellites in 1988, building a constellation that included both polar-orbiting and geostationary satellites. From the beginning the program has embraced international meteorological data sharing norms, making data freely available to international users via direct broadcast. The China Meteorological Administration has also developed a data portal that freely provides data online for noncommercial purposes.

Sharing outside of the meteorological areas is more mixed. Most data from China's ocean observing satellites is made available online for nonmilitary, noncommercial uses, but redistribution is restricted. Data from the CBERS satellite series was made freely available online as of 2010, but data from the majority of China's land remote sensing satellites is not made readily available outside of China or is subject to a data policy that emphasizes commercial efforts and market pricing.

Meteorological Satellites
Since the 1960s, China had been receiving cloud images from foreign satellites, and officials recognized their value to operational meteorological forecasting. In 1977, China initiated its own weather satellite program, the Fengyun (FY) series, jointly coordinated by the Ministry of Aerospace and the China Meteorological Administration (CMA). Polar-orbiting weather satellites, FY-1A and -1B, were launched in 1988 and 1990, respectively. Each satellite was equipped with an imager similar to that carried by US

meteorological satellites, and data collected by the satellites was distributed to ground stations within China using data formats compatible with US systems. Unfortunately, FY-1A failed after only 39 days and FY-1B after six months.[44]

In 1997, China launched FY-2A, its first geostationary meteorological satellite, adopting the convention of designating polar-orbiting Fengyun meteorological satellites with odd numbers and geostationary weather satellites with even numbers. The satellite carried a sensor that collected data useful for a range of meteorological purposes. China made data from the FY-2 satellites openly available to international users via direct broadcasting of high-resolution digital data and low-resolution analog data in formats compatible with those used by other geostationary weather satellites.[45] Five additional FY-2 satellites were launched between 2004 and 2014, with improved imaging and data transmission capabilities.[46]

In 1999, China launched a new polar-orbiting weather satellite, FY-1C. The satellite featured improved sensors as well as an improved overall spacecraft design. FY-1D followed, using the same design, in 2002. Both satellites operated smoothly. Data transmissions from the satellites used a format compatible with US weather satellites, making it possible for international ground stations to receive the data with minimal modifications.[47]

The second generation of polar-orbiting Fengyun satellites incorporated a number of more advanced instruments for broad environmental monitoring. Each satellite included three imaging instruments capable of making useful observations of the oceans, atmosphere, and land. Satellites in the series also include three sounding instruments to collect information about the atmosphere at multiple altitudes. Instruments were added for measuring climate-relevant data, as well, including ozone concentrations and the Earth's radiation budget. FY-3A was launched in 2008, followed by FY-3B and -3C in 2010 and 2013, respectively.[48] The capabilities and orbits of the FY-3 series were similar to that of planned US next-generation Joint Polar Satellite System (JPSS) meteorological satellites. In fact, a report prepared for NOAA stated that use of China's FY-3C and -3D satellites could be "a silver bullet" in addressing expected gaps in US weather data caused by delays in the JPSS program, a recommendation that was rejected by the US Congress.[49]

FY-3 satellites transmitted data according to WMO data formats and were compatible with US and European satellite data transmission characteristics, allowing direct access to satellite data in near real time. The CMACAST and GEONETCAST systems also provided data via telecommunications satellite to users with properly equipped receiving stations. China has emphasized the importance of its meteorological satellite system to the Global Observing System of the World Meteorological Organization (WMO), noting that its satellites not only benefit China, but also provide a valuable contribution to the international community.[50]

Archived data from Chinese meteorological satellites can be accessed online through the China National Satellite Meteorological Center (NSMC) website. The website includes data from all currently operational Chinese meteorological satellites, with both minimally processed data and more highly processed datasets available for free to users. Access requires registration, and users must sign terms of use that require the data only be used for noncommercial purposes.[51]

Earth Observation Satellites

China also pursued a number of other Earth observation projects outside of meteorology. In 2002, China launched the first satellite in its Haiyang (HY) series, focused on ocean observations. HY-1A carried an instrument to measure ocean color. HY-1B, which followed in 2007, had similar capabilities. In 2011, China launched the HY-2 satellite, which carried four sensing instruments able to provide a wide array of ocean-relevant data.[52] Data from the HY-1B and -2 satellites, both of which were operating as of early 2017, was made available to international users through the China National Satellite Ocean Application Service (NSOAS). The data was available at no charge, but use was restricted to nonmilitary, noncommercial use only and redistribution to third parties was forbidden. Only processed data products (level 2 and above) were available and data from China's exclusive economic zone (EEZ) and other sensitive ocean regions was excluded.[53]

NSOAS distinguished between three levels of users: level 1 included individuals or organizations that require small amounts of data with non-real-time delivery. These users can register on the NSOAS website and expect to receive access to the data in two to three business days. Level 2 included organizations that require large amounts of data, but do not need real-time delivery. Level 2 users were required to fill out a more in-depth application that explained their data requirements and the purpose of the project. Level

3 organizations were those that wished to develop operational applications requiring large quantities of data and real-time delivery. These level 3 users were required to provide the same information as level 2 users and also to propose a contract, agreement, or memorandum that would be submitted to the State Oceanic Administration Science and Technology Department for formal approval.[54]

In addition to its weather and ocean satellite programs, China has a land remote sensing program, managed by the China Center for Resource Satellite Data and Application (CRESDA). The program includes five distinct series of satellites: China-Brazil Earth Resource Satellite (CBERS), HuanJing (HJ) environmental and disaster monitoring satellites, ZiYuan (ZY) resource satellites, experimental ShiJian (SJ) systems, and high-resolution GaoFen (GF) imaging satellites. CBERS-1, launched in 1999, was China's first land remote sensing satellite. It was followed by CBERS-2, -2B, and -4 in 2003, 2007, and 2013, respectively. HJ satellites focused on environmental and disaster monitoring. HJ-1A and -1B were launched in 2008. They carry advanced imagers and can revisit the same area within two days. HJ-1C, launched in 2012, was China's first synthetic aperture radar (SAR) satellite, capable of acquiring data in all weather conditions. Two ZY resource satellites have been launched, one each in 2011 and 2012. They carry an instrument that allows collection of three-dimensional geographic information. The SJ satellites are experimental systems designed primarily to test new instruments. Two of these satellites launched in 2012. The GF is the newest series, but also the largest, with five satellites launched between 2013 and 2015. They provide high-resolution imaging capabilities as part of the China High-resolution Earth Observing System (CHEOS).[55]

Distribution of land remote sensing data within China is done through a three-part system. Departments that rely heavily on the data, such as the Ministry of Land Resources or the Ministry of Environmental Protection, have an optical fiber network link to the data. Other government users can access the data through the e-government network. The general public can access data through the Internet. In the early 2000s, China was distributing thousands of images a year. By 2013, it was distributing hundreds of thousands of images a year, and had distributed 4.1 million scenes total.[56] A 2015 CRESDA presentation said that nearly all Chinese demand for medium-resolution satellite imagery, up to 2.5 meters, is fulfilled by Chinese satellites, and prices for high-resolution imagery are dropping.[57] In

2016, the head of the China State Administration for Science, Technology and Industry for National Defense announced that Chinese satellites were providing 80 percent of the satellite data used in China.[58]

China has established partnerships for international distribution of its data, generally on an ad hoc basis. In 2007, it joined the International Charter on Space and Major Disasters, pledging to provide free satellite data to those affected by disasters anywhere in the world. In 2008, China and Brazil built a CBERS station in South Africa to facilitate free CBERS data access for African nations. In 2010, China and Brazil announced that CBERS data would be made freely available to all nations. In 2011, an HJ-1A ground station was constructed in Thailand to allow direct access to real-time data.[59] Also in 2011, the China Center for Earth Observation and Digital Earth implemented its Earth Observation Data Sharing Plan, which made 23,000 scenes of medium-resolution satellite remote sensing data freely available online. Interestingly, however, the data included was exclusively from foreign satellites, such as Landsat and ERS.[60] China has also expressed an interest in international cooperation related to its new CHEOS system.[61]

Military, Intelligence, and Commercial Satellites

As in many other parts of the world, commercial remote sensing efforts are taking place in China. In 2005, China launched Beijing-1, built by SSTL, as part of the international Disaster Management Constellation. Marketing of the satellite was done by the Beijing Landview Mapping Information Technology Company (BLMIT).[62] In July, 2015, the Twenty First Century Aerospace Technology Co. Ltd., a subsidiary of BLMIT, purchased the full capacity of three 1-meter-resolution "Beijing-2" satellites from SSTL, also known as the TripleSat constellation. A few months later, the Chang Guang Satellite Technology Co., a commercial spinoff of the Chinese Academy of Sciences, launched four Jilin-1 satellites. Two of the satellites provided ultrahigh-definition video imagery, one was a technology demonstrator, and the fourth carried a 72-centimeter-resolution imager.[63]

In addition to its civil and dual-use remote sensing systems, China has launched more than 30 reconnaissance satellites, beginning in 1975. The earliest satellites were based on film that needed to be recovered and processed on Earth; hence the program was referred to as the film recovery photographic series in China.[64] The more recent Yaogan (YG) series included more than 20 satellites launched between 2006 and 2013. The three Tianhui (TH) mapping satellites launched between 2010 and 2015 are also controlled by the PLA and used primarily for reconnaissance. As with reconnaissance satellites in other nations, data from these systems has not been made publicly available.

South Africa

Through the use of satellite ground stations, South Africa has been utilizing remote sensing data for more than 30 years, and it has launched two Earth observation satellites of its own. The first, SunSat, was built almost entirely by postgraduate students at the University of Stellenbosch. Launched in 1999, the satellite carried an imager, but data was never made widely available. SumbandilaSat, developed by the same group for the South African National Space Agency (SANSA), was launched in 2009. The satellite carried a 6.5-meter-resolution imager and collected more than 1,000 images during its two-year operation.[65] The data can be accessed by contacting SANSA.

South Africa has been active in efforts to coordinate remote sensing satellite efforts among African nations, including the African Association of Remote Sensing of the Environment (AARSE). In 2014, AARSE called on African governments to support a pan-African space policy. They have also supported data sharing through development of an African Resource Management (ARM) Constellation or through development of a joint data portal, but these actions have not yet been completed.[66]

Satellite Data Sharing in BRICS Countries

All of the BRICS nations share at least some satellite data, and in many cases the level of data availability is quite high. China's meteorological agency has historically embraced open data sharing and continues to provide users with access to free data online. Chinese oceanographic data is also shared fairly extensively. Brazil, in its joint program with China, is a leader in this area, making all of its satellite data freely available online since 2010. After a long history of restricting access to data, particularly related to national security concerns, India has moved to make a significant portion of its data freely available. Russia does not restrict access to its scientific data, but the data is typically not made readily available online. Similarly, South Africa makes its data available upon request. China, India, and Russia have all developed systems to sell remote sensing data, particularly high-resolution imagery, and thus some portion of their data falls under a more restrictive data policy or commercial licensing scheme to enable these efforts.

Part III   Data Sharing Trends

Having developed a model for data sharing policy development and applied it to a wide variety of agencies, it is now possible to solve the two mysteries laid out in the beginning of the book. This part returns to those two questions, illustrating how the key elements of the model can explain the enduring challenges, open data leadership, and intriguing patterns in Earth observation satellite data sharing. The final chapter then explores how the lessons learned here can help predict and shape future data sharing efforts, both in the space sector and for publicly collected data as a whole.

16 Sharing Satellite Data

This chapter begins by explaining why agencies that operated Earth observation satellites were often leaders in making their data openly available. In doing so, it explains the reasons for the patterns seen across this historical development of many of the agency case studies. The chapter then turns to the remaining gaps in satellite data sharing, using the model framework to better understand why some countries continue to restrict access to data from their environmental satellites.

Mystery 2: Open Data Leaders

Why were the agencies that operate Earth observation satellites often ahead of the curve in terms of open data sharing—developing open policies years or even decades before they were announced by national leaders? And why did many of these agencies follow the same pattern—from free data provision to restricted policies and then back to free and open sharing? The answers to both questions lie primarily in the interaction between agency- and national-level decision-makers, with the former often advocating for free and open data sharing to best achieve their agency goals and the latter typically advocating for data sales and commercialization to reduce pressure on the budget. The economic attributes were particularly important in these debates, as well as to the relative strength of these two positions over time. Experience with different types of data policies, including data sales and commercialization, played an important role in improving understanding of these economic attributes, demonstrating that for most, if not all, Earth observation data, there were broad noncommercial benefits, but no viable commercial market. Technological development, particularly the spread of the Internet, also affected these economic attributes in important ways, significantly decreasing the marginal cost of sharing data, and other actors and data attributes reinforced the positions and arguments of each of these groups.

All of the agencies featured in the case studies favored free and open data sharing at the outset of their programs. For the meteorological agencies—NOAA, EUMETSAT, and JMA—free international exchange of data directly contributed to their ability to achieve agency goals of providing high-quality weather forecasts, warnings, and other meteorological information to the public. It was also in keeping with long-standing cultural norms in the international meteorological community. Data sharing helped the space agencies—NASA, ESA, and JAXA—evaluate the capabilities of new technologies and support the scientific research community. Since the United States, Europe, and Japan all began their programs with cooperation between the space and meteorological agencies, these goals reinforced each other.

For both types of agencies, sharing data also helped to build the agency's reputation, domestically and internationally, providing further incentive for sharing. These activities demonstrated international technological leadership and increased national prestige, leading national-level policy-makers to actively support international data sharing. This is seen in statements made by US presidents and Congress people, and in decisions by Europe and Japan to begin their programs with the development of meteorological satellites that would provide a highly visible international contribution.

By the early 1980s, it had become clear that satellite data was valuable. The agencies interpreted this value as reinforcing the importance of building the satellites and sharing the data to produce more research and knowledge or further improve meteorological forecasting capabilities. However, it also became clear that maintaining these programs would be expensive. Some, particularly national-level policy-makers concerned about the budget, believed this cost should be offset by commercialization. Agencies often spoke of applications for agriculture, mining, and many other industries. If data could be sold to these private customers, it could reduce the cost of these systems for the government. If sales were sufficient to sustain a commercial market, the programs could be operated efficiently without any government involvement, driving innovation and returning tax revenues. There was an incentive for nations to act quickly in enabling commercialization: remote sensing companies that were first-movers in the nascent global commercial remote sensing market would likely have an advantage.

Some agencies and other organizations, such as NASA and the Office of Technology Assessment in the United States, warned that although a robust commercial market for satellite data may someday exist, commercialization of satellite systems was not yet feasible. But with tightening budgets, the potential benefits of commercialization, and the perceived importance of being a first-mover in this new industry, many national-level decision-makers felt it was worth pursuing, even given the uncertainties. They put pressure on agencies to implement cost recovery or commercialization of their satellite systems. The impact of this pressure on agencies' data sharing policies depended on the importance of data sharing to the agency's mission and on the estimated economic value of the data produced by the agency.

The first effort at commercialization among the case studies, regarding the Landsat program, was also the most extreme. This occurred because Landsat was perceived to have the greatest commercial potential of any unclassified government satellite operating at that time, and because no agency saw the Landsat program as essential to its mission. Landsat had begun as a research program at NASA, and during that time its data had been freely shared, but NASA was not interested in maintaining an operational satellite series. Its mission was oriented toward developing new technologies to answer specific research questions. NOAA, which took over control of Landsat after NASA, was a more natural fit for an operational satellite series, but that agency did not have a mission related to land monitoring. The perceived large commercial potential and the lack of agency pushback allowed the US Congress to play a dominant role in determining Landsat's future—including its data sharing policy. In fact, Landsat is the only program, across all of the case studies, for which the full data sharing policy was articulated by national-level policy-makers. The Carter and Reagan administrations had both pushed for commercialization of the Landsat program, and Congress required this transition by law, leading directly to significant increases in data access fees for all users.

The Landsat case is an extreme example in terms of the amount of influence, but national-level government actors drove the adoption of more restrictive data sharing policies in other agencies, as well. In the same 1984

Table 16.1

Agency Mission, and Economic Attributes of Data, and Data Sharing Policy

	Agency Mission Focus Area(s)	Perceived Commercial Value of Satellite Data	Degree of Commercialization in Data Sharing Policy
NOAA	Meteorology and oceanography	Medium: sustained data, but narrow commercial applications, mostly in commercial meteorology sector	Given legal authority to charge market prices, NOAA established prices on some data
USGS/ Landsat	No strong agency ownership over early Landsat program	High: sustained data, broad commercial applications	US Congress implemented full privatization
NASA	Science and technology	Low: short-term data, few commercial applications	Legally required data to be provided to users only under negotiated bulk data buys, NASA declined to engage with commercial sector
EUMETSAT	Meteorology and oceanography	Medium: sustained data, but narrow commercial applications, mostly in commercial meteorology sector	Policy restricted access to member nations, tiered policy implemented
ESA	Science and technology	High: sustained data, broad commercial operations	Tiered policy implemented, with advantages for member nations
JMA	Meteorology	Medium: sustained data, but narrow commercial applications, mostly in commercial meteorology sector	Data freely available, but low emphasis on provision to research users
JAXA	Science and technology	High: sustained data, broad commercial applications	Tiered policy implemented

law that commercialized Landsat, Congress required NASA to provide data to commercial users only through the negotiation of bulk marketing rights. However, NASA's situation differed from Landsat in two important ways. First, NASA tended to develop one-off satellites designed to answer particular scientific questions. Data from these types of satellites was not likely to have significant commercial value. Further, NASA saw data sharing as central to its mission of promoting scientific research. These two conditions gave NASA a relative advantage in maintaining its more open data sharing practices. NASA technically abided by the law, but it did so by restricting data to noncommercial use and did not actively pursue commercial partnerships. When the Government Accountability Office called attention to NASA's lack of engagement with the private sector, NASA responded that they "recognize[d] that some data will likely have commercial value but believe the highest priority for the program is science and global climate change research rather than commercial applications."[1]

NOAA's operational satellites produced data that was potentially valuable to some commercial users, particularly commercial meteorological services, but the agency viewed free and open data sharing as central to its mission. When Congress gave NOAA the power to sell data to commercial users at market prices in 1988, NOAA developed a schedule of fees, but prices were lower than would have been necessary to fully recoup the cost of NOAA data collection systems, particularly satellites. Further, NOAA regularly waived even marginal cost fees for research and educational users, providing data for free. The limited commercial value of these products and the perceived conflict between data sales and NOAA's meteorological mission resulted in more limited data sales efforts that Congress intended.

In Europe, national-level policy-makers put significant pressure on their respective meteorological agencies to recover some of their operational costs through the sale of meteorological data and products. This pressure extended to joint activities within EUMETSAT, leading to the establishment of restrictions and fees on EUMETSAT data beginning in 1988. In addition to recovering costs, access fees for developed nations were designed to demonstrate clear benefits to agency members and prevent free riding, a reflection of the fact that EUMETSAT was an international, rather than subnational, agency.

Like NASA, a large part of ESA's mission focused on scientific research, but ESA also faced pressure from national-level policy-makers to promote European leadership in the emerging commercial remote sensing sector. ESA attempted to balance these goals by developing SAR and high-resolution imaging instruments that would produce data of interest to a wide variety of users in the public and private sectors. ERS-1, ERS-2, and Envisat were designed to ensure data continuity, an important factor for commercial viability. In 1992, ESA adopted a tiered policy in an attempt to balance scientific and commercial goals, working directly with European remote sensing firms to market its data to the commercial sector.

Compared to the other two space agencies examined in the case studies, JAXA put relatively more emphasis on technology development than on scientific research, and viewed broad data sharing as less central to its mission. The agency was amenable to national-level policy-makers' desire to develop cutting edge technologies, SAR and high-resolution imagers, that would help to jump-start the Japanese commercial remote sensing sector. Relatively low emphasis on research and scientific data utilization and high perceived commercial value of the data resulted in the adoption of a tiered policy, with data available at marginal cost for researchers and at a market price for commercial users.

JMA also increased its interaction with the private meteorological sector during this period, but did so through the creation of JMBSC, which facilitated data exchange with commercial and research entities, providing data to both groups at marginal cost.

The decisive factor in the transition back toward free and open data policies was the failure of the commercial remote sector to take off. This experience occurred across nearly all of the case studies. With insufficient revenues to sustain the series, commercialization of the Landsat program was officially declared a failure in 1992. Having never utilized its ability to negotiate data purchases with the private sector, NASA did not generate any revenue from data sales. NOAA reports consistently noted the failure to generate significant revenues, to the point that the agency voiced concern that the cost of maintaining the payment collection system was greater than the revenue collected. In Europe, EUMETSAT reported annual revenues equivalent to less than 1 percent of its annual budget. Even with the ability to access all data from ESA at marginal cost, companies marketing ERS and Envisat data did not make significant profits. JAXA does not publicly release information on its revenues from data sales, but anecdotal accounts suggest these are quite low.

The effect of the lack of a viable commercial market and low revenues was augmented by the lobbying of external actors. WMO emphasized the mutual benefits of meteorological data sharing and facilitated the development of Resolution 40, in which states agreed to continue sharing a limited set of essential data on a free and open basis. The meteorological community drew attention to the disparities in the value-added meteorological sector in Europe compared to the United States and Japan. The scientific research community raised awareness of the value of satellite data for research and the negative impact of data fees and restrictions on these efforts. At the same time, the commercial remote sensing companies that were attempting to build a market in data sales advocated for continued support for their activities.

Some organizations raised normative arguments focused on the issues to which data would contribute. This was particularly prominent within the meteorological community, which argued that data should be made freely available because it contributes to saving lives and property. A similar argument was used in the development of the International Charter on Space and Major Disasters, with countries arguing that there is a moral imperative to provide data in the event of a disaster when it may aid in rescue and recovery efforts.

These types of ethical arguments only went so far. Arguably, any data that would be useful for improving the weather forecast would contribute to saving lives and property, but WMO members only agreed to share "essential" data, not all data that proved useful. Similarly, some in the disaster response community argued that recovery is a long process, and free provision of data in only the first days or months of a disaster is insufficient, but restricted data is rarely made available for extended recovery efforts. These arguments, along with more abstract or forward-looking normative arguments about the potential for better understanding of climate change to result in saved lives and property or the importance of open data sharing for transparency and democracy, were raised and contributed to data sharing discussions, but they did not drive changes on their own.

Agencies also wrestled with normative questions about the proper role of the government. There was a desire to maximize the value of the data for citizens, which some agencies argued required free and open data provision for all users. Others argued that if the data was primarily being used by private actors to generate profits, it was more appropriate for the government to require these actors to pay for access to the data, decreasing the burden on the taxpayer. In general, these types of arguments were raised to provide additional support and justification for existing policy preferences, rather than being drivers of policy development themselves. Similarly, security arguments were sometimes raised, but typically did not have a major impact given the availability of similar data from multiple nations.

Altogether, the inability to generate significant revenues, combined with arguments made by the meteorological and research communities, led to a decrease in pressure from national-level policy-makers to engage in commercialization activities, giving the agencies more freedom to move back toward open data sharing. In fact, rather than facing opposition from national-level policy-makers, they often found support. The legal requirement for NASA to provide data to commercial users only through negotiated en bloc sales was repealed as part of the Landsat Act in 1992, and NASA committed to providing data to all users, commercial and noncommercial, at marginal cost or free. The Bromley Principles and OMB Circular A-130 further promoted marginal cost or free data sharing. NOAA noted these policies, gradually increasing the amount of data that it provided for free to both commercial and noncommercial users. Both NASA and NOAA embraced Internet technology early as a way to facilitate low cost or free data access.

EUMETSAT also began to move toward greater data sharing relatively early, in line with its mission and meteorological community norms. After updating its policy to align with WMO 40 in 1995, the agency steadily increased the amount of data that it provided on a free and open basis. When it released its consolidated data policy in 1998, it made all archived data available at no cost for all users, commercial and noncommercial. Eventually, the only restriction remaining was on access to its highest-precision geostationary satellite data by commercial users and medium- and high-income countries, in part a result of its perceived need as an international organization to provide benefits to its members and give disincentive to free riders.

Advocacy by the climate research community, which highly valued Landsat's long history of continuous data collection, played a role in ensuring the program was returned to government control, rather than canceled altogether. Yet the move to open data provision occurred much later. Still without a clear organizational home in 1992, Congress laid out the conditions for accessing data from the existing Landsat 4 and 5 satellites, which would operate under a tiered policy, as well as the upcoming Landsat 7, which would provide marginal cost access to all users. Marginal cost, however, was interpreted in this case to include personnel and capital equipment costs, and prices for a single image were hundreds of dollars. It was only after Landsat found a more permanent home at USGS, a science-focused agency, that serious efforts were undertaken to make Landsat data freely available. In 2008, Landsat data was fully transitioned to an online distribution system, where it was provided free of charge. This transition produced dramatic results in terms of data use, particularly for climate research, and provided further evidence that there were broad noncommercial uses for satellite data.

ESA had begun moving toward a more open data policy in 2002, announcing that data from its science-focused Earth explorer missions would be available to all users at marginal cost or free. European-level support for open data provision in the PSI and INSPIRE Directives lent further support to this trend, and the Landsat policy change provided a dramatic example of how free data provision could impact data use. In 2010, ESA adopted a policy under which nearly all data would be freely available online for all users, with the exception of high volumes of data, which could not be accommodated online, and SAR data, which continued to be treated under a tiered data policy.

In JAXA, reorganization and a national emphasis on climate change issues helped to spur a data policy update in 2013. Under JAXA's new policy, only high-resolution imagery would be subject to fees and sold, while all other data was made freely available online. JMA continues to operate under the same policy it has for decades, continuing to emphasize free provision of data for official use by national meteorological agencies. While the agency has updated its data provision systems serving these entities, less attention has been paid to services for research and other users, who continue to access data primarily through JMSBC and other third parties. There is no automated system for accessing this data, and in some cases, it remains subject to fees.

Overall, the pattern seen across many agencies is explained by the fact that many of these agencies had similar missions, were working with similar technologies, and dealing with data with similar economic attributes. The agencies began by providing data to users to support their meteorological, scientific, and technology-driven missions. Provided with sufficient resources to do so, most of these agencies would have continued to provide data on a free and open basis. Instead, there was significant pressure from national-level policy-makers for agencies to engage in data sales that would offset system costs and to promote the development of a commercial remote sensing sector.

The agencies engaged in commercial activities to various degrees. Even with variations in the exact technologies, timing, and data policy designs, all agencies found that revenues were low, and a robust commercial market was not likely to emerge in the near term. They also found that the fees and restrictions necessary to facilitate these commercial activities had a significant dampening effect on noncommercial use of the data. As agencies began to reduce or eliminate fees and restrictions, the increase in data use provided further evidence of this effect. Seeing that commercialization was not viable, pressure from national-level policy-makers to sell data decreased and in some cases reversed. Seeing the impact that opening access to data could have on data use, an issue central to most agency missions, these agencies increased their efforts to adopt open data policies.

These same trends that decreased national-level pressure to engage in cost recovery and commercialization and increased national-level support for data sharing in the satellite sector continued to evolve into national-level open data movements in general. The space and meteorological agencies were "ahead of the game" on these trends because they had always preferred to make their data available as widely as possible in support of their missions. They were pushing in these areas, rather than being pulled, and adopted open data policies as soon as they were allowed, rather than waiting for national-level initiatives that instructed them to do so.

Mystery 1: Enduring Climate Monitoring Gaps

The first mystery presented in the introduction called attention to the deficiencies in climate monitoring caused by a lack of data sharing. Given high-level attention to the issue on the international stage and efforts on the part of multiple international organizations, why do some nations provide free access to their data while others restrict access to data from unclassified government satellites? To answer these questions, it is useful to look across all countries that have participated in the development of an Earth observation satellite, incorporating data from the BRICS nations and those described in the appendix, in addition to those in the case studies.

The factors that explain the gaps in global satellite data sharing today are largely the same as those that explained trends seen in the case study countries over the past decades: agency mission and understandings of the economic attributes of satellite data. Most agencies that do not make their satellite data available fall into one of two categories: those that do not see data sharing as central to their agency's mission and those who are attempting, often under pressure from national-level policy-makers, to promote cost recovery and growth of the commercial sector. A relatively small portion of data is restricted for alternative reasons, often related to security.

Agency Mission

The first argument for not sharing satellite data, because it is not seen as central to the agency mission, is seen across a wide variety of programs, but is particularly common in nations that operate just a few satellites. Out of the 35 nations that have Earth observation programs, 25 of them have owned five or fewer satellites. Among these are Brazil and South Africa, discussed in chapter 13. The remainder are reviewed in appendix A. Figure 16.1 shows the distribution of data sharing policies for the 57 satellites owned by these nations.

Many of these agencies view their space activities as measured first steps, primarily aimed at capacity building and technological development. Some nations, such as Egypt and Iran, build their programs on their own, typically beginning with relatively simple, experimental Earth observation satellites. Rather than start from scratch, many nations have instead chosen to contract with a company in a nation with a well-established space industry. SSTL, based in the United Kingdom, the Satrec Initiative (SI) in South Korea, and the China Great Wall Industry Corporation all offer programs in which a country can purchase a satellite and training package. As part of the agreement, engineers from the purchasing nation assist in the development of the satellite, receiving hands-on training. The list of foreign

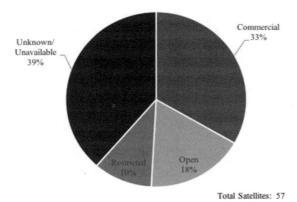

Total Satellites: 57

Figure 16.1
Data Sharing Policies in the 25 Nations with the Smallest Unclassified Earth Observation Satellite Programs

customers that have worked with these companies is long. In more than 20 years of operations, SSTL has developed Earth observations satellites for Portugal, Chile, Thailand, Algeria, Turkey, Nigeria, Kazakhstan, and more. SI customers have included Malaysia, the United Arab Emirates, Singapore, and Turkey, and the China Great Wall Industry Corporation has worked with Pakistan, Venezuela, and others.

For these smaller programs, the goal is to build capacity among national engineers and demonstrate technological capabilities. International sharing of the data is often seen as a secondary issue, if it is considered at all. One way agencies articulate this is by stating that they can't afford to share their data. These agencies aren't against data sharing, per se, it just isn't a high enough priority for their agency, or their national-level policymakers, to justify the funding necessary to set up a data sharing policy and system. The disinclination to share is compounded by a perception that the data is not especially useful. If the satellite is new and experimental, the quality may not be high. Even with the professionally built spacecraft, the focus is often on relatively basic, medium-resolution systems. If nations believe there will be low demand for the data, it is not worth investing time and effort in the development and maintenance of a data distribution system. Similar arguments are often given for not providing data from university satellites and from older satellites, even in countries with very large space programs. Both the United States and Russia do not provide open online access to data from some of their earliest satellites. Once again, there is no strong argument against data sharing, but agencies also don't believe there is a strong enough argument for sharing to make it worth the investment.

It is also quite common for nations with small space programs to sell data from their satellites. Unlike many of the agencies in the case study, these nations are typically not aiming to facilitate the emergence of a new commercial remote sensing market. Instead, commercial sales are seen as an easy, low-effort option for potentially gaining something from an otherwise untapped resource. In many cases, a commercial firm will take on the role of marketing and selling data. Typically, this firm will pay the agency a small fee or a percentage of sales. From the agency's perspective, they spend no time or effort on marketing and sales, but potentially bring in some revenue. From the perspective of the company, they are able to access data at a fraction of the cost of building a satellite. It's a win-win situation.

SSTL actually builds a data sales option into some of their packages, working through their subsidiary, DMCii. Agencies contract with SSTL to develop a standard, medium-resolution imaging satellite. While the satellite is owned by the nation that purchased it, its data is marketed and sold by SSTL as part of the Disaster Management Constellation, and a portion of the revenue is provided to the owner. The ability to sell data from multiple, similar satellites gives the program an advantage, because it is more likely to have the desired data for any particular customer and can offer a better revisit time. The Pan GEO Alliance is designed to offer similar benefits to other small Earth observation programs or companies that own satellites not originally designed to operate as part of a constellation. Like DMCii, the Pan GEO Alliance plans to provide a more comprehensive service by bringing together data from satellites owned by many nations as part of one system. Agencies engaged in these endeavors typically don't see this decision as a trade-off between open data sharing and data sales, but rather between limited or no data sharing and data sales. Viewed in this way, there is essentially no downside to engaging in these commercial activities.

In the larger space programs, this issue of agency priorities is typically referred to as technical or resource challenges. Figure 16.2 shows the distribution of data sharing policies for the 401 satellites owned by the 10 nations and regions with the largest unclassified Earth observation

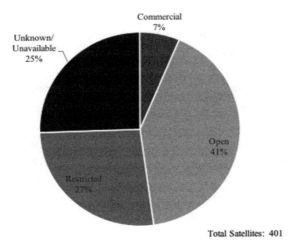

Figure 16.2
Data Sharing Policies in the 10 Nations with the Largest Unclassified Earth Observation Satellite Programs

programs. This group includes the United States, Europe, and Japan, as well as Russia, India, and China, for which case studies and summaries were provided. South Korea, France, Italy, and the United Kingdom are also among the largest programs. Summaries of their activities are provided in appendix A.

Although it is a leader in open data sharing, as of 2016, data from some of NASA's early satellites is not readily available and would likely require a lengthy process to access, if access was possible at all. This is not due to any economic, normative, or security concerns, and NASA does not have any restrictions that would prohibit access to this data. Instead, it reflects a belief that this data is not of the highest utility, and not worth the investment of resources needed to make it available. In developing its open data portal, JAXA has similarly prioritized inclusion of data that is most frequently requested, with other data planned to be added later, or not at all. ESA explains in its data policy that large volumes of data are not provided online due to technical constraints. JMA similarly cites bandwidth issues as the reason for limiting access to its Himawari Cloud system to official NMS use only. Access to Russian satellite data is provided to government users through internal systems and to others through special agreements. New users wishing to access data can contact Roshydromet or Roscosmos, which are often willing to provide the data. However, none of the data from Russia's 84 unclassified government satellites is made openly available online for those outside of Russia, and this technical restriction is a significant barrier to access. In these countries with large space programs, as with those with smaller programs, the problem is not that data portals or technical systems to accommodate larger data traffic are impossible to build. Instead, these agencies do not have sufficient resources to do so, or, in other words, they do not see the activity as a high enough priority to justify redirecting existing resources. In some cases, these agencies may be correct—some data may have limited utility and building a system capable of supporting nontraditional data users or very large volume downloads could potentially be very expensive without providing value to a large number of users, but in many cases, it seems likely that if the data was made more readily available, use would significantly increase.

Some agencies have argued that the era of big data and cloud computing has resulted in a situation in which they are no longer technically able to provide data to users for free. NOAA, for example, is experimenting with a program in which cloud computing companies ingest satellite data at their own cost and then provide that data to users on commercial cloud computing platforms, potentially charging fees to end users. NOAA argues that this type of commercial arrangement is necessary, because its existing data storage and distribution systems are not capable of dealing with the volumes of data produced by the newest generation of satellites. It is true that these technical limitations do currently exist, and it is also likely that it would be expensive and inefficient for the government to build its own cloud computing platforms, given the availability of high-quality commercial options provided by the private sector. However, this does not necessarily imply that the costs of data provision and access should now be borne by the data users rather than producers. For example, whatever funding NOAA would have previously invested in developing and maintaining a data portal, it could now put toward funding an arrangement with these companies under which users could access and use satellite data for free on commercial cloud computing platforms.

Economic Arguments

Although many of the countries have moved toward greater data sharing and more open data policies, government efforts to support commercial remote sensing or engage in cost recovery remain, particularly among the nations with larger space programs. In the United States, although the civil government agencies that operate satellites have adopted open data policies, the defense and intelligence agencies are actively supporting commercial remote sensing. Japan, Europe, France, Germany, Italy, Russia, India, and China all support satellite systems that have a data sales component. Israel, which has a much smaller program, has put significant effort into international sales and marketing.

These efforts persist because officials are still the debating the same question that has intrigued government leaders for decades: Can satellite data efficiently be treated as a commodity and sold, leading to a vibrant commercial market, or is it most efficient for the government to collect the data and give it away for free? Getting the right answer to this question is of significant importance for getting the most out of the investment in satellite technology. As described in chapter 2, the two key elements for determining this answer are as follows: is there a viable commercial market for the data, and are there broad noncommercial uses for the data? Experience with a variety of Earth observation technologies and data sharing policies over the past five decades helps to answer both of these questions.

Landsat provided the most dramatic evidence of the consequences of answering these questions incorrectly and adapting a data sharing policy that is not matched to economic reality. National-level policy-makers in the 1980s believed that Landsat may be commercially viable, and aimed to create a fully commercial entity with a commercially driven data policy, essentially positioning Landsat in quadrant three in Table 4.1. After Landsat was privatized, low revenues meant that EOSAT was reliant on the government to provide sufficient funding to maintain the system, through direct budgetary contributions and data purchases, to cover the costs of satellite development and operations. At the same time, the high cost of data led to a significant reduction in use of the data by government agencies and researchers. The researcher that complained, "We have the worst of all possible worlds: we are both spending the money and making sure that we get nothing out of it," was exactly correct.[2] When Landsat data was made freely available online in 2008, data access and use skyrocketed. Numerous economic analyses supported the intuition that the increased benefits from data use far outweighed the loss of revenue from data sales.[3]

Nations have learned from Landsat and other examples like it. Significantly more government Earth observation data is made freely available than is sold. Existing commercial efforts in these large space programs all focus on very high-resolution radar and visible imagery. After decades of experimentation with data sales, these are the markets that seem the closest to being viable or offer the greatest potential for cost recovery. They are also assumed to have relatively narrow noncommercial applications, primarily benefiting national security and intelligence users. (Whether this is a good assumption is discussed in the next chapter.) In the framework presented in the model in chapter 4, these programs fall into quadrant 1: not commercially viable and narrow noncommercial uses. In this case, public-private partnerships may be economically efficient, depending on the specific circumstances. As the primary customer, the government should expect to shoulder the majority of the cost of developing and operating the satellites. However, if the commercial market for data is sizable (even if not viable on its own), the cost savings for the government that result from sharing the fixed costs of development and operation with these commercial purchasers could be significant. Further, if there truly are only narrow noncommercial applications, the loss of benefits caused by restrictions on access to and redistribution of data would be minimal.

Existing commercial remote sensing companies also fall into this category, relying on sales to government customers for the majority of their revenue. The National Geo-Spatial Intelligence Agency (NGA) has long been the anchor tenant for US commercial remote sensing companies, providing large contracts that include funding for satellite development and guaranteed data purchases. The acquisition of Space Imaging and the merging of GeoEye and Digital Globe, both immediately following losses or significant decreases in NGA contracts, further demonstrate the dependence of this sector on government funding. Yet these commercial partnerships provide the US government with significant savings: nearly 40 percent of Digital Globe revenue came from foreign governments and other customers in 2015. NGA has also mitigated the loss of benefits caused by restrictions on data access by negotiating a license that allows the data to be accessed and used by other US federal agencies without any additional cost. This has allowed agencies such as NASA to take advantage of the data for research conducted by their scientists.

Other nations also provide significant contributions to their commercial remote sensing sectors, negotiating agreements that provide some benefits to the government. France originally planned for Spot Image, launched in 1986, to become self-supporting by 1996.[4] Airbus Defense and Space has made progress toward this goal—it developed SPOT 6 and 7, launched in 2012 and 2014 respectively, without government support, and with no guarantee from French authorities of imagery purchases. However, the company still receives significant support from the French government. It was given exclusive commercial use of the French 50-centimeter resolution, civil-military Pleiades satellites launched in 2011 and 2012.[5] Germany developed the TerraSAR-X satellite, launched in 2007, as part of a public-private partnership with the goal of creating a sustainable market that would allow the commercial partner to finance the follow-on system on its own. Due to lower-than-expected profits, the German government funded the majority of the follow-on TanDEM-X satellite, launched in 2010.[6]

These governments receive benefits in return for their investments. France typically gets a discount on access to Airbus Defence and Space data, and research users can apply to Germany for free or low-cost access to data from its high-resolution radar satellites, TerraSAR-X and TanDEM-X. Similar arrangements were used for the Israeli EROS satellite and Canadian RADARSAT-2 satellite: government subsidized development was provided in exchange for free or low-cost access to data.[7] In some cases, these benefits extend to the larger user community. In 2014, France announced plans to open access to the SPOT satellite data archive for research use. Raw data that is at least five years old, excluding SPOT-5 data with resolution sharper than 10 meters, will be made widely available.[8]

Overall, although commercial remote sensing and cost recovery efforts affect a much smaller portion of satellites than they once did, these arguments remain one of the key reasons that access to satellite data is restricted.

National Security

About 15 percent of Earth observation satellite data is subject to restrictions other than those necessary to allow commercial sales. As mentioned above, many of these restrictions can be traced to technical and financial challenges that reflect agency mission priorities and allocation of resources. Among this subset of restricted access data, security issues are another common justification for restrictions. India has long been concerned with security issues related to its satellite data, limiting access even to its meteorological satellite data for this reason. Security concerns weren't completely unfounded—as India argued, imagery collected by a geostationary satellite could potentially show ship movements or other sensitive information. Even if this data was available from other sources, it would be possible that in some instances the Indian satellite would provide better quality information, or help to confirm information gathered through another method. The security risks of releasing information are never zero. However, most nations, including, eventually, India, came to the determination that the benefits of making the data available outweighed the risks of doing so.

There are also policies that are not directly affected by security restraints on a regular basis, but that have the potential to be restrained in particular circumstances. All nations require companies to apply for a license to operate, ensuring compliance with the United Nations Remote Sensing Principles. The licensing process allows governments to restrict data sales for national security reasons. In developing the data policy for the Copernicus system, the European Commission laid out a relatively complex process for reviewing systems for potential security sensitivities. India, Germany, and France also have in place processes for those wishing to sell or access high-resolution data.

The United States has typically limited the spatial resolution of data that can be sold by the commercial remote sensing sector, adjusting the limits based on an evaluation of other imagery available on the international market. The United States has also put in place measures that would allow it to further restrict access to data in the event of a national security emergency. Provisions in commercial licenses allow the United States to limit data collection or dissemination of commercial remote sensing firms for national security reasons, effectively giving the government "shutter control." The agreement with EUMETSAT regarding the joint international polar-orbiting meteorological system gives the US government the right to restrict access to data from NOAA instruments flying on the Metop satellites in the case of a national security emergency. The United States has never made use of either of these special capabilities, but retains the legal right to do so.

For the most part, however, national security concerns have not been a major factor in the development of data sharing policies covering unclassified government satellites. In many cases, particularly with respect to scientific satellites, this is because the data does not have any major security implications. For others, such as meteorological satellites, and weather forecasting information in general, there are clear defense or military applications of the data, but the benefits of sharing the information outweigh the security risks. For data with high spatial resolution or other attributes that could pose security issues, the data is not restricted because it is not unique. Many nations operate satellites with similar capabilities, and often data from these high-resolution systems is sold commercially. Efforts to limit the distribution of data from domestically owned satellites would likely lead to a loss of market share for that nation, without actually preventing access to data by the would-be customer. In fact, the decision to limit access to data can sometimes spur others to develop systems to collect it on their own. This was the case with regard to classified systems, with multiple nations citing a lack of willingness on the part of the United States to share reconnaissance data as the major driver in the decision to develop their own reconnaissance systems.

Better understanding of the reasons for which nations decide to limit access to data—whether it's related to agency mission and priorities, a desire to promote commercial data sales, or concerns about national security—makes it possible to predict how data sharing in this sector may evolve in future years. The final chapter looks at this issue in depth.

17   Future of Data Sharing

The future of satellite data sharing policies will depend on whether the current trends continue and on which new trends may emerge. Will nations that currently make data freely available continue to do so, or will this data become more restricted? Will nations that restrict access to data begin to instead make it open? How will new actors entering this sector treat their data? Understanding these trends is not only of academic interest—it also shapes the options that interested actors have to influence these changes.

Changing Agency Priorities

One of the reasons that agencies don't share data is that they don't view data sharing as a key part of their mission, and hence don't view it as a priority in requesting and allocating resources. This is important—it means that there are not overriding security, economic, or normative arguments for restrictions or a strong cultural bias within the agency against open data. In fact, if asked directly why they do not share data, these agencies are likely to respond that they do share (just in ad hoc or technically limited ways) or that they would like to share their data, but can't afford it.

This has important implications for the future of data sharing in these agencies, and the types of developments or arguments that might prompt changes. Academic arguments about the economic or normative effects of satellite data sharing, as have been highlighted in multiple reports and meetings in the past, are unlikely to be convincing on their own, as these agencies often do not disagree with these assessments; they just "can't afford" to act on them. Instead, actors who wish to increase data sharing would need to focus on showing these agencies how data sharing can support their mission and/or on providing the resources needed to undertake these activities.

Reputational Effects

The agency priorities and culture that underlie these arguments do not change easily. For agencies to view data sharing as a priority worthy of an investment of time and resources there need to be clear benefits to doing so. For example, reputational benefits gained through national and/or international recognition of open data sharing policies, by government leaders or in organizations such as GEO or WMO, can help provide an incentive for sharing. Negative attention brought to the lack of data sharing may also be effective, but would be risky, as nations receiving criticism may reduce their involvement in these voluntary organizations, rather than change their data sharing policies.

Systems that allow agencies to document and receive credit for the ways in which their data is used—in research, value-added products, or other programs—can also provide reputational benefits and incentives. International organizations and interest groups can help to make this easier by promoting technical methods and professional norms that enable this, such as the use of persistent identifiers that ensure the original data source can be tracked, and expectations for data sources to be properly cited. Collection and presentation of metrics showing which countries are contributing most to global science, perhaps on a per capita basis, could provide additional recognition and benefit.

Argentina provides an interesting model for small space programs that wish to maximize their international impact and reputation. The country has only has only developed three Earth observation satellites, but each was designed to fulfill a particular scientific need not already addressed by the international community. Partnership with larger space programs, particularly NASA, made it possible to design and launch advanced Earth observing systems. Argentina made data from these satellites freely available, providing an important contribution to global environmental monitoring, and increasing its standing in international organizations, with only a relatively small investment.

Demand for Data

Some of the agencies do not make data available because they don't believe the data would be useful. If there was a demonstrated demand for the data, they would be more inclined to invest time and resources in providing it. The lack of demand may be due to low utility of the data, but it may also be due to a lack of awareness of the data's existence, or perceived difficulties of accessing the data. International organizations and interest groups could help to determine potential users of these datasets. They can publicize the existence of the data to appropriate user groups and even assist as a conduit in requesting and accessing the data by providing appropriate contact information.

Lower Cost/Resource Assistance

The decrease in the cost of computing technology and increasing spread of the Internet can also be factors in increasing satellite data sharing among smaller programs. As this technology becomes less expensive and more widely available, it lowers the barriers to data sharing. Combined with reputational benefits and other trends, decreasing costs for making data available will increase the likelihood that it is shared. For agencies that are interested in sharing data, but still feel they can't afford it, resource assistance can be a relatively simple answer. If national-level decision-makers are supportive of open data sharing, this could come in the form of additional national funding directed towards data dissemination systems. If not, international organizations and others can provide resource assistance in a variety of direct and indirect ways, as described below.

Technical and Logistical Assistance

International organizations can share technical and logistical solutions that lower the costs and other barriers to data sharing for these nations. Sharing data portal technology, design, and lessons learned could help lower the time and resources needed to develop a new portal. Sharing best practices for collecting and presenting usage statistics is another area of potential benefit. Already, GEO has helped to provide some guidance to agencies on how to implement standard creative commons usage licenses. These types of assistance improve the situation for data users, who gain access to more data with systems that are more standardized, and data producers, who face reduced challenges to implementing a data sharing system. In the case of developing nations, this assistance could go beyond knowledge sharing, to include provision of ground station or computing equipment and hands-on training, perhaps in partnership with aid organizations such as USAID or the World Bank. These types of programs would help build capacity in developing nations while also benefiting users worldwide by increasing data availability.

International Technical Cooperation

Agencies could also undertake direct cooperation or assistance in developing data sharing technologies and policies. Following the example of the DMCii and the PanGeo Alliance in the commercial sector, a nonprofit or international organization could develop a data portal that would host data from multiple countries—focused around small satellites, satellites from a particular region, satellites of developing nations, or university satellites, for example. These cooperative data portals would provide an easier experience for data users, who only need to visit one location to access data from many countries, and provide a cost effective solution for satellite operators, who could take advantage of economies of scale in data distribution systems, even if they operate only one or two small satellites. The satellite-owning agencies would only need to transmit the data to this central body, with no need to invest in the development or maintenance of a user-friendly data portal of their own. If the central body was equipped with an adequate receiving station, data could be collected directly from the satellites, with no effort on the part of the satellite-owning nation. With multiple agencies funding this central portal, the costs of data distribution would be quite low for each agency, while each would still gain the reputational benefits associated with sharing their satellite data.

International Data Hosting

Rather than building a new data portal, agencies such as NASA, ESA, or JAXA with well-developed data portals could offer to host data collected by foreign satellites in their existing portals, acting as a worldwide host for smaller agencies that would like to make their data available but feel they can't afford it. ESA already hosts external, third-party satellite data, although this data is typically acquired by ESA through special purchases or agreements and access is limited in accordance with these arrangements. Data hosted as a free service for a foreign nation would be provided on an open basis, following the data policy of the hosting agency.

This type of cooperation would provide many of the same benefits as the international consortia described above, with almost no resource investment required on the part of the satellite-owning agency, and very little required on the part of the larger agency, as well. It would benefit the satellite-owning agency, which could promote awareness of its capabilities, and its open data contribution. The system would also benefit the reputation of the agency hosting the data, as it would increase their overall offerings. The Landsat Global Archive Consolidation (LGAC) program provides a successful example of this type of arrangement. International Landsat ground station operators were willing to provide their data free of charge for inclusion in the US Landsat archive in exchange for the United States' willingness to recover the data from outdated media and process it into standard products.

Commercial Data Hosting

The organizations hosting data would not need to be governmental or even nonprofit organizations. Google, Amazon, and others have demonstrated an interest in hosting environmental satellite data on their cloud computing forums, typically providing free access to users (who then have an incentive to purchase further storage or processing capabilities through these providers). Smaller space agencies could potentially work with these organizations to increase the amount of data available online. Once again, such an agreement would require a smaller investment than developing a dedicated data portal and would increase the range of offerings on these commercial sites.

Ensuring Economic Efficiency

The second major reason that nations choose not to share is based on economic attributes of the data. Nations hope to benefit from data sales that offset government costs or to promote the development of a commercial remote sensing sector that will contribute to national innovation and economic growth. However, after decades of experimentation with a variety of remote sensing technologies, to date, no commercial remote sensing endeavor has been able to survive without government support. By the definition in chapter 4, there is no viable commercial market for satellite data. Further, there are broad noncommercial uses for Earth observation satellite data. The framework in chapter 4 suggests, and nations have found in practice, that under these conditions, economic benefits are maximized by providing the data on a free and open basis. This realization largely explains the trend toward free and open policies seen among space programs in the past two decades.

However, this trend is not absolute. Existing free and open policies are regularly questioned, particularly by national-level policy-makers wondering if these expensive activities can be conducted more efficiently. As noted above, many nations, even those that have largely adopted free and open policies, continue to attempt data sales and commercialization. Some new commercial remote sensing companies are advocating for government to turn certain types of data collection over to the private sector. How can the government address these emerging issues?

Documenting Economic Benefits of Open Data Policies

Economic theory and decades of experience with data sales and commercialization efforts have shown that in practice, just as in theory, open data provision is the most economically efficient method of distributing data for all, or nearly all, Earth observation satellite data. Government data collection and free and open distribution maximize net social benefits. Governments that make their Earth observation satellite data openly available will generate the greatest benefits for themselves, for their citizens and for people around the world.

When data is made openly available, without restrictions on access or redistribution, its use is maximized. It can be used by scientists, nonprofits, entrepreneurs, or the general public. Use of the data results in improved scientific understanding that benefits everyone, in more efficient or effective programs to serve the public, and in the creation of new value-added products that increase offerings for consumers and generate tax revenue for the government. Often the full breadth of applications is not realized by the original developers, and may not be discovered until years after the data has been collected, as new technologies, new scientific questions, or new markets create new opportunities for data use. Open data is more easily combined with other data, making it even more likely that the data will be used in new and productive ways.

Experience over the past two decades has shown that this is the case in the Earth observation satellite sector. However, efforts to further study, document, and publicize these effects will be important to sustaining support for these policies among agencies, and the national-level policy-makers that influence them. The benefits of commercial activity are easy to demonstrate in monetary terms and descriptions of international sales that everyone can understand. It is important for metrics on the benefits of open data to also be collected, shared, and explained to make their benefits clear, as well.

Increasing Benefits from Commercial Efforts

As mentioned above, none of the existing commercial remote sensing efforts can survive without government support. However, public-private partnerships for data collection may be more economically efficient than open data, even if the commercial market for the data is not viable on its own, if the decrease in costs achieved through these arrangements exceeds the decrease in benefits caused by restrictions on data access and redistribution. Therefore, it is useful to examine the extent to which this is the case for the existing public-private partnerships.

Any policy that places restrictions on data access and redistribution, as is necessary to facilitate commercial sales, will decrease data use compared to an open data policy. However, this decrease in use is likely to be relatively low if there are only narrow noncommercial uses of the data. For example, if high-resolution SAR and visible imagery is primarily of use for intelligence and defense purposes, it can be purchased by these agencies at a cost lower than would be necessary to develop the satellites themselves. Since meteorologists, for example, do not have a need for this data, the fact that they cannot access the data does not cause a loss in benefits. However, there are some indications that this data may have broader uses. Under the current NGA-negotiated license, other federal agencies can access the data. Some NASA scientists have taken advantage of this, finding the high-resolution imagery to be particularly useful in studies of the cryosphere. NASA's short-lived Science Data Buy program also showed that the research community would find the high-resolution imagery useful, particularly with some adjustments to ensure careful calibration. However, with low-cost access to the data only available for five years under the program, researchers did not find it prudent to undertake a long-term research portfolio that would depend on this data. In 2001, the National Research Council review stated that SAR data was "one of the

most exciting remote-sensing technologies" for scientists.[1] Such a statement does not imply narrow, security-related uses.

Given this situation, governments may be able to increase economic
efficiency by adopting arrangements that increase access to the data, particularly for noncommercial users. Licenses that allow the data to be shared
freely with other government agencies, as is the case with the US data purchases, are a good start. Governments should look to expand this access,
perhaps negotiating for a license that allows government-purchased data
to be used for noncommercial purposes by any user—not just those in federal agencies. This is unlikely to have a significant effect on commercial
revenues, as nongovernment, noncommercial users typically make up a
very small portion of their customer base. If the license conditions on the
data could be made clear—perhaps with persistent identifiers—it may even
be possible to allow free redistribution of the data among noncommercial
users. ESA has done this to some extent. In 2006, the agency signed a three-
year contract that would give it access to 12,000 images from the SPOT-2
and SPOT-4 satellites, which could then be provided for free to research and
education users.

Data sharing agreements in public-private partnerships can also differentiate other attributes of the data. In the case of remote sensing satellites,
timeliness can be particularly important. Many commercial uses of the data
rely on it being acquired quickly, and very old data is typically less commercially viable. France's arrangement to provide archived SPOT data will
provide a good test of the potential benefits of this type of arrangement,
demonstrating whether there is significant demand for this type of data
and the range of noncommercial uses that may be possible.

Evaluating Cost Savings from Commercial Efforts

In considering the economic efficiency of these endeavors, governments
should remember that the cost savings are those stemming from general
commercial efficiency—the idea that a commercial entity can develop
and operate a satellite more efficiently than the government—and from
decreased reliance on general taxation, and the subsequent reduction in
deadweight loss, needed to fund the satellite system. The first benefit may
not be as large in the commercial remote sensing sector as in some other
cases. With high fixed costs to build and launch a satellite, and low marginal costs of data distribution, commercial remote sensing is a natural

monopoly. This has led to significant consolidation in the remote sensing sector, resulting in one, or very few, commercial remote sensing companies in each country or region. The lack of competition decreases normal market incentives to innovate and keep prices low. However, it is still true that commercial entities can operate with less bureaucracy and red tape than the government, and although governments tend to favor domestic industry, international commercial remote sensing companies do compete with one another to some extent.

With respect to efficiency benefits from decreased reliance on general taxation, it is important to make a distinction between savings that occur due to sales to the commercial sector and savings due to sales to foreign governments. If the vast majority of data sales are to governments, this does not reduce the reliance on general taxation—just shifts it from one country to another. In these cases, collaboratively built and operated satellites might be more efficient. Taking high-resolution imagery in the United States as an example, if the United States were to pay 60 percent of the cost, and two or three other governments paid the remaining 40 percent, the cost of the program would be the same from the perspective of the governments involved, but as a fully government-owned program, the data could be provided on a free and open basis, maximizing social benefits. From a global economic perspective, this would likely be a more efficient arrangement. This type of shift—from commercial provision to government provision—is rare, however. The Copernicus program in Europe provides the only example, with its plans to collect and openly share medium-resolution imagery of a type currently collected and sold by some private companies, and even that plan has yet to be implemented.

Transition from commercial to government control is also not without drawbacks. It may be challenging for the United States to convince other nations to participate—the government would essentially be taking on the marketing role currently undertaken by the commercial entity. Returning to the theoretical example above, if raising the remaining 40 percent of costs requires the participation of many other governments, rather than two or three, this is an even greater challenge. Further, due to the need for international negotiations and government participation, a cooperatively developed international satellite is likely to be more expensive than one built by the commercial sector. Nations will need to be convinced that it is worthwhile to increase their investment, at least slightly, to get the benefits of free data provision. Effort will be required to maintain this international effort, particularly if it is to continue to support a series of satellites, rather than a single satellite. If any nation decides to decrease or end its support for the program, sustainability is in question for all others. For the example given, which included a 60/40 cost-sharing arrangement, as time goes on, agency officials and policy-makers not involved in the original planning may ask why the United States pays more than half of the cost of a system that benefits all partners equally (since the data is freely available to all). These practical considerations and transaction costs must be factored into efforts to increase economic efficiency.

One possible compromise between international cooperation and commercial provision would be an international open data buy. Rather than cooperating to jointly build a satellite, nations could negotiate to jointly purchase data from a commercial entity under an open license. The result would be the same—data could be freely shared with all, maximizing social benefits, but countries would not need to reinvent the wheel, and could benefit from existing commercial efficiencies. There may even be some opportunities for offsetting costs raised by general taxation if the company also sells specialized datasets or value-added products to commercial users, although governments should expect such commercial cost sharing to be minimal.

Future Commercial Remote Sensing

This type of open data buy procurement would be the most efficient approach for some emerging commercial remote sensing activities, as well. GPS Radio Occultation (GPS-RO) data provides the most immediate candidate. GPS-RO data has proven to be very valuable for meteorology, significantly improving the accuracy of weather forecasts when combined with data from traditional weather satellites and other weather observation systems. Currently, the only dedicated GPS-RO constellation is the COSMIC-1 system, developed as a joint effort between the United States and Taiwan and launched in 2006. Data from this constellation has been freely shared internationally, and has been incorporated into weather forecasting systems in many countries. Planning for the follow-on COSMIC-2 system is underway.

GPS-RO technology is well understood and relatively inexpensive. In recent years, a number of companies have announced plans to develop commercial GPS-RO constellations. These organizations have lobbied the US government to purchase GPS-RO data from commercial entities rather than continuing government development. US decision-makers, particularly in Congress, are eager to promote the growth of US commercial remote sensing activities and to benefit from cost decreases promised by these commercial entities. However, NOAA officials question whether the US government will truly save money by purchasing commercial data, given that Taiwan is paying more than half the cost of the existing and planned government systems. They also worry about the impact of ending free international sharing of this data that has proven to be of such value to weather forecasting, and the implications of doing so for US commitments to share data under WMO Resolution 40 and on the norm of free international exchange of meteorological data in general. A move by the United States from free data distribution to a commercial model that restricts access could lead other nations to do the same. Right now, the United States (and every other nation) receives much more free data from the WMO World Weather Watch than it puts in. If nations begin restricting access to data to allow commercial purchases, this situation could change, and all nations would be worse off than before.

The framework from chapter 4 further illustrates why such an arrangement would be unlikely to provide significant benefits to the United States. GPS-RO data is primarily useful to governments for use in numerical weather forecasting models—there is not significant commercial demand for the data. As such, this data falls into quadrant I or II in Table 17.1, and the government should not expect to achieve significant cost savings or economic efficiency gains from commercial sales. If GPS-RO data is narrowly useful for the meteorological sector, it may belong best in quadrant I. In this case, a data purchase that allowed the data to be freely shared for official meteorological use and perhaps also for research uses may be efficient. However, it seems likely that the data would also have important uses for climate and other environmental studies, in which case it would more appropriately be placed in quadrant II. In this instance, free and open data provision would maximize net social benefits. If the US government could arrange to share the costs of purchasing the data under a global open data license with other nations—Taiwan, at least, and perhaps others—it could capture the savings of commercial development and operation while maintaining the significant benefits of open data sharing. At least one of the emerging

Table 17.1
Commercial Remote Sensing and Economic Efficiency

	Narrow Noncommercial Uses *Less benefit to open data*	Broad Noncommercial Uses *Greater benefit to open data*
Nonviable Commercial Market (government funding required) *Less savings from data sales*	I Open data policy or tiered data policy *Existing remote sensing companies (e.g., DigitalGlobe, Airbus Defence and Space, e-Geos)*	II Open data policy *Most Earth observation satellites*
Viable Commercial Market (no government funding required) *Greater savings from data sales*	III Data sales *Possibly emerging remote sensing companies (e.g., Planet Labs, Terra Bella)*	IV Open data policy or tiered data policy *Possibly emerging remote sensing companies (e.g., Planet Labs, Terra Bella)*

GPS-RO companies has expressed a willingness to engage with the government in this type of arrangement.

There has also been significant attention to the rise of commercial remote sensing companies, particularly in the United States, that are using new small satellite technology and traditional Silicon Valley techniques to provide new types of data and lower the cost of data access. Top among these are Planet Labs and Terra Bella. These companies are focusing primarily on temporal resolution—imaging the same area once a day, or even providing persistent video, rather than spatial resolution, as other commercial remote sensing companies have done. They argue that there is a commercial demand for this type of information that is not currently being met. While both companies have launched satellites, neither has achieved full operational status, so the reality of these claims cannot yet be assessed. As of yet, neither company has required or requested significant government assistance in the form of development funding or guaranteed data purchases. It is possible that these will be the first remote sensing companies to be commercially viable in the traditional sense.

Governments should assess the value of the data for noncommercial uses, as well. The NGA has already announced a plan to examine the value of the data for its own national security purposes. If purchased using licensing agreements similar to those used in its contracts with other remote sensing companies, this may provide an opportunity for other US agencies to evaluate the data, as well.

If there is a commercially viable market, and the government assessment shows that there are relatively narrow noncommercial uses, the government can purchase the data at the same prices and under the same conditions as other users—just as it would any other commodity, such as a laptop or an office chair. To the extent that the data has broad noncommercial uses, the government could investigate options for purchasing a license that allows free access to the data for noncommercial uses or making a bulk purchase of archived data, to be made freely available under an open license. The most beneficial arrangement will depend on the specific attributes of the data.

If these systems do need government subsidization to continue operations, the government will need to consider the value of noncommercial uses to determine whether these benefits outweigh the requirement for government support. Government leaders should consider whether this support is needed only in the short term or for a longer period, and whether provision of this "start-up funding" may be a good investment for the government, given the potential benefits of a vibrant commercial remote sensing sector. This echoes the decision that nations have made, and continue to make, with existing, high-resolution imaging companies. Some countries argue that while the "start-up" support for these existing companies has lasted longer than many nations had hoped or projected, it does not necessarily mean these efforts at commercialization have failed. The government should continue to realistically evaluate the likelihood that these activities will become commercially viable in the future.

Future of Satellite Data Sharing

Overall, what does all of this mean for the future of satellite data sharing? The economic realities of satellite data mean that, as a rule, it is most efficiently treated as a public good and shared freely. Failed efforts at commercialization and dramatic examples of the impact of free data policies have helped to clarify these economic attributes and have helped to drive a trend toward free and open data policies. To maintain this trend, governments and others will need to actively document and demonstrate these benefits, which are not always clearly visible to those outside the Earth observation satellite community, or even everyone within it.

Organizations and nations that hope to increase the availability of satellite data and decrease gaps in climate data collection and sharing should actively work to increase the reputational benefits of data sharing, through organizations such as GEO and WMO. They should consider direct investments of time, technology, and funding to make archived data or data collected by smaller programs that do not or cannot prioritize data sharing more openly available. Governments should promote interactions with the private sector and the growth of commercial remote sensing activities, but this should be done in a way that maximizes long-term, global economic benefits over private returns and short-sighted cost savings. They must carefully consider whether new technological trends—small satellites, big data, and cloud computing—are truly changing the economics of the remote sensing sector, or just providing new ways of operating within the same basic structure. There is great potential to increase the benefits from existing and future Earth observation activities, if the right actions are taken to do so.

Future of Open Data Movement

Finally, we can ask what all of this means for the open data movement in general. How can the lessons learned from the Earth observation satellite community inform policy-making in other areas? For individuals interested in these questions the greatest takeaway from this book is the validation of the model of data sharing policy development presented in part I. Application to the issue of satellite Earth observation helped to demonstrate the value of using this model, which places the agency as the central actor, and considers the impact of both people—national government, international, and nongovernment actors—and ideas—the economic, normative, security, and technical attributes of the data—in understanding the development of data sharing policies. This model can be used to understand why open data sharing initiatives have been successful in some cases and not others, and it can help to direct the development of future policies.

The model demonstrates that the mission and culture of the agency are key to understanding how and why particular data sharing policies are adopted and implemented the way they are. Free data sharing isn't free from the perspective of the data producer, so the agency needs to have a reason, and an ability, to undertake these activities. Just as seen in the Earth observation satellite community, if an agency doesn't believe data sharing is an important part of its mission, it is unlikely to make its data available. In these cases, national-level policy-makers that direct agencies to share data, through initiatives or laws, are unlikely to see these directives implemented as enthusiastically as they'd like. Similarly, if agencies view open data sharing as an important element of their mission, they are likely to push back against efforts to commercialize or otherwise restrict data sharing.

National-level policy-makers can attempt to affect agencies by better understanding the role data sharing plays in the agency's mission. Sharing can be incentivized by pairing open data initiatives with resources needed to undertake these activities and with efforts to highlight how the data is being used, providing reputational benefits for the agency. Laws that prohibit data sharing are typically effective, but more nuanced directives to share in particular ways or with particular groups require resources and reputational incentives, just as in the case of open data sharing.

Actors that are external to the government—interest groups and international organizations, for example—can play an important role in reinforcing agency culture, providing reputational benefits, and providing information about the effects of relevant data sharing policies in other nations that allow more informed policy-making. These organizations can help bring attention to the distributional effects of data sharing policies, keeping in mind that one of the benefits of open policies is the ability to use data in ways not originally intended or predicted.

Considering these actors and their political interactions without including the ideas and issues that they are debating will paint an incomplete picture. It is important to understand the range of interpretations regarding the security and privacy aspects of the data. Can sharing the data contribute to national security, or does it pose significant security or privacy risks? Are there alternative versions of the data or types of policies that could manage these risks? Normative arguments can also be very strong. Are there ethical reasons that the data must be either shared or restricted? Data that has broad noncommercial uses is likely to result in significant economic benefits if shared widely. If there is a viable commercial market, there may be opportunities to benefit from public-private partnerships. Finally, it is important to understand the technologies that are used to collect, produce, and distribute the data. Changes in technology can affect the other attributes of the data, changing the overall conditions for sharing. The table at the end of part I can act as a guide to consider these issues.

The model and the example of the space sector can also help to provide a general understanding of where open data policies will be more or less likely to be implemented. For example, open data provision is unlikely in cases where such sharing is seen as unrelated or even contrary to the agency mission or culture. This may be the case in agencies that don't typically deal directly with the public, or in agencies—focused on national security or working with personal information—that value secrecy and discretion. National-level policy-makers who decline to provide resources needed to support data sharing systems or who fail to recognize the benefits of agency data sharing will inhibit the adoption of open data policies. Legislatures can also prohibit data sharing by making distribution of the data illegal. In cases where there are significant national security or privacy risks of sharing data, this may be appropriate, although opportunities to aggregate or otherwise transform data into a form that could be shared can also be explored. Further, if the data has not been shared in the past, there is unlikely to be a large group of users advocating for data sharing or looking for these types of opportunities.

When the technologies for collecting and/or providing the data are relatively expensive, and when the primary users of data outside the government are private actors, data sales will be attractive, and potentially more economically efficient, than open data policies. If the agency does not have a strong preference for data sharing based on its mission or culture, it may favor cost recovery efforts or commercial partnerships to decrease pressure on its budget, and national-level policy-makers concerned with budget issues are likely to share these views. Organizations that can afford these data purchases may advocate for status quo pricing if it gives existing commercial actors an advantage over new entrants. Data collection technology that is sufficiently well understood can also lead to commercial involvement and even commoditization of the data.

Agencies that have a culture of openness and see data sharing as part of their mission—perhaps because of an affiliation with the science community or a broad public information or service-oriented responsibility—will be more amenable to data sharing. In these cases, there may be existing users who understand the value of the data and are interested in using it. These groups can advocate for sharing and make the value of the data clear to others, including national-level decision-makers. This makes national-level support for sharing more likely, and can lead to provision of resources and reputational benefits that further promote sharing by the agency. This will be made even easier if the technologies for data collection and/or provision are relatively inexpensive and the uses are primarily noncommercial. Strong normative arguments for data sharing, related to uses of the data and/or to issues of government transparency, can reinforce these trends.

As a whole, the spread of Internet and computing technology is making data more easily accessible and resulting in a more information-savvy citizenry. Users are more likely to recognize the benefits of open data sharing and have the skills needed to use the data—as scientific researchers, as journalists, as entrepreneurs, or as members of the informed public. Interest groups, agencies, and legislatures, and all other elements of the model of data sharing policy development are affected by these developments. The trend toward open sharing of government data will not be smooth and it may never extend to all government data, but it is a trend that will continue moving forward.

Appendix A Global Satellite Data Sharing

By early 2016, 39 countries had developed or operated at least one remote sensing satellite, including government, university, and commercial activities. This chapter reviews remote sensing activities in each of these countries, and documents the state of satellite data sharing as of the beginning of 2016.

Algeria

The Algerian National Space Agency, created in 2002, owns and operates the AlSat-1 and -2A remote sensing satellites, launched in 2002 and 2010, respectively. Remote sensing offers Algeria the ability to monitor desertification, coastal regions, urbanization, and other environmental issues.

AlSat-1 was developed in cooperation with Surrey Satellite Technology Limited (SSTL) in the United Kingdom as the first contribution to the Disaster Monitoring Constellation (DMC). The goal of the DMC, conceived by SSTL, was to take advantage of low-cost, small satellite technology to create a constellation of imaging satellites with medium spatial resolution, but high temporal resolution. This would help to support disaster monitoring and response, which requires the frequent collection of imagery. To further reduce the investments needed for the system, SSTL envisioned individual ownership for each satellite and ground station pair.[1]

Data collected by each DMC satellite is the property of the satellite owner, and nations can develop their own data sharing policy, although a provision of the contract with DMCii ensures data will be made available free of charge for major international disasters through the International Charter for Space and Major Disasters. Data from the constellation as a whole, including AlSat-1, is marketed and sold by DMC International

Imaging (DMCii), a subsidiary of SSTL, with a portion of the revenues going to DMC member nations.[2]

AlSat-2A was developed by Airbus Defence and Space in an arrangement that included hands-on training for Algerian engineers. It carries a high-resolution optical imager. The data from this instrument has been used for a variety of environmental applications within Algeria, but there is no system for foreign users to access the data.[3]

Argentina

The Comisión Nacional de Actividades Espaciales (CONAE), the Argentinian Space Agency, was formed in 1991. CONAE has launched three remote sensing satellites, all in partnership with NASA. The Satellite for Scientific Applications A, or Satelite de Aplicaciones Cientificas A (SAC-A), mission, launched in 1998, was primarily designed to test the physical infrastructure for satellite control and communications prior to the launch of SAC-C. However, it carried a CCD camera capable of taking digital images of the Earth and a magnetometer that was used for investigations of Earth's magnetic field.[4]

SAC-C, launched in 2000, carried 10 instruments from 6 countries, including the United States and Argentina. The satellite included multiple imaging sensors, a GPS occultation experiment, a magnetic mapping payload, and an experiment related to space radiation. Use of the data was reportedly a priority for Argentina. CONAE issued an Announcement of Opportunity for the use of data from the SAC-C satellite in 2000, and approved 200 projects from a wide range of countries.[5] Data from two of the SAC-C instruments, the High Resolution Tracking Camera (HRTC) and the Multispectral Medium Resolution Scanner (MMRS), is available for free online through CONAE. The remaining four instruments are not listed in the CONAE database.[6] Data from the French ICARE instrument is hosted on the University of Lille Cloud-Aerosol-Water-Radiation Interactions (ICARE) website, and made available subject to registration.[7]

The successful launch of SAC-C was followed by the development of the Aquarius/SAC-D satellite, designed primarily to monitor sea surface salinity, a key climate variable. The satellite was launched in 2011, carrying instruments from the United States, Argentina, France, and Italy. The Aquarius sea surface salinity data is available free of charge from NASA and

CONAE.[8] CONAE also makes data from its High Sensitivity Camera (HSC), Microwave Radar (MWR), and the New Infrared Sensor Technology (NIRST) instruments freely available online.[9] Data from the ICARE and Radio Occultation Sounder for Atmosphere (ROSA) instruments are made available through France, and Italy, respectively.

Brazil

Refer to chapter 15: Brazil, Russia, China, India, South Africa.

Canada

Canada's RADARSAT series, which began with the launch of RADARSAT-1 in 1995, made Canada a leader in global Synthetic Aperture Radar (SAR) capabilities. From the beginning, the satellite was envisioned as a multiuse system that could provide important information to both civil and military leaders.[10] Canada's Earth Observation Data Sets (EODS) program gave researchers the opportunity to submit research proposals for access to data, and the United States received free access to RADARSAT-1 data in exchange for providing the launch vehicle.[11]

The mission also had a commercial component: in 1989, RADARSAT International was established to process, market, and distribute RADARSAT-1 data. Royalties from sales were used to offset the cost of satellite operations. By 1999, revenues from RADARSAT data sales totaled about C$3 million, and revenues from other products and services reached C$14 million. Together these revenues accounted for about 3 percent of the C$620 million cost of satellite development.[12]

After proving that there was commercial interest in the RADARSAT products with RADARSAT-1, Canada hoped to engage more closely with the Canadian remote sensing sector with the follow-on system. CSA chose to develop RADARSAT-2 as part of a public-private partnership with Canadian company MacDonald Dettwiler Associates (MDA). Under the agreement, CSA was to pay C$225 million and MDA C$80 million toward development of the system, and MDA was responsible for ongoing operations and commercialization. MDA would own the satellite and ground stations to allow for full commercial exploitation. The government of Canada would receive free access to data products and services valued up to the total amount

it contributed to system development. Canadian academics and researchers could submit a research proposal to gain access to RADARSAT-2 data
through the Science and Operational Applications Research (SOAR) program.[13] The decision to privatize RADARSAT-2 led NASA to renege on an
agreement to launch the satellite in exchange for access to data, leading
to increases in development costs for Canada. According to Canadian officials, the United States cited concerns that the enhanced performance of
the system would make it a competitor to US industry.[14]

To address potential security issues and ensure compliance with the
Outer Space Treaty, Canada passed the Remote Sensing Space Systems Act
in 2005. The law required remote sensing satellite operators to apply for a
government license to operate, and gave the government the right to limit
the types of data or products that can be sold as well as the individuals or
groups to whom they may be sold.[15] This ensured that commercial efforts
would be adequately regulated, in accordance with international law, and
provided Canada assurance that it would have the power to prevent any
potential security issues related to data distribution.

By the time RADARSAT-2 was launched in 2007, six years later than originally planned, the investment by the government of Canada had nearly
doubled, to C$437.1 million. MDA's contribution was C$91.6 million. The
cost overrun triggered an official review of the project. The review acknowledged that the program had dual goals: it was designed not only to provide
data continuity from RADARSAT-1, but also to build the capacity of the private Earth observing sector in Canada. Reviewers noted that MDA saw the
program as a success. However, in the words of the report authors, "Some
government stakeholders believe that the P3 [Public-Private Partnership]
was not successful for the Government of Canada (GoC), primarily because
the CSA had insufficient control over the project, the GoC absorbed all of
the project risk and paid for most of the system, and in the end the GoC
does not own the system." There were also concerns that due to delays
and competition, MDA profits would not meet expectations, and the value-
added sector would also not benefit as much as originally foreseen. The
report concluded that the next generation radar satellite should adopt a
different development model.[16]

In addition to the review of the RADARSAT-2 program, in 2011, the
Canadian Space Agency carried out a comprehensive evaluation of its

Earth Observation Data and Imagery Utilization Program. The goals of the program were to promote use of satellite data by other government departments, support the Canadian value-added sector, and maximize the benefits realized from CSA projects. The review identified data access and sharing as the major obstacles to greater use of satellite data and data cost as the main impediment to commercialization of value-added products. The report recommended increasing the use of grants and contributions to fund university research and suggested supplementing the support provided under the SOAR program. One of the specific recommendations was the development of a data policy for the follow-on RADARSAT system that would facilitate data access and sharing.[17]

In the same year, Canada launched its Open Government strategy and announced its intention to join the international Open Government Partnership. Its first open government action plan, released in 2012, identified open data as one of the three key activity streams.[18] A 2014 update to the policy emphasized the global nature of open data sharing.[19] The data utilization evaluation and the open data plan did not lead to a change in the RADARSAT-1 and -2 data access policies; the data is still sold commercially by MDA.[20] However, the draft policy for the RADARSAT Constellation Mission, scheduled to be launched in 2018, has taken into account the trend toward open data, aiming to balance these goals with concerns related to the commercial sector and national security.[21]

As of 2015, the draft data policy for the RADARSAT Constellation Mission states that government requirements in support of sovereignty, security, and safety have top priority. The data should be used in a way that fuels prosperity and advances foreign policy objectives. The policy aims to strengthen the Canadian industry's capacity to commercialize value-added products, to enable cooperation with allies and partners to meet socioeconomic and security objectives, and to support international organizations with safety, humanitarian programs, and other initiatives of benefit to Canadians. Finally, the policy states that Canada will aim to enable commercial distribution of RADARSAT data while remaining compliant with the Open Government Strategy.[22] The policy does not provide details on how this would be achieved.

In addition to its RADARSAT series, Canada has launched two scientific Earth observation missions. SCISAT-1, launched in 2003 in partnership

with NASA, was designed to study atmospheric chemistry in the strato-
sphere.[23] Data from the satellite is freely available through the University of
Waterloo and the ESA Earth Online portal.[24] In 2013, Canada launched the
CASSIOPE satellite. While this mission focused primarily on space weather
effects, it also collected data on the ionosphere and magnetosphere that
may be relevant to climate studies.[25] Minimally processed data from the
mission is freely available online through the University of Calgary, but
due to limited manpower, highly processed, publication-quality data is not
available.[26] Users can contact the principal investigator for the relevant
instrument directly regarding this type of data.[27]

Chile

Chile has launched two Earth observation satellites. The first, the Fuerza
Aerea Satellite–Bravo (FASat-B),* was launched in 1998. FASat-B was built
by SSTL in cooperation with the Chilean Air Force primarily as a capacity-
building effort. It included instruments to monitor ozone, but data was
never made widely available.[28]

   In 2011, the Chilean Air Force launched the Sistema Satelital para la
Observación de la Tierra (SSOT) satellite, also known as FASat-C. SSOT carried a high-resolution imager suitable for resource management, precision
agriculture, disaster monitoring, and other civil and commercial uses. With
a resolution of 1.5 meters, SSOT is also useful for defense and intelligence
uses. SSOT data can be purchased online from the photogrammetric service
of the Air Force of Chile.[29]

China

Refer to chapter 15: Brazil, Russia, China, India, South Africa.

Denmark

Denmark has only undertaken one satellite mission. The Ørsted satellite,
launched in 1999, was developed by the Danish Meteorological Institute,

---

* FASat-B was a replacement for FASat-A, which never became operational, because
it failed to separate from another craft during launch.

the Danish Space Research Institute, and a number of Danish universities and companies. The satellite took highly accurate measurements of Earth's geomagnetic field. Data from the satellite was made freely available online via the Ørsted Satellite website until 2007. The website is no longer maintained, leaving no current options for accessing data.[30]

Egypt

The Egyptian space program is run by the National Authority for Remote Sensing & Space Sciences (NARSS), established in 1991. Egypt's first remote sensing satellite, EgyptSat-1, was launched in 2007 and carried two imaging instruments. EgyptSat-2 followed in 2014 with a high-resolution imager. Both satellites failed to survive for their full design lives, with EgyptSat-1 failing after three years, and EgyptSat-2 failing after just one year. Still, Egypt views both satellites as important to capacity building and demonstrating Egyptian technical capabilities, and hopes to use data collected for practical applications within the country. While EgyptSat data has been used in some research, it is not made widely available.[31]

Europe

Refer to chapter 11: European Organization for the Exploitation of Meteorological Satellites, and chapter 12: European Space Agency.

France

France is one of the largest contributors to ESA, and it also operates the largest national space program in Europe. The Centre National d'Etudes Spatiales (CNES), the French space agency, was established in 1961. The four earliest satellites launched to support Earth science activities were Diadème 1/2, Starlette, and Stella, launched in 1967, 1975, and 1993, respectively. These were very simple spherical satellites covered in reflectors that allowed detailed measurement of the Earth's gravitational field using satellite laser ranging techniques. Because they are passive and do not transmit data, any nation can make use of these satellites. Starlette and Stella continue to contribute to satellite laser ranging studies.[32] Satellite laser ranging data is available through anonymous ftp on NASA's Crustal Dynamics Data Information System (CDDIS) website.[33]

These satellites were followed by the launch of four small satellite missions. The Detection of Electro-Magnetic Emissions Transmitted from Earthquake Regions (DEMETER) microsatellite, launched in 2004, aimed to improve understanding of earthquakes and volcanic activity on the electromagnetic environment of the atmosphere.[34] Just a few months later, in cooperation with NASA, CNES lofted the Polarization and Anisotropy of Reflectances for Atmospheric Sciences coupled with Observations from a Lidar (PARASOL) mission aimed at better understanding the role of clouds and aerosols in climate change. PARASOL carried a digital CCD camera.[35] The third microsatellite, Picard, was launched in 2010 with three instruments focused on solar irradiance and variability, two of which were developed by Switzerland and Belgium.[36] In 2013, France launched the Satellite for ARgos and ALtiKa (SARAL) satellite in cooperation with India and EUMETSAT to provide continuity of near-real-time ocean topography. The satellite is also designed to improve understanding of climate change by providing better insight into ocean dynamics.[37]

Data from the DEMETER mission is hosted online by the CNES Centre de Données de la Physique des Plasmas (CDPP) archive. Access requires registration and agreement of the CDPP team, though access is generally provided to anyone who requests it.[38] PARASOL data is available through the University of Lille Cloud-Aerosol-Water-Radiation Interactions (ICARE) system. Most ICARE data is available free of charge for research and application development. Access requires registration and an agreement to properly credit ICARE and the original data providers. Access to some data, including selected PARASOL data, requires further approval.[39] Picard data is hosted by the Belgian User Support and Operations Centre (B.USOC) and was made publicly available beginning in 2013.[40] To access SARAL data, users are directed to the AVISO website, which again requires registration, and in some cases, approval, to access data.[41]

In addition to its civil program, France developed a series of reconnaissance satellites, prompted in part by the US decision not to share its own satellite reconnaissance images with the French defense minister during the first Gulf War.[42] Four Helios satellites have been launched, in 1995, 1999, 2004, and 2009. The first two satellites were developed in partnership with Italy and Spain, and this was later expanded to include Belgium and Greece.

A data-exchange agreement was reached with Germany in exchange for data from its SAR-Lupe radar reconnaissance satellite. Data from these satellites is not made publicly available.[43]

Within the Earth observation community, France is perhaps most well known for being home to the first high-resolution commercial remote sensing company: the Spot Image Corporation. Satellite Pour l'Observation de la Terre 1 (SPOT-1) was launched in 1986 and collected imagery with 10-meter resolution, much better than Landsat-5's 30-meter resolution and sufficient for some military and intelligence uses.[44] The launch of the Spot-5 satellite in 2002 improved this further, with 2.5-meter resolution.

The primary customers for SPOT imagery were national governments, particularly defense and intelligence agencies. However, data was purchased for other uses as well. In 1999, the United States purchased data of North America collected between 1986 and 1998. Under the license agreement, the data is freely available for authorized federal users only.[45] In 2000, the United States made 10-meter black-and-white SPOT images taken between 1986 and 1994 freely available to the public through the "Imagery for Citizens" program.[46] Another data buy made images of North America taken by SPOT-4 and -5 between 2009 and 2013 available to US federal, state, and local government users.[47] In 2006, ESA signed a three-year contract that would give it access to 12,000 images from the SPOT-2 and -4 satellites, which could then be provided for free to research and education users.[48]

Spot Image has always had close ties with CNES. The French space agency funded the construction of the first four SPOT satellites and the majority of the fifth, while charging Spot Image only annual system-maintenance fees between 3 and 8 million euros. CNES remained the largest stakeholder in the company until 2008, when European Aeronautic Defence and Space (EADS) Astrium Services (later renamed to Airbus Defence and Space Geo-Information Division) became the majority owner.[49]

When annual revenues reached more than 100 million euros, it was decided that government support should decrease. The Spot-6 and -7 satellites were financed without government funding, and with no guaranteed data purchase agreements provided by the French government. Launched in 2012 and 2014, respectively, the satellites had a resolution of 1.5 meters. Government support did not disappear completely, however. France gave the company exclusive commercial use of the French dual-use civil-military

Pleiades satellites, launched in 2011 and 2012, both with a resolution of 50 centimeters.[50]

At the January, 2014, GEO meeting, France announced that it would be opening access to the SPOT satellite data archive, marking the first major contribution to GEOSS from the private sector. Access to data will not be fully open, with licenses specifying that imagery is for research use only. The release is limited to data that is at least five years old, and SPOT-5 data with resolution sharper than 10 meters is not included. Only raw data is being made available, although the French government has committed to processing the first 100,000 images free of charge. Additional images will be subject to a nominal processing fee. With more than 30 million images produced since 1986, the release has the potential to provide great benefits, particularly in studying change over time.[51] In addition, ESA is making a limited amount of Spot-6 and -7 and Pleiades imagery available for free, but access requires the submission and acceptance of a full project proposal.[52] And, of course, data can be purchased from Airbus Defence and Space.[53]

Germany

Along with France, Germany is one of the largest contributors to ESA and also maintains a relatively large national program for satellite Earth observations that includes civil, military, and commercial activity. The German space agency, the Deutsches Zentrum für Luft- und Raumfahrt (DLR), launched a series of small Earth observation satellites, beginning with a laser ranging satellite, GeoForschungsZentrum-1 (GFZ-1) in 1995. This was followed by DLR-TUBSAT, a cooperative project with the Technical University of Berlin (TUB) that launched in 1999 and carried two CCD cameras. The following year, DLR launched the Challenging Minisatellite Payload (CHAMP) satellite to study the Earth's magnetic field and do atmospheric sounding. Finally, the Bispectral and Infrared Remote Detection (BIRD) satellite launched in 2001 to demonstrate the possibility of detecting and monitoring forest fires from space.[54]

There is no uniform data policy governing the use of data from German satellite systems. Instead, each program develops its own terms and conditions.[55] Data from the experimental DLR-TUBSAT project as well as the BIRD mission is not made publicly available. CHAMP data, from raw instrument data to highly processed scientific data products, is available for free through the CHAMP Information System and Data Centre at the German Research Centre for Geosciences in Potsdam, but access requires registration and administrator approval.[56]

DLR's civil small-satellite program led directly to its support for a commercial constellation of five small Earth observation satellites, called RapidEye.[57] Rather than extremely high resolution, RapidEye focused on quick revisit times: the satellites had a resolution of 5 meters, but could revisit the same area once per day, allowing them to guarantee fast access to desired imagery. The company saw this guaranteed access as the key to gaining customers in agriculture, insurance, and other fields.[58] RapidEye originally hoped to finance development of the 160 million euro system without the need for direct aid, but ultimately required a 15 million euro investment from DLR as well as funds from local government. In return for its investment, DLR received free access to a limited amount of RapidEye data for scientific use. The full constellation was launched in 2008.[59]

Data sales in 2009 and 2010 did not meet forecasts, and RapidEye declared bankruptcy in 2011. The company was purchased by an imagery distributer in Canada and renamed RapidEye Canada.[60] In 2013, the company changed its name to BlackBridge. Then, in 2015, the company and its RapidEye constellation were acquired by a US commercial remote sensing satellite company, Planet Labs.[61] The Planet Labs website notes that researchers and other noncommercial users in Germany can apply to access RapidEye data for free through the RapidEye Science Archive (RESA). RESA, operated by Planet Labs on behalf of DLR, includes 20 million square kilometers of archived imagery. Researchers also have limited access to new data acquisitions.[62]

At the same time it was investing in RapidEye, DLR was engaged in development of another public-private system: TerraSAR-X, which featured a Synthetic Aperature Radar (SAR) instrument. The primary goal of the satellite was scientific, supporting studies in a range of fields, including climatology. The second goal was to create a sustainable commercial Earth observation market in Europe, sufficiently profitable to allow industry to privately fund follow-on satellite systems on its own, a goal that was later reconsidered due to lower than expected profits from radar data sales. Under an agreement with EADS Astrium (later Airbus Defence and Space), DLR owned the satellite and all of its data and was responsible for scientific utilization of the data.[63] Airbus Defence and Space paid 20 percent of the satellite cost, and in return received exclusive commercial data exploitation rights.[64] Nonscientific use, including administrative use by the German government, requires commercial purchase of the data.[65] The TerraSAR-X add-on for Digital Elevation Measurement (TanDEM-X) satellite was developed under a similar public-private arrangement with Airbus Defence and Space. The satellite was synchronized with TerraSAR-X to develop a very accurate global digital elevation model and test new SAR techniques and applications.[66] TerraSAR-X and TanDEM-X launched in 2007 and 2010, respectively.

Recognizing the sensitive nature of these high-resolution radar satellites, Germany passed a law in 2007 laying out a data security policy for space-based Earth remote sensing systems, the Satellitendatensicherheitsgesetz (SatDSiG). The law applies to primary distribution of data from nonmilitary German satellites, satellites operated by German citizens, and satellites operated from Germany. Rather than limiting data distribution based on spatial resolution, as is done in France and the United States, the German law requires that the data provider carry out an automated sensitivity check on every data request. The algorithm for this check considers the information content of the product, the target area, the time period, and the individual customer, as well as other information. If the sensitivity check reports that dissemination is sensitive, the data provider must apply to the German Federal Office of Economics and Export Control for a permit to release the data.[67]

Users can apply for access to TerraSAR-X and TanDEM-X data for scientific uses by registering on the appropriate portal and submitting a project proposal. If accepted, data is provided online at the cost of fulfilling the user request.[68] In the case of TerraSAR-X, this ranges from 160 to 200 euros per product.[69] For TanDEM-X prices range from 60 to 100 euros per SAR data product.[70] The TerraSAR-X and TanDEM-X science service system data license specifies that data is exclusively for scientific use and exclusively for the purpose of the approved project. If unauthorized transfer or sale of the data occurs, the project leader is responsible for paying compensation at commercial rates. Further, DLR would have the right to exclude the leader, their coinvestigators, and their institution from future access to TerraSAR-X and TanDEM-X data.[71] Data provision is subject to additional restrictions related to security, including delayed provision, requirement of special user authorization, and restrictions based on the geographic location of the data.[72]

Although TerraSAR-X and TanDEM-X provide data of high utility for defense and intelligence users, DLR also developed a dedicated military radar satellite constellation, SAR-Lupe. The decision was reportedly prompted by US reticence to share reconnaissance satellite data with allies during NATO action in Kosovo.[73] SAR-Lupe consists of five identical small satellites, launched between 2006 and 2008, and is Germany's first satellite reconnaissance system. Germany has signed an agreement to share data with France in return for access to the French satellite reconnaissance system, but data from the system is not made publicly available.[74]

India

Refer to chapter 15: Brazil, Russia, China, India, South Africa.

Indonesia

Indonesia, an archipelago with over 17,000 islands, saw the value of space technology early, creating the National Institute of Aeronautics and Space (LAPAN) in 1963. In 1973, Indonesia began acquiring NOAA weather satellite data in Jakarta, eventually expanding its ground station capabilities to receive and process data from a wide variety of foreign remote sensing satellites.[75]

In 2003, Indonesia began development of its first Earth observation satellite, the LAPAN-TUBSAT microsatellite. LAPAN-TUBSAT was built in cooperation with the Technical University Berlin (TUB). Launched in 2007, LAPAN-TUBSAT carried two cameras and used an experimental control technique to decrease revisit time over equatorial regions. LAPAN engaged in the mission primarily to gain experience in developing, launching, and operating a satellite, although the satellite also carried a camera that provided data useful for applications monitoring natural resources and environmental conditions.[76] A follow-on system, LAPAN-A2, was launched in 2015, using a similar design, but carrying a higher-resolution camera.[77] Indonesia does not make data from these satellites publicly available.

Iran

In 2011, Iran launched a small satellite, Rasad-1, carrying an Earth observation camera. A second small satellite, Navid, was developed by students at the Iran University of Science and Technology and launched in 2012. Very little information was released about the two satellites, and the data from them is not publicly available.[78]

Israel

Civil remote sensing activities in Israel are quite limited, but the country is well known for its commercial and intelligence remote sensing systems. Plans for an indigenous reconnaissance satellite began in the early 1970s, following what Israel saw as a lack of adequate access to US satellite intelligence data during the Yom Kippur war. Israel established its space agency in 1983 and launched the Ofeq 1 (Horizon 1) satellite in 1988.[79] Launches of the series continued steadily, with technology improving over time. Ofeq 10 launched in 2014 carrying both a high-resolution imager and a synthetic aperture radar.[80] In 2006, Israel considered partnering with a commercial distributor for sales of some Ofeq images, but chose not to pursue this option. Data from the reconnaissance satellites is not shared.[81]

In 2000, Israel launched its first commercial remote sensing satellite, the Earth Resources Observation System A (EROS A). The launch of EROS B followed in 2006. The satellites were developed by the government-owned Israel Aerospace Industries (IAI) based on Ofeq technology, and IAI subsidiary ImageSat International is in charge of marketing and sales of the imagery. With a resolution of 2 meters for EROS A and 70 centimeters for EROS B, the system both augmented Israel's reconnaissance capabilities and marked Israel's entry into the global commercial remote sensing market.[82] ImageSat offers a number of options for procuring imagery. Customers can request specific images in real time, either working through ImageSat or directly tasking the satellite themselves. ImageSat offers exclusive and confidential purchase of all imagery in a given footprint or long-term lease of the entire satellite for real-time operation over a given geographical area.[83]

In 1998, Israel launched Gurwin-Techsat, a small satellite carrying an imager and a radiation sensor, in partnership with the Technion-Israel

Institute of Technology. The satellite survived for nine years on orbit, but as it was a largely experimental and educational effort, the data was not made publicly available.[84]

Italy

Italy contributes to the development of Earth observation satellites primarily within ESA, maintaining a relatively small domestic program. Italy has launched three satellites that contribute to international laser ranging services.[85] However, its largest contribution to Earth observations is through its COnstellation of small Satellites for Mediterranean basin Observation (COSMO-SkyMed). COSMO-SkyMed is managed by the Italian Space Agency, Agenzia Spaziale Italiana (ASI), in cooperation with the Ministry of Defense. It is designed to serve civil, commercial, and military users. The four satellites in the constellation were launched between 2007 and 2010, with each carrying a high-resolution synthetic aperture radar that allows the system to collect global imagery during the night and day and in all weather conditions.[86]

Italy shares COSMO-SkyMed intelligence data with France in exchange for access to its optical satellite reconnaissance imagery.[87] Marketing and distribution of COSMO-SkyMed imagery to commercial users is carried out by e-GEOS, a company jointly owned by ASI and the Italian company Telespazio. The company advertises the main applications of the data as maritime domain awareness, defense and intelligence, natural resource monitoring, and emergency response.[88] Prices range from 825 euros for a low-resolution, archived product to 6,000 euros for new acquisition of a high-resolution "spotlight" product. Prices are lower if a larger number of scenes is purchased, although there are additional fees for near-real-time service or image reprocessing.[89] The majority of e-GEOS revenue comes from applications and value-added services, with only 15 percent from data sales.[90]

Noncommercial users can apply for access to COSMO-SkyMed data by registering and submitting a project proposal to ASI. Costs for noncommercial users range from 50 euros for archived products more than six months old to 860 euros per product for bulk purchase of real-time high-resolution "spotlight" products. Data received by those whose proposals are approved can only be used for the purpose expressed in the project proposal, and cannot be disclosed or transferred to a third party. Researchers are not permitted to retain copies of the products for other uses after completion of the project.[91]

Japan

Refer to chapter 13: Japan Meteorological Agency, and chapter 14: Japan Aerospace Exploration Agency.

Kazakhstan

Kazakhstan procured two remote sensing satellites, both launched in 2014. KazEOSat-1 was a 1-meter-resolution imaging satellite built by Airbus Defence and Space, and KazEOSat-2 was a 6.5-meter-resolution satellite built by Surrey Satellite Technology (SSTL).[92] The satellites were part of Kazakhstan's national space plan, which envisions use of the imagery for civil and defense purposes. Kazakhstan Gharysh Sapary (KGS), working together with the Kazakh Space Agency, KazCosmos, is responsible for operating the satellites and marketing the images to commercial and other customers.[93]

Malaysia

The first Malaysian remote sensing satellite, RazakSat, was launched in 2009 and is operated by the Malaysian space agency (ANGKASA). RazakSat includes a medium-resolution imager and operates in an orbit that allows it to pass over Malaysia more frequently than most other Earth observation satellites.[94] RazakSat data over Malaysia is classified in the interest of national security. Data of other areas can be accessed by submitting a data application form and nondisclosure agreement. Data is then provided free of charge via DVD.[95]

Morocco

Maroc-TUBSat was jointly developed between Morocco's Centre Royale de Télédétection Spatiale (CRTS) and the Technical University of Berlin (TUB).

Launched in 2001, the satellite carried a medium-resolution imager, but was primarily seen as an experimental effort. Data from the project was not made publicly available.[96]

Nigeria

Nigeria has developed three Earth observation satellites: NigeriaSat 1, 2, and X, in cooperation with Surrey Satellite Technology Limited (SSTL) in the United Kingdom. Each of the satellites included a training and transfer agreement that allowed engineers from the Nigerian National Space Research and Development Agency (NASRDA) to participate in satellite development and gain hands-on training. NigeriaSat 1 and NigeriaSat X, launched in 2003 and 2011, respectively, were based on the same satellite platform, each carrying a 22-meter-resolution imager. NigeriaSat 2, also launched in 2011, was capable of collecting imagery with up to 2.5-meter resolution. All three satellites were part of the Disaster Management Constellation. Like all satellites in the constellation, imagery is provided for free in the case of natural disasters. In fact, the first satellite image of New Orleans after Hurricane Katrina was captured by NigeriaSat 1.[97] Imagery is also made freely available to Nigerians for noncommercial use.[98] For commercial users within Africa, imagery is available for purchase from GeoApps Plus Limited, the commercial arm of NASRDA.[99] DMCii has responsibility for sales outside of Africa.[100]

Pakistan

The Space and Upper Atmosphere Research Commission (SUPARCO) was established in 1991. In 2001, SUPARCO launched Badr-B, a small satellite carrying a CCD camera built by Rutherford Appleton Laboratory in the United Kingdom.[101] Data from the satellite was not made publicly available.

Portugal

The first Portuguese satellite, PoSAT-1 was built by SSTL on behalf of a consortium of universities and companies in Portugal in a deal that included training for a team of Portuguese engineers. The satellite carried two imagers, and data was made freely available online through an image archive. However, the website has not been maintained and the archive is no longer accessible.[102]

Russia

Refer to chapter 15: Brazil, Russia, China, India, South Africa.

Saudi Arabia

Remote sensing activities in Saudi Arabia are managed by the King Abdulaziz City for Science and Technology (KACST) Satellite Technology Center (STC) and the Saudi Center for Remote Sensing (SCRS). In 2002, Saudi Arabia developed a National Policy for Science and Technology, which included space and aeronautics as one of its 11 strategic priorities. The plan included development of a fleet of remote sensing satellites to provide data for national security and domestic needs including emergency monitoring and response, environmental monitoring, mapping, and other applications. Saudi Arabia also aims to become a leading provider of commercial Earth observation products in the Middle East and support defense and national security needs.[103]

Efforts on these first two objectives began with the development of experimental satellites carrying basic imaging equipment, including SaudiSat 1C and SaudiSat 2 launched in 2002 and 2004, respectively. SaudiSat 3, launched in 2007, carried a more advanced 2.5-meter-resolution imager, and is considered Saudi Arabia's first remote sensing satellite.[104] While the country has plans to provide satellite data to both civil and commercial users, a system for international online distribution is not yet available.[105]

Singapore

In 2011, Singapore launched X-Sat, a small satellite carrying a 10-meter-resolution imager. The project was a collaboration between the Singapore Centre for Research in Satellite Technology (CREST), the Nanyang Technological University (NTU), and the Singapore National Defence (DSO)

National Laboratories, carried out primarily for technology demonstration purposes, though the satellite does produce data relevant to environmental monitoring.[106] X-Sat images are made available online by the Singapore Centre for Remote Imaging, Sensing and Processing (CRISP).[107]

In December 2015, Singapore launched TeLEOS-1, its first commercial Earth observation satellite, capable of providing 1-meter-resolution imagery. The satellite is in an equatorial orbit that allows it to pass over Singapore and other equatorial nations frequently, improving the chances of collecting cloud-free images quickly. AgilSpace was created to market data and products from the satellite.[108]

With the support of the Singapore government, Nanyang Technical University (NTU) and the National University of Singapore (NUS) have also been active in developing and launching satellites. In addition to X-Sat, NTU has developed six satellites. Velox-I, launched in 2014, carried a basic imager designed at the university. In 2015, NTU launched Velox-CI, a small satellite that will use radio occultation techniques to obtain data useful for weather and climate studies. NUS's Kent Ridge 1 (KR-1) microsatellite, launched on the same rocket, was equipped with an advanced imaging sensor that can be used to detect changes in temperature on the Earth's surface.[109] Data is not made publicly available online by either university.

South Africa

Refer to chapter 15: Brazil, Russia, China, India, South Africa.

South Korea

South Korea's first remote sensing satellite was the Korea Institute of Technology Satellite 1 (KitSat-1), developed by the Korea Advanced Institute of Science and Technology (KAIST) and launched in 1992. This was followed by KitSat-2 in 1993 and KitSat-3 in 1999. The Science and Technology Satellite (STSat) series followed, with four satellites launched between 2003 and 2013, this time in cooperation with the Korea Aerospace Research Institute (KARI). All three KitSat satellites as well as STSAT-2 and -3 carried Earth observing sensors.[110] However, data from these experimental satellite programs is not made widely available.

Developed by KARI, the KOrea Multi-Purpose SATellite (KOMPSAT) satellites were more advanced and demonstrated a range of capabilities. KOMPSAT-1, launched in 1999, carried two Earth observation instruments. A sample of data from the Electro Optical Camera is available through ESA.[111] Data from the Ocean Scanning Multi-spectral Imager (OSMI) was distributed to noncommercial users by KARI while the satellite was operating, but the data is no longer accessible.[112]

KOMPSAT-2, -3 and -3A, launched in 2006, 2012, and 2015, respectively, focused on high-resolution imaging. KOMPSAT-5, launched in 2013, carried a synthetic aperture radar, providing all-weather day and night monitoring capabilities.[113] Data from KOMPSAT-2, -3, -3A, and -5 is sold commercially by Satrec Initiative (SI) Imaging Services and a range of licensed distributers.[114] Scientific users can apply to access KOMPSAT-2 data through ESA.[115]

South Korea's first geostationary satellite, the Communication, Ocean, and Meteorological Satellite (COMS), was developed by the Korea Meteorological Administration (KMA). In addition to a traditional imager for meteorological purposes, COMS carries the Geostationary Ocean Color Imager (GOCI) instrument, which is useful for a variety of oceanographic applications.[116] Minimally processed data from GOCI is made freely available online by the Korea Ocean Satellite Center (KOSC) within hours of collection. Access to the data does not require registration, but users have to sign an agreement stating they will not redistribute data to third parties.[117] Imagers from the Meteorological Imager (MI) on COMS can be viewed freely online. Users can also apply to receive free access to encrypted weather satellite transmissions via ground station.[118]

Spain

Spain contributes to ESA's Earth observation program and has plans for domestic development of civil and military remote sensing systems. However, the only Spanish Earth observation satellites developed as of 2015 were privately built and operated. Deimos-1 and -2 were launched in 2009 and 2014, respectively. Imagery from the two satellites is marketed and sold by Deimos Imaging. As Deimos-1 was originally designed as part of the Disaster Monitoring Constellation (DMC), it also contributes data for free in the case of natural disasters.[119] In 2014, Deimos joined a number of other commercial remote sensing satellite operators around the world to form the PanGeo Alliance, allowing the imagery from the full constellation of satellites to be jointly marketed and sold.[120]

Sweden

The Swedish National Space Board (SNSB), in cooperation with Canada, Finland, and France, launched the Odin satellite in 2001. One of the instruments included on the satellite was designed to study the depletion of the ozone layer and the effects of global warming.[121] Minimally processed data from this instrument is freely available from the project website.[122] More highly processed data from this instrument, as well as data from the imaging sensor, is available without charge by ESA but requires registration and approval by the mission scientist.[123]

Taiwan

Taiwan's first remote sensing satellite, FORMOSAT-1, was launched in 1999. It carried an ocean color imager and an instrument to measure conditions in the ionosphere. Archived data from these instruments is available online through the Taiwan National Space Organization (NSPO) and NASA.[124] FORMOSAT-2, launched in 2004, was a high-resolution imaging satellite. Data from the satellite was sold commercially through approved distributers.[125] The Taiwan National Space Organization (NSPO) also provided the opportunity for global researchers to access up to 10 FORMOSAT-2 images for free by submitting a proposal.[126]

In 2006, Taiwan launched the COSMIC constellation, also referred to as FORMOSAT-3, in cooperation with the United States. COSMIC was the first constellation dedicated to GPS radio occultation, a technique that can be used to collect data about the atmosphere. Originally designed as an experimental program, COSMIC data proved so useful that it was incorporated into operational weather forecasting models in multiple nations. Access to COSMIC data is free for science use, but access requires approval of NSPO, and data cannot be redistributed without permission.[127]

Thailand

In 1998, Thailand launched its first satellite. Originally called the Thai Microsatellite (TMSAT), the King renamed it Thai-Paht (Freedom Development) shortly after launch. The satellite was built by Mahanakorn University of Technology (MUT) in collaboration with the University of Surrey in the UK. The primary goal for MUT was technology transfer and training for its students. Imagery from the two onboard cameras was used for educational purposes, but never made widely available.[128]

Thailand created a space agency, the Geo-Informatics and Space Technology Development Agency (GISTDA), in 2000. In 2008, it launched Thailand's second satellite, the Thailand Earth Observation System (THEOS), also known as Thaichote, to give Thailand a sustainable source for environmental data relevant to development. THEOS was designed primarily to meet the needs of the Thai government, but data was made available to other users for purchase online.[129] In sales of GISTDA imagery, Thailand distinguishes between domestic users, users in the Association of Southeast Asian Nations (ASEAN), and other international users.[130] GISTDA noted that they only expected to regain about 10 percent of the cost of the satellite through data sales, and were more focused on using the satellite data in many applications than in gaining the highest possible profit margins.[131]

Turkey

Turkey's first satellite was BILSAT, built in cooperation with SSTL. The goal of the small satellite program was to build within Turkey the capability to develop and operate a satellite independently, but it also carried an imaging instrument and was designed to contribute to the Disaster Management Constellation.[132] Data is available for purchase through DMCii.[133]

Turkey's next satellite, RASAT, was built entirely by engineers at TÜBİTAK-UZAY, Turkey's space agency. In its first four years of operation, RASAT collected over seven million square kilometers of imagery.[134] Turkish citizens can download RASAT data imagery free of charge on the Gezgin national satellite imagery portal. Registration and verification of citizenship is required.[135]

GÖKTÜRK-2 is a dual-use system, designed to provide high-resolution imagery useful for intelligence, reconnaissance, and surveillance for the Turkish armed forces as well as civilian uses. The satellite was launched in 2012.[136] Data is not made publicly available.

Ukraine

Ukraine has developed three Earth observing satellites, Sich-1, -1M, and -2, launched in 1995, 2004, and 2011, respectively. Sich-1 was built by the Ukrainian Yuzhnoye State Design Office using the same design they had used for the Soviet Okean-O1 satellite series prior to the breakup of the Soviet Union. Sich-1M was a cooperative project with the Russian Space Agency, Roscosmos. Sich-2 was a 2-meter-resolution imaging satellite.[137] Sich-2 data is available via a GeoPortal set up by the Ukranian National Space Research Institute (NSAU). Access requires registration and no information is publicly available on the criteria for access, prices, or other restrictions.[138]

United Arab Emirates

DubaiSat-1 was developed by the United Arab Emirates under an agreement with the Satrec Initiative (SI) company in South Korea. The satellite was launched in 2009, carrying a 2.5-meter-resolution imager. This was followed by the launch of DubaiSat-2 in 2013, developed under a similar agreement with SI and featuring a 1-meter-resolution imager.[139] Data from both satellites can be purchased through the Mohammed Bin Rashid Space Centre (MBRSC).[140] Imagery is also marketed and sold as part of the PanGeo Alliance.[141]

United Kingdom

The University of Surrey and later, SSTL, built nearly 50 satellites, working with more than 15 different nations over the past 35 years. The first satellites they developed were for themselves. UoSAT-1, -2, and -5, launched in 1981, 1984, and 1991, respectively, all carried CCD imagers. However, these early satellites were developed primarily as technology test beds and for education purposes. As such, the data collected was not made publicly available.[142]

In 2000, the British National Space Agency (now the UK Space Agency) undertook the Micro Satellite Applications in Collaboration (MOSAIC) program. Under this program, the United Kingdom procured the UK-DMC-1 satellite from SSTL. UK-DMC-1 was the second confirmed satellite for the Disaster Monitoring Constellation after AlSat-1, and the sale helped to provide the confidence needed for additional countries to sign on to the DMC concept. The launch of UK-DMC-1 also marked the United Kingdom's entry into the International Charter for Space and Major Disasters.[143] The United Kingdom later invested in an additional UK-DMC satellite, launched in 2009. Data collected by these satellites is sold through DMCii and other approved distributers.[144]

The MOSAIC program also provided support for the development of the TopSat satellite, also built by SSTL. Cofunded by the UK Ministry of Defence, the satellite demonstrated the ability of SSTL to develop small satellites capable of collecting militarily-relevant high-resolution imagery.[145] TopSat data is available for purchase through Airbus Defence and Space Geo-Information Services.[146]

United States

Refer to chapter 7: US National Oceanic and Atmospheric Administration; chapter 8: US Geological Survey; chapter 9: US National Aeronautics and Space Administration; and chapter 10: US Defense, Intelligence, and Commercial Satellites.

Venezuela

Venezuela's space agency, the Bolivarian Agency for Space Activities (ABAE) was formed in 2007. In 2012, the country launched the first Venezuelan Remote Sensing Satellite (VRSS-1), built in cooperation with the China Great Wall Industry Corporation. The satellite houses four cameras, and is designed for use for a variety of natural resource and environmental monitoring tasks.[147] Data from the satellite is not made publicly available.

Vietnam

The Vietnam Natural Resources, Environment, Disaster Satellite (VNREDSat-1) was launched in 2013 by the Vietnam Academy of Science and Technology Space Technology Institute.[148] Images from the satellite are reportedly useful for both civil remote sensing and national security uses.[149] Data from the satellite is not made publicly available.

Appendix B Satellite Data Sharing Database

In support of this book, I developed a comprehensive dataset contain-
ing the data sharing policies for all unclassified remote sensing satellites
that were successfully launched between 1957 and 2016. This appendix
describes in detail the process for creating the dataset and coding the data
sharing policies.

The database builds on the World Meteorological Organization (WMO)
Observing Systems Capability Analysis and Review Tool (OSCAR) updated
on July 5, 2016.[1] Additional unclassified Earth observation satellites were
added to the database based on a comparison with the NASA Space Science
Data Coordinated Archive (NSSDCA) Master Catalog, the FAA compilation
of commercial remote sensing satellites in the 2015 Commercial Space
Transportation Forecasts report, and additional outside research.[2] Reconnaissance satellites, satellites that failed during or shortly after launch, and
those that did not carry Earth observing instruments—for example, data
collection and relay satellites—were removed from the database.

The database includes the designation of a "Lead Country" and "Participating Countries." Information for these variables was determined based
on the lead agency listed by WMO and additional research on the specific
mission. In some cases, satellites were developed as joint projects, with no
country officially taking a leadership role. In these cases, leadership was
randomly assigned to one of the two countries, with the same leader kept
for all satellites in the same series. When satellites were developed by commercial or university groups, this designation was placed in the "participating countries" column. This makes it possible to identify the country in
which the company operates.

The database also includes information on the data sharing policy,
official or unofficial, governing access to data from each satellite in the

database. In cases in which the data could not be located and inquiries to agencies or affiliated researchers were unanswered, the data policy is listed as "unknown." Data was coded as "unavailable" if it was confirmed that data is not made publicly available. For data that was available, the data policy is coded as commercial, restricted, or open. This information was gathered through analysis of documents, websites, Internet portals, and personal communications related to each satellite and instrument.

The data sharing policy covering the satellite is coded as "commercial" if the data is sold at prices above the marginal cost of fulfilling the user request. It should be noted, however, that some of these satellites provide low-cost or free access to data for selected users, typically as a result of public-private partnerships or special programs.

The data sharing policy for any satellite for which access, use, or redistribution of data is significantly restricted for purposes other than commercial sales was coded as "restricted." Examples of restricted policies are those that require submission of a data access request or detailed proposal and approval by agency or mission officials, those for which access to the data is restricted to particular uses (e.g., noncommercial), or users (e.g., citizens), and those that significantly limit reuse or redistribution of the data.

Data classified as "open" follows the definition used by the Group on Earth Observations. It includes data that is available without charge or restrictions on reuse, but may be subject to conditions of registration and attribution when the data are reused.

For a small number of cases (fewer than 10), two or more instruments on the same satellite are subject to different data sharing policies. In these cases, the satellite is categorized as commercial if data from at least one instrument is sold; otherwise, it is categorized as restricted.

Notes

1   Two Mysteries

1. IPCC, "Climate Change 2014 Synthesis Report. Contribution of Working Groups I, II, and III to the Fifth Assessment Report of the Intergovernmental Panel on Climate Chagne," ed. Intergovernmental Panel on Climate Change Core Writing Team, P K Pachauri, and L A Meyer (Geneva, Switzerland: IPCC, 2014).

2. Bruce R Barkstrom, "The Earth Radiation Budget Experiment (ERBE)," *Bulletin of the American Meteorological Society* 65, no. 11 (1984).

3. GCOS, "Status of the Global Observing System for Climate" (Geneva, Switzerland: Global Climate Observing Group, 2015).

4. Intergovernmental Panel on Climate Change, *Climate Change: The 1990 and 1992 IPCC Assessments: IPCC First Assessment Report, Overview and Policymaker Summaries, and 1992 IPCC Supplement* (WMO, 1992).

5. United Nations, "United Nations Framework Convention on Climate Change (UNFCCC)" (1992).

6. D Brent Smith, David F Williams, and Akihiro Fujita, "Satellite Missions, Global Environment, and the Concept of a Global Satellite Observation Information Network: The Role of the Committee on Earth Observation Satellites (CEOS)," *Acta Astronautica* 34 (1994).

CEOS, "1998 CEOS Consolidated Report" (Committee on Earth Observations Satellites, 1998).

7. GCOS Data and Information Mangement Panel, "Data and Information Management Plan Version 1.0" (Global Climate Observing System, 1995).

8. Paul Wapner, "World Summit on Sustainable Development: Toward a Post-Jo'burg Environmentalism," *Global Environmental Politics* 3, no. 1 (2003).

9. G8, "Science and Technology for Sustainable Development: A G8 Action Plan" (Evian, France, 2003).

10. GEO, "The Global Earth Observation System of Systems (GEOSS) 10-Year Implementation Plan as Adopted 16 February 2005," in *Earth Observation Summit III* (Brussels, Belgium, 2005).

11. Jonathan T Overpeck et al., "Climate Data Challenges in the 21st Century," *Science (Washington)* 331, no. 6018 (2011).

12. Sandra Cabrera et al., "The Progressive Use of Satellite Technology for Disaster Management Relief: Challenges to a Legal and Policy Framework" (paper presented at the 64th International Astronautical Congress, 2013).

13. GEO, "GEO Strategic Plan 2016–2025: Implementing GEOSS: MS4 Approved by GEO-XII Plenary," in *GEO-XII Plenary & Mexico City Ministerial Summit* (Mexico City, Mexico: Group on Earth Observations, 2015).

14. Peter R. Orszag, "Open Government Directive," ed. US Office of Management and Budget (OMB) (Washington, DC, 2009).

15. Alon Peled, "When Transparency and Collaboration Collide: The USA Open Data Program," *Journal of the American Society for Information Science and Technology* 62, no. 11 (2011).

16. NASA, "Data and Information Policy," National Atmospheric and Space Administration, http://science.nasa.gov/earth-science/earth-science-data/data-information -policy/.

NOAA, "Policy on Partnerships in the Provision of Environmental Information," ed. National Oceanic and Atmospheric Administration (NOAA) (2007).

USGS, "Landsat Data Distribution Policy," ed. United States Geological Survey (2008).

17. US Department of State, "Memorandum of Understanding between the Brazilian Instituto De Pesquisas Espaciais and the United States National Aeronautics and Space Administration," in *United States Treaties and Other International Agreements*, ed. United States Department of State (Department of State, 1973).

Harlan Yu and David G Robinson, "The New Ambiguity of 'Open Government'," *UCLA Law Review* (2012).

18. Rashmi Krishnamurthy and Yukika Awazu, "Liberating Data for Public Value: The Case of data.gov," *International Journal of Information Management* 36, no. 4 (2016).

19. Gilberto Câmara (INPE) and Guo Jian Ning (CRESDA), "CBERS Data Policy Version 1.8," ed. CBERS JPC (2004).

Hilcéa Ferreira, "Benefits of Data Sharing: The CBERS Programme," in *Symposium on the Data Sharing Action Plan for GEOSS and the Benefits of Data Sharing* (Beijing, China: National Institute for Space Research (INPE) Brazil, 2010).

20. EUMETSAT, "Resolution on EUMETSAT Principles on Data Policy Adopted at the 28th Meeting of the EUMETSAT Council" (1998).

21. ESA, "ESA Data Policy for ERS, Envisat and Earth Explorer Missions (Simplified Version)" (ESA, 2012).

22. Osamu Ochiai, "Jaxa's Earth Observation Data and Information System," in *Copernicus Big Data Workshop*, ed. Japan Aerospace Exploration Agency (JAXA) (Brussels, Belgium, 2014).

Part I   A Model of Data Sharing Policy Development

1. Marijn Janssen, Yannis Charalabidis, and Anneke Zuiderwijk, "Benefits, Adoption Barriers and Myths of Open Data and Open Government," *Information Systems Management* 29, no. 4 (2012).

Jane E Fountain, "Challenges to Organizational Change: Multi-Level Integrated Information Structures (MIIS)," *Governance and information technology: from electronic government to information government* (2007).

Alon Peled, "When Transparency and Collaboration Collide: The USA Open Data Program," *Journal of the American Society for Information Science and Technology* 62, no. 11 (2011).

2. Chris Martin, "Barriers to the Open Government Data Agenda: Taking a Multi-Level Perspective," *Policy and Internet* 6, no. 3 (2014).

Emily Barry and Frank Bannister, "Barriers to Open Data Release: A View from the Top," *Information Polity* 19, nos. 1–2 (2014).

2   Defining Data Sharing

1. Russell L Ackoff, "From Data to Wisdom," *Journal of Applied Systems Analysis* 16, no. 1 (1989).

Charlotte Hess and Elinor Ostrom, "Introduction: An Overview of the Knowledge Commons," eds. Charlotte Hess and Elinor Ostrom *Understanding Knowledge as a Commons. From Theory to Practice* (Cambridge, MA: The MIT Press, 2009).

2. Harlan Yu and David G Robinson, "The New Ambiguity of 'Open Government'" (2012).

Antti Halonen, "Being Open About Data: Analysis of the UK Open Data Policies and Applicability of Open Data," *London: The Finnish Institute in London* (2012).

3. Bryn Nelson, "Data Sharing: Empty Archives," *Nature* 461 (2009).

Jane Kaye et al., "Data Sharing in Genomics—Re-Shaping Scientific Practice," *Nature Reviews Genetics* 10, no. 5 (2009).

4. Tung-Mou Yang and Terrence A Maxwell, "Information-Sharing in Public Organizations: A Literature Review of Interpersonal, Intra-Organizational and Inter-Organizational Success Factors," *Government Information Quarterly* 28, no. 2 (2011).

5. Jeni Tennison, "Being Open About Data: Analysis of the UK Open Data Policies and Applicability of Open Data," https://theodi.org/blog/data-sharing-is-not-open -data (2014).

6. Chris Martin, "Barriers to the Open Government Data Agenda: Taking a Multi-Level Perspective," *Policy & Internet* 6, no. 3 (2014).

Mireille Van Eechoud and Brenda Van Der Wal, "Creative Commons Licensing for Public Sector Information—Opportunities and Pitfalls," available at SSRN 1096564 (2008).

3 People

1. Thomas H Hammond, "Veto Points, Policy Preferences, and Bureaucratic Autonomy in Democratic Systems," eds. G A Krause and K J Meier, *Politics, Policy, and Organizations: Frontiers in the Scientific Study of Bureaucracy* (University of Michigan Press, 2003).

2. Daniel P Carpenter, *The Forging of Bureaucratic Autonomy: Reputations, Networks, and Policy Innovation in Executive Agencies, 1862–1928* (Princeton University Press, 2001).

3. Karine Nahon and Alon Peled, "Data Ships: An Empirical Examination of Open (Closed) Government Data" (paper presented at the 48th Annual Hawaii International Conference on System Sciences (HICSS 48), 2014).

4. Sandra Braman, *Change of State: Information, Policy, and Power* (The MIT Press, 2009).

Robert O Keohane and Joseph S Nye, Jr., "Power and Interdependence in the Information Age," *Foreign Affairs*– 77 (1998).

Sharon S Dawes, "Interagency Information Sharing: Expected Benefits, Manageable Risks," *Journal of Policy Analysis and Management* 15, no. 3 (1996).

Barry and Bannister, "Barriers to Open Data Release: A View from the Top."

Alon Peled, "When Transparency and Collaboration Collide: The USA Open Data Program," *Journal of the American Society for Information Science and Technology* 62 (2011).

5. Francis Harvey and D Tulloch, "Local-Government Data Sharing: Evaluating the Foundations of Spatial Data Infrastructures," *International Journal of Geographical Information Science* 20, no. 7 (2006).

Alon Peled, "Re-Designing Open Data 2.0" (paper presented at the Conference for E-Democracy and Open Governement, 2013).

Zorica Nedovic-Budic and Jeffrey K Pinto, "Understanding Interorganizational GIS Activities: A Conceptual Framework," *URISA Journal* 11 no. 1 (1999).

6. Bijan Azad and Lyna L Wiggins, "Dynamics of Inter-Organizational Geographic Data Sharing: A Conceptual Framework for Research," eds. Harlan J Onsrud and Gerard Rushton, *Sharing Geographic Information* (Center for Urban Policy Research, 1995).

7. Chris Martin, "Barriers to the Open Government Data Agenda: Taking a Multi-Level Perspective," *Policy and Internet* 6, no. 3 (2014).

8. Zorica Nedović-Budić and Jeffrey K Pinto, "Interorganizational GIS: Issues and Prospects," *The Annals of Regional Science* 33, no. 2 (1999).

9. Emily Barry and Frank Bannister, "Barriers to Open Data Release: A View from the Top," *Information Polity* 19, nos. 1–2 (2014).

Jane E Fountain, "Challenges to Organizational Change: Multi-Level Integrated Information Structures (MIIS)." *National Center for Digital Government Working Paper Series 15* (2006).

10. Daniel P. Carpenter, *The Forging of Bureaucratic Autonomy: Reputations, Networks, and Policy Innovation in Executive Agencies, 1862–1928* (Princeton University Press, 2001).

11. Karine Nahon and Alon Peled, "Data Ships: An Empirical Examination of Open (Closed) Government Data," *48th Annual Hawaii International Conference on System Sciences* (2015).

Bruce Joffe, "Ten Ways to Support GIS without Selling Data," *URISA Journal* 16, no. 2 (2005).

12. Nahon and Peled, "Data Ships: An Empirical Examination of Open (Closed) Government Data."

13. Bastiaan van Loenen, "Developing Geographic Information Infrastructures: The Role of Access Policies," *International Journal of Geographical Information Science* 23, no. 2 (2009).

14. Joffe, "Ten Ways to Support GIS without Selling Data."

15. van Loenen, "Developing Geographic Information Infrastructures: The Role of Access Policies."

Harlan J Onsrud, "In Support of Cost Recovery for Publicly Held Geographic Information," *GIS Law* 1, no. 2 (1992).

16. Jeffrey K Pinto and Harlan J Onsrud, "Sharing Geographic Information across Organizational Boundaries: A Research Framework," eds. Harlan J Onsrud and Gerard Rushton, *Sharing Geographic Information* (Center for Urban Policy Research, 1995).

David L Tulloch and Francis Harvey, "When Data Sharing Becomes Institutionalized: Best Practices in Local Government Geographic Information Relationships," *URISA Journal* 19, no. 2 (2007).

17. Joffe, "Ten Ways to Support GIS without Selling Data."

18. Tulloch and Harvey, "When Data Sharing Becomes Institutionalized: Best Practices in Local Government Geographic Information Relationships."

Martin, "Barriers to the Open Government Data Agenda: Taking a Multi-Level Perspective."

19. Theresa A Pardo and Giri Kumar Tayi, "Interorganizational Information Integration: A Key Enabler for Digital Government," *Government Information Quarterly* 24, no. 4 (2007).

Peled, "When Transparency and Collaboration Collide: The USA Open Data Program."

20. Barry and Bannister, "Barriers to Open Data Release: A View from the Top."

21. Halonen, "Being Open About Data: Analysis of the UK Open Data Policies and Applicability of Open Data."

22. Tulloch and Harvey, "When Data Sharing Becomes Institutionalized: Best Practices in Local Government Geographic Information Relationships."

23. Ibid.

24. Nahon and Peled, "Data Ships: An Empirical Examination of Open (Closed) Government Data."

25. National Research Council, "Resolving Conflicts Arising from the Privatization of Environmental Data," ed. Freeman Gilbert and William L Chameides (2001).

26. Joffe, "Ten Ways to Support Gis without Selling Data."

27. Halonen, "Being Open About Data: Analysis of the Uk Open Data Policies and Applicability of Open Data."

28. Ben Worthy, "Making Transparency Stick: The Complex Dynamics of Open Data," *Available at SSRN 2497659* (2014).

29. Tom Christensen et al., *Organization Theory and the Public Sector: Instrument, Culture and Myth* (Routledge, 2007).

Barry and Bannister, "Barriers to Open Data Release: A View from the Top."

30. James G March and Johan P Olsen, "Elaborating the "New Institutionalism"," *The Oxford Handbook of Political Institutions* 5 (2006).

31. Angela Schäfer et al., "Baseline Report on Drivers and Barriers in Data Sharing," *Opportunities for Data Exchange (ODE)* (2011).

Tulloch and Harvey, "When Data Sharing Becomes Institutionalized: Best Practices in Local Government Geographic Information Relationships."

32. David Constant, Sara Kiesler, and Lee Sproull, "What's Mine Is Ours, or Is It? A Study of Attitudes About Information Sharing," *Information Systems Research* 5, no. 4 (1994).

33. David B Drake, Nicole A Steckler, and Marianne J Koch, "Information Sharing in and across Government Agencies the Role and Influence of Scientist, Politician, and Bureaucrat Subcultures," *Social Science Computer Review* 22, no. 1 (2004).

Martin, "Barriers to the Open Government Data Agenda: Taking a Multi-Level Perspective."

34. Dawes, "Interagency Information Sharing: Expected Benefits, Manageable Risks."

35. W Craig, "White Knights of Spatial Data Infrastructure: The Role and Motivation of Key Individuals," *URISA Journal* 16, no. 2 (2005).

36. Christensen et al., *Organization Theory and the Public Sector: Instrument, Culture and Myth.*

Harvey and Tulloch, "Local-Government Data Sharing: Evaluating the Foundations of Spatial Data Infrastructures."

37. Janssen, Charalabidis, and Zuiderwijk, "Benefits, Adoption Barriers and Myths of Open Data and Open Government."

38. Ibid.

Barry and Bannister, "Barriers to Open Data Release: A View from the Top."

Halonen, "Being Open About Data: Analysis of the UK Open Data Policies and Applicability of Open Data."

39. Barry and Bannister, "Barriers to Open Data Release: A View from the Top."

40. Richard McDermott and Carla O'dell, "Overcoming Cultural Barriers to Sharing Knowledge," *Journal of Knowledge Management* 5, no. 1 (2001).

41. D. R. Kim, "Political Control and Bureaucratic Autonomy Revisited: A Multi-Institutional Analysis of Osha Enforcement," *Journal of Public Administration Research and Theory* 18, no. 1 (2006).

Hammond, "Veto Points, Policy Preferences, and Bureaucratic Autonomy in Democratic Systems."

42. Kutsal Yesilkagit and Sandra van Thiel, "Political Influence and Bureaucratic Autonomy," *Public Organization Review* 8, no. 2 (2008).

43. T. H. Hammond and J. H. Knott, "Who Controls the Bureaucracy?: Presidential Power, Congressional Dominance, Legal Constraints, and Bureaucratic Autonomy in a Model of Multi-Institutional Policy-Making," *Journal of Law, Economics, and Organization* 12, no. 1 (1996).

44. Nahon and Peled, "Data Ships: An Empirical Examination of Open (Closed) Government Data."

Paul F Uhlir and Peter Schröder, "Open Data for Global Science," *Data Science Journal* 6 (2007).

Worthy, "Making Transparency Stick: The Complex Dynamics of Open Data."

45. Nahon and Peled, "Data Ships: An Empirical Examination of Open (Closed) Government Data."

Uhlir and Schröder, "Open Data for Global Science."

46. Peled, "When Transparency and Collaboration Collide: The USA Open Data Program."

47. Hammond and Knott, "Who Controls the Bureaucracy?: Presidential Power, Congressional Dominance, Legal Constraints, and Bureaucratic Autonomy in a Model of Multi-Institutional Policy-Making."

48. Halonen, "Being Open About Data: Analysis of the Uk Open Data Policies and Applicability of Open Data."

49. Ibid.

50. Ibid.

51. Hammond and Knott, "Who Controls the Bureaucracy?: Presidential Power, Congressional Dominance, Legal Constraints, and Bureaucratic Autonomy in a Model of Multi-Institutional Policy-Making."

52. Barry and Bannister, "Barriers to Open Data Release: A View from the Top."

53. Ibid.; Peled, "When Transparency and Collaboration Collide: The USA Open Data Program."

54. Fountain, "Challenges to Organizational Change: Multi-Level Integrated Information Structures (Miis)."

Barry and Bannister, "Barriers to Open Data Release: A View from the Top."

55. Hammond and Knott, "Who Controls the Bureaucracy?: Presidential Power, Congressional Dominance, Legal Constraints, and Bureaucratic Autonomy in a Model of Multi-Institutional Policy-Making."

56. James Q Wilson, *Bureaucracy: What Government Agencies Do and Why They Do It* (Basic Books, 1989).

57. Ibid.

58. Peled, "Re-Designing Open Data 2.0."

Anneke Zuiderwijk et al., "Socio-Technical Impediments of Open Data," *Electronic Journal of eGovernment* 10, no. 2 (2012).

Janssen, Charalabidis, and Zuiderwijk, "Benefits, Adoption Barriers and Myths of Open Data and Open Government."

59. Uhlir and Schröder, "Open Data for Global Science."

60. Worthy, "Making Transparency Stick: The Complex Dynamics of Open Data."

61. Deepti Agrawal, William Kettinger, and Chen Zhang, "The Openness Challenge: Why Some Cities Take It on and Others Don't" (2014).

62. Stephen E Fienberg, Margaret E Martin, and Miron L Straf, *Sharing Research Data* (National Academies, 1985).

63. Halonen, "Being Open About Data: Analysis of the Uk Open Data Policies and Applicability of Open Data."

64. Sébastien Martin et al., "Risk Analysis to Overcome Barriers to Open Data," *Electronic Journal of e-Government* 11, no. 1 (2013).

65. US National Committee for CODATA. Committee on Issues in the Transborder Flow of Scientific Data, *Bits of Power: Issues in Global Access to Scientific Data* (Haworth Press, 1997).

66. Joseph E Stiglitz, "Knowledge as a Global Public Good," *Global Public Goods* 1, no. 9 (1999).

67. Robert O Keohane, "International Institutions and State Power: Essays in International Relations Theory" (1989).

68. Martha Finnemore and Kathryn Sikkink, "International Norm Dynamics and Political Change," *International Organization* 52, no. 4 (1998).

4   Ideas and Technology

1. Peter Arzberger et al., "Promoting Access to Public Research Data for Scientific, Economic, and Social Development," *Data Science Journal* 3 (2004).

Janssen, Charalabidis, and Zuiderwijk, "Benefits, Adoption Barriers and Myths of Open Data and Open Government."

2. Chris Martin, "Barriers to the Open Government Data Agenda: Taking a Multi-Level Perspective."

3. Antti Halonen, "Being Open About Data: Analysis of the UK Open Data Policies and Applicability of Open Data."

4. Marco Fioretti, "Open Data: Emerging Trends, Issues and Best Practices," in *Open Data, Open Society* (Laboratory of Economics and Management of Scuola Superiore Sant'Anna, Pisa, 2011).

5. Francesco Molinari and Jesse Marsh, "Does Privacy Have to Do with Open Data?" (paper presented at the Conference for E-Democracy and Open Government, 2013).

Frederik Zuiderveen Borgesius, Jonathan Gray, and Mireille van Eechoud, "Open Data, Privacy, and Fair Information Principles: Towards a Balancing Framework," *Berkeley Technology Law Journal* 30 (2015).

6. "Open Data, Privacy, and Fair Information Principles: Towards a Balancing Framework."

7. Chris Clifton et al., "Privacy-Preserving Data Integration and Sharing" (paper presented at the Proceedings of the 9th ACM SIGMOD Workshop on Research Issues in Data Mining and Knowledge Discovery, 2004).

8. Barry and Bannister, "Barriers to Open Data Release: A View from the Top."

9. Harlan Onsrud, "Access to Geographic Information: Openness versus Security," eds. S Cutter, D Richardson, and T Wilbanks, *Geographic Dimensions of Terrorism* (Routledge, 2003).

10. Ann Florini, "Behind Closed Doors," *Harvard International Review* 26, no. 1 (2004).

11. Patricia Salkin, "GIS in an Age of Homeland Security: Accessing Public Information to Ensure a Sustainable Environment," *William & Mary Environmental Law and Policy Review* 30 (2005).

John C Baker et al., *Mapping the Risks: Assessing the Homeland Security Implications of Publicly Available Geospatial Information* (Rand Corporation, 2004).

Federal Geographic Data Committee, "Guidelines for Providing Appropriate Access to Geospatial Data in Response to Security Concerns," *National Spatial Data Infrastructure (NSDI)* (2005).

12. Baker et al., *Mapping the Risks: Assessing the Homeland Security Implications of Publicly Available Geospatial Information.*

13. Committee, "Guidelines for Providing Appropriate Access to Geospatial Data in Response to Security Concerns."

14. Geospatial Information & Technology Association, "Free or Fee: The Governmental Data Ownership Debate," GITA White Paper (2005).

Joseph E Stiglitz, "On Liberty, the Right to Know, and Public Discourse: The Role," eds. Matthew J Gibney, *Globalizing Rights: The Oxford Amnesty Lectures 1999* (2003).

Halonen, "Being Open About Data: Analysis of the UK Open Data Policies and Applicability of Open Data."

Uhlir and Schröder, "Open Data for Global Science."

15. Ibid.

16. Geospatial Information & Technology Association, Association, "Free or Fee: The Governmental Data Ownership Debate," GITA White Paper (2005).

17. Harlan J Onsrud, "Tragedy of the Information Commons," ed. D R F Taylor, *Policy Issues in Modern Cartography* (Elsevier, 1998).

18. Janssen, Charalabidis, and Zuiderwijk, "Benefits, Adoption Barriers and Myths of Open Data and Open Government."

19. Onsrud, "In Support of Cost Recovery for Publicly Held Geographic Information."

20. Martin et al., "Risk Analysis to Overcome Barriers to Open Data."

21. Ibid.

22. Stiglitz, "On Liberty, the Right to Know, and Public Discourse: The Role."

Halonen, "Being Open About Data: Analysis of the UK Open Data Policies and Applicability of Open Data."

23. Ibid.

24. Yu and Robinson, "The New Ambiguity Of 'Open Government'."

25. Martin et al., "Risk Analysis to Overcome Barriers to Open Data."

Florini, "Behind Closed Doors."

Halonen, "Being Open About Data: Analysis of the UK Open Data Policies and Applicability of Open Data."

Janssen, Charalabidis, and Zuiderwijk, "Benefits, Adoption Barriers and Myths of Open Data and Open Government."

26. Janssen, Charalabidis, and Zuiderwijk, ibid.

27. Martin et al., "Risk Analysis to Overcome Barriers to Open Data."

28. Dawes, "Interagency Information Sharing: Expected Benefits, Manageable Risks."

Agrawal, Kettinger, and Zhang, "The Openness Challenge: Why Some Cities Take It on and Others Don't."

29. Barry and Bannister, "Barriers to Open Data Release: A View from the Top."

Halonen, "Being Open About Data: Analysis of the UK Open Data Policies and Applicability of Open Data."

30. Barry and Bannister, "Barriers to Open Data Release: A View from the Top."

Janssen, Charalabidis, and Zuiderwijk, "Benefits, Adoption Barriers and Myths of Open Data and Open Government."

31. Halonen, "Being Open About Data: Analysis of the UK Open Data Policies and Applicability of Open Data."

32. Hess and Ostrom, "Introduction: An Overview of the Knowledge Commons," eds. Charlotte Hess and Elinor Ostrom, Elinor, *Understanding Knowledge as a Commons. From Theory to Practice* (Cambridge, MA: The MIT Press)."

33. Kenneth Arrow, "Economic Welfare and the Allocation of Resources for Invention," in *The Rate and Direction of Inventive Activity: Economic and Social Factors* (Princeton University Press, 1962).

34. Arzberger et al., "Promoting Access to Public Research Data for Scientific, Economic, and Social Development."

35. Arrow, "Economic Welfare and the Allocation of Resources for Invention."

36. Onsrud, "In Support of Cost Recovery for Publicly Held Geographic Information."

37. Peter Weiss and Y Pluijmers, *Borders in Cyberspace: Conflicting Public Sector Information Policies and Their Economic Impacts* (Edward Elger Publishing, 2004).

Tulloch and Harvey, "When Data Sharing Becomes Institutionalized: Best Practices in Local Government Geographic Information Relationships."

38. Jonathan T Overpeck et al., "Climate Data Challenges in the 21st Century," *Science(Washington)* 331, no. 6018 (2011).

39. Kaye et al., "Data Sharing in Genomics—Re-Shaping Scientific Practice."

Henry Rodriguez et al., "Recommendations from the 2008 International Summit on Proteomics Data Release and Sharing Policy: The Amsterdam Principles," *Journal of Proteome Research* 8, no. 7 (2009).

Heidi L Williams, "Intellectual Property Rights and Innovation: Evidence from the Human Genome," *The Journal of Political Economy* 121, no. 1 (2010).

40. Pira International and European Commission. Information Society DG., *Commercial Exploitation of Europe's Public Sector Information: Executive Summary* (Office for Official Publications of the European Communities, 2000).

41. Janssen, Charalabidis, and Zuiderwijk, "Benefits, Adoption Barriers and Myths of Open Data and Open Government."

Uhlir and Schröder, "Open Data for Global Science."

Weiss and Pluijmers, *Borders in Cyberspace: Conflicting Public Sector Information Policies and Their Economic Impacts*.

42. Martin Feldstein, "Tax Avoidance and the Deadweight Loss of the Income Tax," *Review of Economics and Statistics* 81, no. 4 (1999).

Onsrud, "In Support of Cost Recovery for Publicly Held Geographic Information."

43. The situation is analogous for government purchases of data. If the government would have spent $100 million to build its own satellite, but can instead purchase equivalent data from a commercial provider for $90 million, this will be a $10 million savings to the government, but just a $3 million (or less) savings for society as a whole.

44. Nathan Rosenberg, "Science, Invention and Economic Growth," *The Economic Journal* 84, no. 333 (1974).

Robert G King and Sergio Rebelo, "Public Policy and Economic Growth: Developing Neoclassical Implications" (National Bureau of Economic Research, 1990).

Joseph E Stiglitz, "Capital Market Liberalization, Economic Growth, and Instability," *World Development* 28, no. 6 (2000).

Halonen, "Being Open About Data: Analysis of the UK Open Data Policies and Applicability of Open Data."

Gregor Eibl and Brigitte Lutz, "Money for Nothing—Data for Free" (paper presented at the Conference for E-Democracy and Open Governement, 2013).

Martin et al., "Risk Analysis to Overcome Barriers to Open Data."

Fienberg, Martin, and Straf, *Sharing Research Data*.

45. Philip B Evans and Thomas S Wurster, "The New Economics of Information," *Harvard Business Review* 5 (1997).

46. Zuiderwijk et al., "Socio-Technical Impediments of Open Data."

Agrawal, Kettinger, and Zhang, "The Openness Challenge: Why Some Cities Take It on and Others Don't."

47. John Carlo Bertot et al., "Big Data, Open Government and E-Government: Issues, Policies and Recommendations," *Information Polity* 19, nos. 1–2 (2014).

48. Janssen, Charalabidis, and Zuiderwijk, "Benefits, Adoption Barriers and Myths of Open Data and Open Government."

Uhlir and Schröder, "Open Data for Global Science."

49. Dawes, "Interagency Information Sharing: Expected Benefits, Manageable Risks."

Zuiderwijk et al., "Socio-Technical Impediments of Open Data."

Arzberger et al., "Promoting Access to Public Research Data for Scientific, Economic, and Social Development."

50. "Promoting Access to Public Research Data for Scientific, Economic, and Social Development."

51. Nahon and Peled, "Data Ships: An Empirical Examination of Open (Closed) Government Data."

Arzberger et al., "Promoting Access to Public Research Data for Scientific, Economic, and Social Development."

Agrawal, Kettinger, and Zhang, "The Openness Challenge: Why Some Cities Take It on and Others Don't."

Janssen, Charalabidis, and Zuiderwijk, "Benefits, Adoption Barriers and Myths of Open Data and Open Government."

Martin et al., "Risk Analysis to Overcome Barriers to Open Data."

52. Arzberger et al., "Promoting Access to Public Research Data for Scientific, Economic, and Social Development."

Barry and Bannister, "Barriers to Open Data Release: A View from the Top."

53. Zuiderwijk et al., "Socio-Technical Impediments of Open Data."

54. Ibid.

Janssen, Charalabidis, and Zuiderwijk, "Benefits, Adoption Barriers and Myths of Open Data and Open Government."

5 World Meteorological Organization

1. World Meteorological Organization (WMO), "IMO: The Origin of the WMO," http://www.wmo.int/pages/about/wmo50/e/wmo/history_pages/origin_e.html.

2. "Convention of the World Meteorological Organization," in *Basic Documents No. 1* (2012).

WMO, "Convention of the World Meteorological Organization" (1947).

3. Susan Barr and Cornelia Lüdecke, *The History of the International Polar Years (IPYS)* (Springer, 2010).

4. International Council of Scientific Unions (ICSU) Comité International de Géo-physique, "Guide to International Data Exchange through the World Data Centers" (London, 1963).

5. ICSU, "The International Council for Science and Climate Change, 60 Years of Facilitating Climate Change Research and Informing Policy" (International Council for Science [ICSU], 2015).

6. Arthur Davies and Oliver M Ashford, *Forty Years of Progress and Achievement: A Historical Review of WMO* (World Meteorological Organization, 1990).

7. NASA, "TIROS," http://science.nasa.gov/missions/tiros/.

8. United Nations General Assembly, "Resolution Adopted by the General Assembly 1721 (XVI). International Co-Operation in the Peaceful Uses of Outer Space," ed. United Nations General Assembly (1961).

9. World Meteorological Organization, *First Report on the Advancement of Atmospheric Sciences and Their Application in the Light of Developments in Outer Space [with maps]* (Secretariat of the World Meteorological Organization, 1961).

10. United Nations General Assembly, "Resolution Adopted by the General Assembly 1802 (XVII), International Cooperation in the Peaceful Uses of Outer Space," ed. United Nations General Assembly (1962).

11. World Meteorological Organization (WMO), "World Weather Watch" (Geneva, Switzerland, 1966).

12. Gerald S. Schatz, *The Global Weather Experiment: An Informal History* (National Academy of Sciences, 1978).

13. US National Oceanic and Atmospheric Administration (NOAA), "Satellite Activities of Noaa 1978" (Washington, DC, 1979).

14. United States Committee for the Global Atmospheric Research Program et al., *The Global Weather Experiment, Perspectives on Its Implementation and Exploitation: Report of the FGGE Advisory Panel to the U.S. Committee for the Global Atmospheric*

Research Program, Assembly of Mathematical and Physical Sciences, National Research Council (National Academy of Sciences, 1978).

15. Tillmann Mohr, "International Cooperation of Meteorological/Earth Observing Satellites," *The Richard Hallgren Symposium* (2008).

16. World Meteorological Organization (WMO), "Exchanging Meteorological Data Guidelines on Relationships in Commercial Meteorological Activities: WMO Policy and Practice" (Geneva, Switzerland: World Meteorological Organization, 1996).

17. WMO, "Eleventh World Meteorological Congress Abridged Report with Resolutions" (Geneva, Switzerland: World Meteorological Organization, 1991).

18. "Resolution 40 (Cg-XII) WMO Policy and Practice for the Exchange of Meteorological and Related Data and Products Including Guidelines on Relationships in Commercial Meteorological Activities," ed. World Meteorological Organization (1995).

19. Ibid.

20. Ibid.

21. "Resolution 25 (Cg-XIII) Exchange of Hydrological Data and Products," ed. World Meteorological Organization (Geneva, 1999).

22. R J Fleming, T M Kaneshige, and W E McGovern, "The Global Weather Experiment 1. The Observational Phase through the First Special Observing Period," *Bulletin of the American Meteorological Society* 60, no. 6 (1979).

23. International Council for Science (ICSU), "The International Council for Science and Climate Change, 60 Years of Facilitating Climate Change Research and Informing Policy" (2015).

World Meteorological Organization (WMO), "Declaration of the World Climate Conference" (1979).

24. Intergovernmental Panel on Climate Change, *Climate Change: The 1990 and 1992 IPCC Assessments: IPCC First Assessment Report, Overview and Policymaker Summaries, and 1992 IPCC Supplement* (WMO, 1992).

25. Global Climate Observing System (GCOS), "Memorandum of Understanding between the World Meteorological Organization, the Intergovernmental Oceanographic Commission of the United Nations Educational, Scientific and Cultural Organization, the International Council of Scientific Unions, and the United Nations Environment Program" (1992).

26. United Nations, "United Nations Framework Convention on Climate Change (UNFCCC)" (1992).

27. EUMETSAT, "Amendments to the Convention" (EUMETSAT, 1991).

28. National Research Council, "Ensuring the Climate Record from the NPOESS and GOES-R Spacecraft: Elements of a Strategy to Recover Measurement Capabilities Lost in Program Restructuring" (2008).

29. GCOS Data and Information Mangement Panel, "Data and Information Management Plan Version 1.0" (Global Climate Observing System, 1995).

30. J T Houghton and Intergovernmental Panel on Climate Change, *Climate Change 1995: The Science of Climate Change: Contribution of Working Group I to the Second Assessment Report of the Intergovernmental Panel on Climate Change* (Cambridge University Press, 1996).

R T Watson et al., *Climate Change 2001: Synthesis Report* (World Meteorological Organization, 2001).

31. P J Mason et al., "The Second Report on the Adequacy of the Global Observing Systems for Climate in Support of the UNFCCC," *GCOS Report* 82 (2003).

32. Ibid.

33. Mariel Borowitz, "International Cooperation in Global Satellite Climate Monitoring," *Astropolitics* 13, nos. 2–3 (2015).

34. Ibid.

35. GCOS, "Systematic Observation Requirements for Satellite-Based Products for Climate" (Global Climate Observing System [GCOS], 2006).

36. CEOS, "Satellite Observation of the Climate System: The Committee on Earth Observation Satellites (CEOS) Response to the Global Climate Observing System (GCOS) Implementation Plan (IP)" (Committee on Earth Observation Satellites [CEOS] 2006).

37. Ibid.

38. GCOS, "Implementation Plan for the Global Observing System for Climate in Support of the UNFCCC (2010 Update)" (Geneva, Switzerland: Global Climate Observation System [GCOS], 2010).

39. "Systematic Observation Requirements for Satellite-Based Data Products for Climate 2011 Update" (Geneva, Switzerland: Global Climate Observation System [GCOS], 2011).

CEOS, "The Response of the Committee on Earth Observation Satellites (CEOS) to the Global Climate Observing System Implementation Plan 2010 (GCOS IP-10)" (Committee on Earth Observation Satellites [CEOS], 2012).

40. GFCS, "Implementation Plan of the Global Framework for Climate Services" (Geneva, Switzerland: Global Framework for Climate Services, 2014).

41. WMO, "Resolution 60 (Cg-17) WMO Policy for the International Exchange of Climate Data and Products to Support the Implementation of the Global Framework for Climate Services," ed. World Meteorological Organization (Geneva, 2015).

42. Ibid.

6   Group on Earth Observations

1. United Nations Centre for Disarmament, United Nations Department for Disarmament Affairs, and United Nations, *The Implications of Establishing an International Satellite Monitoring Agency: Report of the Secretary-General, Department For Disarmament Affairs* (United Nations, 1983).

2. Ibid.

3. Ann M Florini, "The Opening Skies: Third-Party Imaging Satellites and US Security," *International Security* 13, no. 2 (1988).

4. John H Gibbons, "US Congress Office of Technology Assessment, Commercial Newsgathering from Space: A Technical Memorandum" (Washington, DC: Government Printing Office, 1987).

5. J I Gabrynowicz et al., *The United Nations Principles Relating to Remote Sensing of the Earth from Space: A Legislative History—Interviews of Members of the United States Delegation* (National Remote Sensing & Space Law Center, 2002).

6. Ibid.

7. Ibid.

United Nations, "Principles Relating to Remote Sensing of the Earth from Outer Space" (1986).

8. Committee on Earth Observation Satellites (CEOS), "Applications of Satellite Earth Observations Serving Society, Science, & Industry" (CEOS, 2015).

9. D Brent Smith, David F Williams, and Akihiro Fujita, "Satellite Missions, Global Environment, and the Concept of a Global Satellite Observation Information Network: The Role of the Committee on Earth Observation Satellites (CEOS)," *Acta Astronautica* 34 (1994).

CEOS, "CEOS Consolidated Report 1992" (London, United Kingdom: Committee on Earth Observation Satellites, 1992).

10. Smith, Williams, and Fujita.

CEOS, "1998 CEOS Consolidated Report" (Committee on Earth Observation Satellites, 1998).

11. CEOS, "1998 CEOS Consolidated Report."

12. Paul Wapner, "World Summit on Sustainable Development: Toward a Post-Jo'burg Environmentalism," *Global Environmental Politics* 3, no. 1 (2003).

13. Jose Achache, "Statement by Prof. Jose Achache, CEOS Chairman, Committee on Earth Observation Satellites (CEOS)," in *World Summit for Sustainable Development* (Johannesburg, South Africa, 2002).

14. Jana Goldman, "NOAA Administrator Promotes Role of Global Observations to Sustainable Development at WSSD; Calls for More International Cooperation," news release, 30 August 2002, http://www.publicaffairs.noaa.gov/releases2002/aug02/noaa02109.html.

15. United Nations Department of Public Information, *Johannesburg Declaration on Sustainable Development and Plan of Implementation of the World Summit on Sustainable Development: The Final Text of Agreements Negotiated by Governments at the World Summit on Sustainable Development, 26 August–4 September 2002, Johannesburg, South Africa* (United Nations Department of Public Information, 2003).

16. G8, "Science and Technology for Sustainable Development: A G8 Action Plan" (Evian, France, 2003).

17. Bob Hopkins, "U.S.-Hosted Summit Brings Nations Together to Take the Pulse of Planet Earth," news release, 2003, http://www.publicaffairs.noaa.gov/releases2003/jul03/noaa03091.html.

18. Ibid.

19. Conrad Lautenbacher, "Three Imperatives for an Integrated Earth Observation System: Speech by Vice Admiral (Ret.) Conrad C. Lautenbacher Jr., U.S. Navy under Secretary of Commerce for Oceans & Atmosphere, NOAA Administrator" (paper presented at the 2003 WMO Congress, Geneva, Switzerland, 2003).

20. "Declaration of the Earth Observation Summit" (2003).

21. Hopkins.

"Earth Observation Summit," https://web.archive.org/web/20070913054152/http://www.earthobservationsummit.gov/.

22. Earth Observation Summit II, "From Observation to Action—Achieving Comprehensive, Coordinated, and Sustained Earth Observations for the Benefit of Humankind: Framework for a 10-Year Implementation Plan as Adopted by Earth Observation Summit II, 25 April 2004," in *Earth Observation Summit II* (Tokyo, Japan, 2004).

23. Participants of the Second Earth Observation Summit, "Communiqué of the Second Earth Observation Summit as Adopted 25 April 2004," in *Earth Observation Summit II* (Tokyo, Japan, 2004).

Earth Observation Summit II, "From Observation to Action—Achieving Comprehensive, Coordinated, and Sustained Earth Observations for the Benefit of Humankind: Framework for a 10-Year Implementation Plan as Adopted by Earth Observation Summit II, 25 April 2004," in *Earth Observation Summit II* (Tokyo, Japan, 2004).

24. Participants of the Second Earth Observation Summit, "Communiqué of the Second Earth Observation Summit as Adopted 25 April 2004," in *Earth Observation Summit II* (Tokyo, Japan, 2004).

Earth Observation Summit II, "From Observation to Action—Achieving Comprehensive, Coordinated, and Sustained Earth Observations for the Benefit of Humankind: Framework for a 10-Year Implementation Plan as Adopted by Earth Observation Summit II, 25 April 2004," in *Earth Observation Summit II* (Tokyo, Japan, 2004).

25. Conrad Lautenbacher, "The Global Earth Observation System," in *The Washington Roundtable on Science and Public Policy* (Washington, DC: The George Marshall Institute, 2004).

26. Participants of the Earth Observation Summit III, "Resolution of the Third Earth Observation Summit as Adopted 16 February 2005," in *Third Earth Observation Summit* (Brussels, Belgium, 2005).

27. GEO, "The Global Earth Observation System of Systems (GEOSS) 10-Year Implementation Plan as Adopted 16 February 2005," in *Earth Observation Summit III* (Brussels, Belgium, 2005).

28. Ibid.

29. GEO, "2006 General Report on GEOSS Progress" (Group on Earth Observations, 2006).

30. CODATA, "Furthering the Practical Application of the Agreed GEOSS Data Sharing Principles" (2006).

31. GEO, "Report on Progress 2007" (Group on Earth Observations, 2007).

32. Janez Potocnik, "Earth Observations for Sustainable Growth and Development," in *GEO Ministerial Summit: GEO Ministerial Summit* (Cape Town, South Africa, 2007).

33. Egypt, "Statement by Egypt," in *GEO Ministerial Summit: GEO Ministerial Summit* (Cape Town, South Africa, 2007).

34. Heikki Tuunanen, "EO Summit IV Statement by Finland," in *GEO Ministerial Summit: GEO Ministerial Summit* (Cape Town, South Africa, 2007).

35. Germany, "Statement of Germany on Behalf of State Secretary Jorg Hennerkes," in *GEO Ministerial Summit: GEO Ministerial Summit* (Cape Town, South Africa, 2007).

36. GEO, "GEO-IV Report" (Group on Earth Observations, 2008).

37. "Report of GEO-V" (2009).

38. "Implementation Guidelines for the GEOSS Data Sharing Principles as Accepted at GEO-VI 18 November 2009" (Group on Earth Observations, 2009).

39. "GEO-IV Report."

40. G8, "G8 Hokkaido Toyako Summit Leaders Declaration" (Hokkaido Toyako, Japan, 2008).

41. "Responsible Leadership for a Sustainable Future" (L'Aquila, Italy, 2009).

42. GEO, "Mid-Term Evaluation of GEOSS Implementation" (2010).

43. "GEOSS Data Sharing Action Plan as Accepted at GEO-VIII 4 November 2010" (2010).

44. Ibid.

45. "Report of GEO-VII" (2011).

46. David J. Hayes, "United States Remarks to the GEO 2010 Ministerial " in *GEO 2010 Ministerial Summit* (Beijing, China, 2010).

47. Estonia, "Estonia Statement at ESA Ministerial," in *GEO 2010 Ministerial Summit* (Beijing, China, 2010).

48. Antonio Tajani, "European Commission Statement: Observe, Share Inform," in *GEO 2010 Ministerial Summit* (Beijing, China, 2010).

49. ESA, "ESA Statement for Geo Summit," in *GEO 2010 Ministerial Summit* (Beijing, China, 2010).

50. GEO, "Second Evaluation of GEOSS Implementation" (Group on Earth Observations, 2011).

51. "GEO 2012–2015 Work Plan" (Group on Earth Observations, 2011).

52. "Report of Data Sharing Working Group" (Group on Earth Observations, 2012).

53. Peter R Orszag, "Open Government Directive," ed. US Office of Management and Budget (OMB) (Washington, DC, 2009).

54. Rashmi Krishnamurthy and Yukika Awazu, "Liberating Data for Public Value: The Case of Data. Gov," *International Journal of Information Management* 36, no. 4 (2016).

55. G8, "G8 Open Data Charter and Technical Annex" (Group of Eight, 2013).

56. GEO, "GEO Strategic Plan 2016–2025: Implementing GEOSS: MS4 Approved by GEO-XII Plenary," in *GEO-XII Plenary & Mexico City Ministerial Summit* (Mexico City, Mexico: Group on Earth Observations, 2015).

57. Ibid.

58. GEO and CODATA, "The Value of Open Data Sharing" (Geneva, Switzerland: Group on Earth Observations, 2015).

7   US National Oceanic and Atmospheric Administration

1. Patrick Hughes, *A Century of Weather Service: A History of the Birth and Growth of the National Weather Service, 1870–1970* (Gordon & Breach Science Pub, 1970).

2. NASA, "TIROS," http://science.nasa.gov/missions/tiros/.

3. US Weather Bureau, "Catalogue of Meteorological Satellite Data—TIROS I Television Cloud Photography," ed. United States Department of Commerce, Key to Meteorological Records Documentation No. 5.31 (Washington, DC, 1961).

4. World Meteorological Organization (WMO), "IMO: The Origin of the WMO," http://www.wmo.int/pages/about/wmo50/e/wmo/history_pages/origin_e.html.

5. Arthur Davies and Oliver M Ashford, *Forty Years of Progress and Achievement: A Historical Review of WMO* (World Meteorological Organization, 1990).

6. John F. Kennedy, "Annual Message to the Congress on the State of the Union" (1961).

7. "Address at the U.N. General Assembly" (1961).

8. United Nations General Assembly, "Resolution Adopted by the General Assembly 1721 (XVI). International Co-Operation in the Peaceful Uses of Outer Space," ed. United Nations General Assembly (1961).

9. Harry Wexler and D S Johnson, "Meteorological Satellites," *Bulletin of the Atomic Scientists* 17, nos. 5–6 (1961).

10. US Weather Bureau, "National Operational Meteorological Satellite System" (Washington, DC: US Department of Commerce, 1962).

11. Ibid.

12. NASA.

13. R A Stampfl, W G Stroud, and Goddard Space Flight Center, *The Automatic Picture Transmission (APT) TV Camera System for Meteorological Satellites* (National Aeronautics and Space Administration, 1963).

14. Morris Tepper, "Meteorological Satellites: Statement by Dr. Morris Tepper, Director of Meteorological Systems, National Aeronautics and Space Administration," in *Second National Conference on Peaceful Uses of Outer Space* (Seattle, Washington,1962).

15. Francis Wilton Reichelderfer, "Meteorological Satellite Systems in Weather Research and Services," *Aerospace Engineering* (1961).

16. R L Pyle, "Archiving of TIROS Data" (paper presented at the Proceedings of the International Meteorological Satellite Workshop, Washington, DC, 13–22 November, 1961).

17. Bureau, "Catalogue of Meteorological Satellite Data—TIROS I Television Cloud Photography."

18. Pyle.

19. World Meteorological Organization, *First Report on the Advancement of Atmospheric Sciences and Their Application in the Light of Developments in Outer Space [with maps]* (Secretariat of the World Meteorological Organization, 1961).

United Nations General Assembly, "Resolution Adopted by the General Assembly 1802 (XVII), International Cooperation in the Peaceful Uses of Outer Space," ed. United Nations General Assembly (1962).

20. World Meteorological Organization (WMO), "World Weather Watch" (Geneva, Switzerland, 1966).

21. John F. Kennedy, "Letter from President Kennedy to Chairman Khrushchev" (1962).

22. United States Department of State Historical Office, United States Congress Senate Committee on Aeronautical, and Space Sciences, *Documents on International Aspects of the Exploration and Use of Outer Space, 1954–1962: Staff Report* (US Government Printing Office, 1963).

23. Nikita Khruschev, "Letter from Chairman Khrushchev to President Kennedy" (1962).

Lani Hummel Raleigh, "Soviet Application of Space to the Economy 1971–1975," http://www.globalsecurity.org/space/world/russia/meteor-system-series.htm.

NASA, "Cosmos 122," National Aeronautics and Space Administration, http://nssdc.gsfc.nasa.gov/nmc/spacecraftDisplay.do?id=1966-057A.

24. J Gordon Vaeth, "Establishing an Operational Weather Satellite System," in *Advances in Space Science and Technology*, ed. F I Ordway (Elsevier Science, 2014).

US Congress, "Supplemental Appropriation Act 1962" (1961).

25. United States House of Representatives Committee on Science and Astronautics, "National Meteorological Satellite Program" (Washington, DC, 1961).

26. US Weather Bureau, "TIROS Operational Satellite System Fact Sheet" (1965).

27. US Department of Commerce Weather Bureau, "National Operational Meteorological Satellite System Fact Sheet" (1962).

28. US Weather Bureau, "TIROS Operational Satellite System Fact Sheet" (1965).

29. Executive Office of the President, "Reorganization Plan No. 2 of 1965: Environmental Science Services Administration, Department of Commerce" (1965).

30. Environmental Science Services Administration, "Satellite Activities of ESSA 1968" (Washington, DC, 1969).

31. US Environmental Science Services Administration, "Catalogue of Meteorological Satellite Data—ESSA-1 Television Cloud Photography Part 1," in *Key to Meteorological Records Documentation No. 5.311* (1966).

32. Richard Nixon, "Reorganization Plan No. 3 of 1970" (Executive Office of the White House, 1970).

33. Executive Office of the President, "Reorganization Plan No. 4 of 1970 National Oceanic and Atmospheric Administration (NOAA)" (1970).

34. United States National Oceanic and Atmospheric Administration (NOAA), "Satellite Activities of NOAA 1970" (Washington, DC, 1971).

35. Gerald S Schatz, *The Global Weather Experiment: An Informal History* (National Academy of Sciences, 1978).

36. US National Oceanic and Atmospheric Administration (NOAA), "Satellite Activities of NOAA 1978" (Washington, DC, 1979).

37. United States Committee for the Global Atmospheric Research Program et al., *The Global Weather Experiment, Perspectives on Its Implementation and Exploitation: Report of the FGGE Advisory Panel to the U.S. Committee for the Global Atmospheric Research Program, Assembly of Mathematical and Physical Sciences, National Research Council* (National Academy of Sciences, 1978).

38. US National Oceanic and Atmospheric Administration, "Satellite Activities of NOAA 1979" (Washington, DC, 1980).

39. US National Oceanic and Atmospheric Administration, "National Environmental Satellite Service Catalog of Products" (Washington, DC, 1980).

40. O Weiss, D Namian, and A Schwalb, "Analysis of Past Funding for NOAA's Satellite Programs" (Washington, DC: National Oceanic and Atmospheric Administration [NOAA], 1985).

41. Ibid.

42. Ibid.

43. Ed Harper, "Ed Harper, Office of Management and Budget, Memorandum to Craig Fuller/Martin Anderson, "Resolution of Issues Related to Private Sector Transfer of Civil Land Observing Satellite Activities," July 13, 1981," in *Exploring the Unknown Volume III: Using Space*, ed. Roger Launius, John Logsdon, David Onkst, and Stephen Garber (NASA, 1998).

44. US House of Representatives, "Transfer of Civil Meteorological Satellites," House Concurrent Resolution 168, November 14, 1983," in *Exploring the Unknown Volume III: Using Space*, ed. Roger Launius, John Logsdon, David Onkst, and Stephen Garber (NASA, 1998).

45. David Johnson, "Statement of David Johnson, University Corporation for Atmospheric Research and Verner Suomi, Professor of Meteorology, University of Wisconsin" (Washington, DC: United States Senate Committee on Commerce, Science, and Transportation, 1983).

46. US House of Representatives, "Transfer of Civil Meteorological Satellites, House Concurrent Resolution 168, November 14, 1983," in *Exploring the Unknown Volume III: Using Space*, ed. Roger Launius, John Logsdon, David Onkst, and Stephen Garber (NASA, 1998).

47. Ibid.

48. United States Congress, "Land Remote-Sensing Commercialization Act of 1984," ed. United States Congress (1984).

49. Robert G Fleagle, "The Case for a New NOAA Charter," *Bulletin of the American Meteorological Society* 68, no. 11 (1987).

50. Ibid.

51. United States, "National Aeronautics and Space Administration Authorization Act, Fiscal Year 1989," ed. United States Congress (Washington, DC, 1988).

52. "Omnibus Budget Reconciliation Act of 1990," ed. United States Congress (Washington, DC, 1990).

53. National Oceanic and Atmospheric Administration (NOAA), "Policy Statement on the Weather Service/Private Sector Roles," *Federal Register* 56, no. 13 (1991).

54. Ibid.

55. *Global Change Research Act of 1990*, Public Law 101-606, 101st Congress (16 November, 1990).

56. Intergovernmental Panel on Climate Change, *Climate Change: The 1990 and 1992 IPCC Assessments: IPCC First Assessment Report, Overview and Policymaker Summaries, and 1992 IPCC Supplement* (WMO, 1992).

57. D Brent Smith, David F Williams, and Akihiro Fujita, "Satellite Missions, Global Environment, and the Concept of a Global Satellite Observation Information Network: The Role of the Committee on Earth Observation Satellites (CEOS)," *Acta Astronautica* 34 (1994).

CEOS, "CEOS Consolidated Report 1992" (London, United Kingdom: Committee on Earth Observations Satellites, 1992).

58. Smith, Williams, and Fujita.

CEOS, "1998 CEOS Consolidated Report" (Committee on Earth Observation Satellites, 1998).

59. National Oceanic and Atmospheric Administration (NOAA), "NOAA's Data and Information Management Strategy: A Vision for the 1990s and Beyond" (US Department of Commerce, 1992).

60. US Office of Management and Budget (OMB), "OMB Circular No. A-130" (1993).

61. US National Oceanic and Atmospheric Administration, "Schedule of New and Revised Fees for Access to NOAA Environmental Data and Information," *Federal Register* 59, no. 248 (1994).

62. Ken Singer et al., "Building a COTS Archive for Satellite Data" (1994).

63. NOAA, "The Nation's Environmental Data: Treasures at Risk, Report to Congress on the Status and Challenges for NOAA's Environmental Data Systems" (2001).

64. World Meteorological Organization (WMO), "Exchanging Meteorological Data Guidelines on Relationships in Commercial Meteorological Activities: WMO Policy and Practice" (Geneva, Switzerland: World Meteorological Organization, 1996).

65. WMO, "Resolution 40 (Cg-XII) WMO Policy and Practice for the Exchange of Meteorological and Related Data and Products Including Guidelines on Relationships in Commercial Meteorological Activities," ed. World Meteorological Organization (1995).

66. "Resolution 25 (Cg-XIII) Exchange of Hydrological Data and Products," ed. World Meteorological Organization (Geneva, 1999).

67. Lisa R Shaffer and Peter Backlund, "Towards a Coherent Remote Sensing Data Policy," *Space Policy* 6, no. 1 (1990).

68. US National Research Council Committee on Geophysical and Environmental Data, "On the Full and Open Exchange of Scientific Data" (Washington, DC, 1995).

69. US National Oceanic and Atmospheric Administration, "Schedule of Fees for Access to NOAA Environmental Data and Information and Products Derived Therefrom," *Federal Register* 62, no. 88 (1997).

70. NOAA.

71. Ibid.

72. United States, "Agreement between the United States National Oceanic and Atmospheric Administration and the European Organization for the Exploitation of Meteorological Satellites on an Initial Joint Polar-Orbiting Operational Satellite System" (Washington, DC, 1998).

73. National Research Council, "Resolving Conflicts Arising from the Privatization of Environmental Data," ed. Freeman Gilbert and William L Chameides (2001).

74. NOAA.

75. "NOAA Environmental Data: Foundation for Earth Observations and Data Management System" (2003).

76. American Meteorological Society (AMS), "Free and Open Exchange of Environmental Data," *Bulletin of the American Meteorological Society* 83 (2002).

77. George Bush, "U.S. Commercial Remote Sensing Policy Fact Sheet" (2003).

78. National Research Council, "Fair Weather: Effective Partnerships in Weather and Climate Services" (Washington, DC: The National Academies Press, 2003).

79. NOAA, "Policy on Partnerships in the Provision of Environmental Information," ed. National Oceanic and Atmospheric Administration (NOAA) (2007).

80. Conrad Lautenbacher, "Three Imperatives for an Integrated Earth Observation System: Speech by Vice Admiral (Ret.) Conrad C. Lautenbacher Jr., U.S. Navy Under Secretary of Commerce for Oceans & Atmosphere, NOAA Administrator" (paper presented at the 2003 WMO Congress, Geneva, Switzerland, 2003).

81. GEO, "The Global Earth Observation System of Systems (GEOSS) 10-Year Implementation Plan as Adopted 16 February 2005," in *Earth Observation Summit III* (Brussels, Belgium, 2005).

82. L Cucurull and J C Derber, "Operational Implementation of Cosmic Observations into NCEP's Global Data Assimilation System," *Weather and Forecasting* 23, no. 4 (2008).

83. TACC, "Formosat-3/Cosmic Data Distribution Policy," Taiwan Analysis Center for COSMIC, http://tacc.cwb.gov.tw/service/policy.htm.

84. Daniel Karlson, Email, 27 September 2016.

85. NOAA, "NOAA National Data Centers Free Data Distribution Policy" (2009).

86. "Policy on Access and Distribution of Environemntal Satellite Data and Products" (2011).

87. NOAA CLASS Help Desk, Email, 20 July 2012.

88. NOAA, "Schedule of New and Revised Fees for Access to NOAA Environmental Data and Information," *Federal Register* 74, no. 49 (2009).

"Schedule of New and Revised Fees for Access to NOAA Environmental Data and Information," *Federal Register* 75, no. 247 (2010).

"Schedule of New and Revised Fees for Access to NOAA Environmental Data and Information," *Federal Register* 77, no. 244 (2012).

"Schedule of New and Revised Fees for Access to NOAA Environmental Data and Information," *Federal Register* 80, no. 204 (2015).

"Schedule of New and Revised Fees for Access to NOAA Environmental Data and Information," *Federal Register* 81, no. 6 (2016).

89. Karlson.

90. Mariel Borowitz, "Is It Time for Commercial Weather Satellites? Analyzing the Case of Global Navigation Satellite System Radio Occultation," *New Space* 4, no. 2 (2016).

91. NOAA, "NOAA Commercial Space Policy Draft" (2015).

92. "NOAA Commercial Space Policy" (2016).

93. Steven VanRoekel, Sylvia Burwell, Todd Park, and Dominic Mancini, "Open Data Policy—Managing Information as an Asset" (Washington, DC: Executive Office of the President, 2013).

94. Stefaan Verhulst, Andrew Young, and Christina Rogawski, "NOAA's Open Data Portal: Creating a New Industry through Access to Weather Data" (Open Data Impact, 2016).

8   US Geological Survey

1. Roger Launius, 23 December 2009, http://blog.nasm.si.edu/history/the-whole -earth-disk-an-iconic-image-of-the-space-age/.

2. Richard Nixon, "Address before the 24th Session of the General Assembly of the United Nations," The American Presidency Project, http://www.presidency.ucsb.edu/ ws/index.php?pid=2236.

3. Department of the Interior (DOI), "Office of the Secretary, U.S. Department of the Interior, 'Earth's Resources to Be Studied from Space,' news release, September 21, 1966," in *Exploring the Unknown Volume III: Using Space*, ed. Roger Launius, John Logsdon, David Onkst, and Stephen Garber (1998).

Charles F. Luce, "Charles F. Luce, Under Secretary, U.S. Department of the Interior, to Dr. Robert C. Seamans, Jr., Deputy Administrator, NASA, October 21, 1966, with

Attached: "Operational Requirements for Global Resource Surveys by Earth-Orbital Satellites: EROS Program," in *Exploring the Unknown Volume III: Using Space*, ed. Roger Launius, John Logsdon, David Onkst, and Stephen Garber (1998)..

4. Ralph Bernstein, "Digital Image Processing of Earth Observation Sensor Data," *IBM Journal of Research and Development* 20, no. 1 (1976).

5. R S Williams and W D Carter, *ERTS-1: Earth Resources Technology Satellite 1: A New Window on Our Planet* (1976).

6. US General Accounting Office (GAO), "Land Satellite Project" (1976).

7. Williams and Carter.

8. US Department of State, "Memorandum of Understanding between the Brazilian Instituto De Pesquisas Espaciais and the United States National Aeronautics and Space Administration," in *United States Treaties and Other International Agreements*, ed. United States Department of State (Department of State, 1973).

9. (GAO).

10. Ibid.

11. United Nations General Assembly, "Resolution 3182 (XXVIII). International Cooperation in the Peaceful Uses of Outer Space," ed. United Nations General Assembly (1973).

12. United Nations, "Treaty on Principles Governing the Activities of States in the Exploration and Use of Outer Space, Including the Moon and Other Celestial Bodies," ed. United Nations (1967).

13. US Department of State, "James V. Zimmerman for Arnold W. Frutkin, Assistant Administrator for International Affairs, to Dr. John V. N. Granger, Acting Director, Bureau of International Scientific and Technological Affairs, Department of State, September 12, 1974, with Attached: "Foreign Policy Issues Regarding Earth Resource Surveying by Satellite: A Report of the Secretary's Advisory Committee on Science and Foreign Affairs," July 24, 1974," in *Exploring the Unknown Volume III: Using Space*, ed. Roger Launius, John Logsdon, David Onkst, and Stephen Garber (1998).

14. Ibid.

15. Ibid.

16. (GAO).

17. United Nations Centre for Disarmament, United Nations Department for Disarmament Affairs, and United Nations, *The Implications of Establishing an International Satellite Monitoring Agency: Report of the Secretary-General, Dept. For Disarmament Affairs* (United Nations, 1983).

18. Ann M Florini, "The Opening Skies: Third-Party Imaging Satellites and US Security," *International Security* 13, no. 2 (1988).

19. Government Accountability Office (GAO), "Landsat's Role in an Earth Resources Information System" (1977).

20. Zbigniew Brzezinski, "Presidential Directive/ NSC-54 Civil Operational Remote Sensing" (National Security Council, 1979).

21. US Department of State, "James V. Zimmerman for Arnold W. Frutkin, Assistant Administrator for International Affairs, to Dr. John V. N. Granger, Acting Director, Bureau of International Scientific and Technological Affairs, Department of State, September 12, 1974, with Attached: "Foreign Policy Issues Regarding Earth Resource Surveying by Satellite: A Report of the Secretary's Advisory Committee on Science and Foreign Affairs," July 24, 1974," in *Exploring the Unknown Volume III: Using Space*, ed. Roger Launius, John Logsdon, David Onkst, and Stephen Garber (1998).

22. Christopher C. Kraft, "Christopher C. Kraft, Jr., Director, Johnson Space Center, to Associate Administrator for Applications, NASA Headquarters, "Private Sector Operation of Landsat Satellites," March 12, 1976," in *Exploring the Unknown Volume III: Using Space*, ed. Roger Launius, John Logsdon, David Onkst, and Stephen Garber (1998).

23. (GAO).

24. Satellite Task Force, "Planning for a Civil Operational Land Remote Sensing Satellite System: A Discussion of Issues and Options" (Rockville, MD: US Department of Commerce, 1980).

25. Ed Harper, "Ed Harper, Office of Management and Budget, Memorandum to Craig Fuller/Martin Anderson, "Resolution of Issues Related to Private Sector Transfer of Civil Land Observing Satellite Activities," July 13, 1981," in *Exploring the Unknown Volume III: Using Space*, ed. Roger Launius, John Logsdon, David Onkst, and Stephen Garber (1998).

26. US House of Representatives, "Transfer of Civil Meteorological Satellites, House Concurrent Resolution 168, November 14, 1983," in *Exploring the Unknown Volume III: Using Space*, ed. Roger Launius, John Logsdon, David Onkst, and Stephen Garber (1998).

27. General Accounting Office (GAO), "Effects on Users of Commercializing Landsat and the Weather Satellites" (1984).

28. M Mitchell Waldrop, "Imaging the Earth (I): The Troubled First Decade of Landsat" (1982).

29. (GAO).

30. "Costs and Uses of Remote Sensing Satellites" (1983).

31. "Effects on Users of Commercializing Landsat and the Weather Satellites."

32. Ibid.

33. Office of Technology Assessment, "Remote Sensing and the Private Sector: Issues for Discussion" (Washington, DC: US Congress, 1984).

34. United States Congress, "Land Remote-Sensing Commercialization Act of 1984," ed. United States Congress (1984).

35. Ibid.

36. Ibid.

37. Ibid.

38. Anthony Calio, "Statement of Anthony J. Calio, Deputy Administrator National Oceanic and Atmospheric Administration, U.S. Department of Commerce, before the Subcommittee on Natural Resources, Agriculture, Research, and Environment and Subcommittee on Space Science and Applications" (1985).

39. David Baker, "Remote Future for Third World Satellite Data," *New Scientist* 116 (1987).

40. J I Gabrynowicz et al., *The United Nations Principles Relating to Remote Sensing of the Earth from Space: A Legislative History—Interviews of Members of the United States Delegation* (National Remote Sensing & Space Law Center, 2002).

41. Ibid.

42. Ibid.

United Nations, "Principles Relating to Remote Sensing of the Earth from Outer Space" (1986).

43. USGS, "Historical International Ground Stations," http://landsat.usgs.gov/Historical_IGS.php.

Florini.

Bhupendra Jasani and Christer Larsson, "Security Implications of Remote Sensing," *Space Policy* 4, no. 1 (1988).

44. Congressional Budget Office (CBO), "Encouraging Private Investment in Space Activities" (Washington, DC, 1991).

45. Science Senate Committee on Commerce, and Transportation, "Landsat Amendments Act of 1995" (Washington, DC: US Congress, 1995).

46. Samuel N Goward, "Landsat 1989: Remote Sensing at the Crossroads," *Remote Sensing of Environment* 28 (1989).

47. Myrna Watanabe, "Failure of Landsat 6 Leaves Many Researchers in Limbo," *The Scientist*, 13 December 199.

48. Steven Ashley, "Bringing Launch Costs Down to Earth," *Mechanical Engineering* 120, no. 10 (1998).

49. Government Accountability Office (GAO), "Satellite Data Archiving: U.S. And Foreign Activities and Plans for Environmental Information" (1988).

50. Eliot Marshall, "Landsats: Drifting toward Oblivion?: US "Commercial" Earth Observing Satellites May Be Abandoned in March, Creating a Gap in Coverage That May Last until 1991 or Later," *Science (New York, NY)* 243, no. 4894 (1989).

51. EOSAT, "Shutdown of U.S. Landsat System Halted," news release, 15 March 1989.

52. Executive Office of the President, "Statement by Press Secretary Fitzwater on Landsat Satellite Program Funding" (1989).

53. Robert Corell, "The Value of Landsat to the Global Change Program" (US Intergovernmental Committee on Earth and Environmental Sciences, 1990).

54. Byron D. Tapley, "On Research Uses of Landsat: Letter Report" (Committee on Earth Studies, National Research Council, 1991).

55. Office of Technology Assessment (OTA), "Remotely Sensed Data from Space: Distribution, Pricing, and Applications" (Washington, DC: US Congress, 1992).

56. Ibid.

57. Vipin Gupta, "New Satellite Images for Sale," *International Security* 20, no. 1 (1995).

58. George Bush, "U.S. Commercial Space Policy Guidelines" (1991).

59. "Landsat Remote Sensing Strategy" (Executive Office of the President, 1992).

60. *Land Remote Sensing Policy Act of 1992*, PL 102-555 (28 October 1992).

61. Ibid.

62. Ibid.

63. USGS, "Global Land Information System (GLIS)" (1993).

Office of Technology Assessment (OTA), "Remotely Sensed Data: Technology, Management and Markets" (Washington, DC, 1994).

64. Ted Shelsby, "Successful Launch of Satellite Would Lift EOSAT's Sales," *Baltimore Sun*, 5 October 1993.

65. Watanabe.

66. William Clinton, "PDD/NSTC-3 Land Remote Sensing Strategy" (Executive Office of the President, 1994).

67. Landsat Program Management, "Current Status and Summary of Agreement between Landsat Program Management and EOSAT Corporation on Cost and Reproduction Rights for Landsat 4/5 Thematic Mapper Data" (1994).

68. USGS, "Landsat 7 Data Prices Announced," news release, 31 October 1997, http://web.archive.org/web/19980120090116/http://edcwww.cr.usgs.gov/programs/landsat7price.html.

69. Herbert Satterlee, John R Copple III, and Gilbert D Rye, Letter, 1999.

70. Ibid.

71. Director of the Office of Science and Technology Policy, Dr. Neal Lane, Letter, 27 August 1999.

72. NASA Procurement Office, "Landsat Continuity Mission (LCM) Request for Information (RFI)" (1999).

73. Lawrence R. Pettinger, Email, 7 July 1999.

74. NASA Administrator Daniel Goldin, Charles Groat (USGS Director), James Baker (NOAA Administrator), Letter, 25 January 1999.

William Clinton, "Amendment to Presidential Decision Directives/MSTC-3" (2000).

75. Ray Byrnes, "Landsat 4/5 Operations to End," news release, 8 May 2001,.

76. Barbara Ryan, interview by Mariel Borowitz, 20 February 2013.

77. EROS Data Center, "Annual Report of Data Services" (USGS, 2001).

Tim Johnson, Letter, 21 January 2005.

78. USGS, "Ramifications of the Landsat 7 Failure: Short-Term Funding Strategies" (2004).

79. Ron Beck, "USGS Reduces Price for Landsat 7 Scenes," news release, 7 May 2004, http://eros.usgs.gov/sites/all/files/external/eros/history/2000s/2004%20USGS%20Reduces%20Price%20for%20Landsat%207%20Scenes%20News%20Release.pdf.pdf.

USGS, "Ramifications of the Landsat 7 Failure: Short-Term Funding Strategies."

80. "Ramifications of the Landsat 7 Failure: Short-Term Funding Strategies."

81. INPE, "Brazil and China Set Up Policy for Space Data Distribution," news release, 15 April 2010, http://www.cbers.inpe.br/ingles/news.php?Cod_Noticia=152.

82. Ryan.

83. Ray A Williamson and John C Baker, "Current US Remote Sensing Policies: Opportunities and Challenges," *Space Policy* 20, no. 2 (2004).

84. John Marburger III, "Landsat Data Continuity Strategy Adjustment," news release, 23 December 2005.

85. Executive Office of the President, "Restructuring the National Polar-Orbiting Operational Environmental Satellite System (NPOESS)" (2010).

86. John Marburger III.

87. Future of Land Imaging Interagency Working Group, "A Plan for a U.S. National Land Imaging Program" (Executive Office of the President, 2007).

88. Robert Tetrault, "Access and Availability of Resourcesat-1 AWIFS Data for Agriculture," in *USDA FAS Agriculture Applications Seminar 2006*, ed. USDA Satellite Imagery Archive (2006).

89. Glenn Bethel, "USDA Remote Sensing Programs, Priorities, and Partnerships" (USDA Remote Sensing Advisor, 2008).

90. NASA, "Free Landsat 7 Data Available from USGS," news release, 2007, http://landsat.gsfc.nasa.gov/?p=880.

"USGS Pilot Project Makes High-Quality Landsat Data Available for Download," news release, 2007, http://landsat.gsfc.nasa.gov/?p=866.

91. Barbara Ryan and Michael Freilich, "Landsat Data Distribution Policy" (USGS, 2008).

92. Curtis E Woodcock et al., "Free Access to Landsat Imagery," *Science* 320, no. 5879 (2008).

93. Group on Earth Observations, "GEO Announces Free and Unrestricted Access to Full Landsat Archive," news release, 20 November 2008, http://www .earthobservations.org/documents/pressreleases/pr_0811_bucharest_landsat.pdf.

94. Joan Moody, "Secretary Kempthorne Showcases Free Public Availability of Landsat Satellite Image Archive at ESRI Conference," news release, 4 August 2008, http://landsat.gsfc.nasa.gov/?p=1069.

95. Observations Working Group of the US Climate Change Science Program (CCSP) On Behalf of the United States Government, "The United States National Report on Systematic Observations for Climate for 2008: National Activities with Respect to the Global Climate Observing System (GCOS) Implementation Plan" (Washington, DC, 2008).

96. Peter R. Orszag, "Open Government Directive," ed. US Office of Management and Budget (OMB) (Washington, DC, 2009).

97. Michael A Wulder et al., "Opening the Archive: How Free Data Has Enabled the Science and Monitoring Promise of Landsat," *Remote Sensing of Environment* 122 (2012).

98. Thomas R Loveland and John L Dwyer, "Landsat: Building a Strong Future," ibid.

99. Landsat Advisory Group, "The Value Proposition for Ten Landsat Applications" (National Geospatial Advisory Committee, 2012).

100. Dr. Victoria (Tori) Adams, "Improving the Way the Government Does Business: The Value of Landsat Moderate Resolution Satellite Imagery in Improving Decision-Making" (Booz Allen Hamilton, 2012).

101. National Research Council, "Landsat and Beyond: Sustaining and Enhancing the Nation's Land Imaging Program" (Washington, DC, 2013).

102. Office of Science and Technology Policy, "National Plan for Civil Earth Observations" (Washington, DC: Executive Office of the President, 2014).

103. Loveland and Dwyer.

104. Michael A Wulder et al., "The Global Landsat Archive: Status, Consolidation, and Direction," *Remote Sensing of the Environment 185.* (2016).

105. Ibid.

106. Jon Campbell, "Technical Announcement: USGS Completes Decommissioning of Landsat 5," news release, 19 June 2013

107. Wulder et al.

108. Landsat Advisory Group, "Cloud Computing: Potential New Approaches to Data Management and Distribution" (National Geospatial Advisory Committee, 2013).

109. Ibid.

110. USGS, "More Ways to Get Landsat Data," news release, 26 March 2015, http://landsat.gsfc.nasa.gov/?p=10221.

Amazon Web Services (AWS), "Landsat on AWS," http://aws.amazon.com/public-data-sets/landsat/.

9   US National Aeronautics and Space Administration

1. Susan Barr and Cornelia Lüdecke, *The History of the International Polar Years (IPYS)* (Springer, 2010).

2. *National Aeronautics and Space Act of 1958*, Public Law 85-568 (29 July 1958).

3. United States House of Representatives Committee on Science and Astronautics, "National Meteorological Satellite Program" (Washington, DC, 1961).

4. R S Williams and W D Carter, *ERTS-1: Earth Resources Technology Satellite 1: A New Window on Our Planet* (1976).

5. R L Pyle, "Archiving of TIROS Data" (paper presented at the Proceedings of the International Meteorological Satellite Workshop, Washington, DC, 13–22 November 1961).

6. Roger D Launius, "A Western Mormon in Washington, DC: James C. Fletcher, NASA, and the Final Frontier," *Pacific Historical Review* 64, no. 2 (1995).

7. R J Fleming, T M Kaneshige, and W E McGovern, "The Global Weather Experiment 1. The Observational Phase through the First Special Observing Period," *Bulletin of the American Meteorological Society* 60, no. 6 (1979).

8. United States Committee for the Global Atmospheric Research Program, *Understanding Climatic Change: A Program for Action* (National Academy of Sciences, 1975).

9. John J Quann, "NASA and the US Climate Program—A Problem in Data Management" (1978).

10. Nicholas M Short and Locke M Stuart, Jr., "The Heat Capacity Mapping Mission (HCMM) Anthology, NASA Sp-465," *NASA Special Publication* 465 (1982).

11. Lee-Leung Fu and Benjamin Holt, "Seasat Views Oceans and Sea Ice with Synthetic Aperture Radar" (1982).

G H Born, J A Dunne, and D B Lame, "Seasat Mission Overview," *Science* 204, no. 4400 (1979).

T D Allan, "Seasat's Short-Lived Mission," *Nature* 281 (1979).

Diane L Evans et al., "Seasat—a 25-Year Legacy of Success," *Remote Sensing of Environment* 94, no. 3 (2005).

Steven H Pravdo et al., "Seasat Synthetic-Aperture Radar Data User's Manual" (1983).

NOAA, "National Holdings of Environmental Satellite Data of the National Oceanic and Atmospheric Administration" (Washington, DC, 1981).

12. C Scott Southworth, "General Characteristics and Availability of Landsat 3 and Heat Capacity Mapping Mission Thermal Infrared Data" (US Geological Survey, 1983).

13. ICSU, "The International Council for Science and Climate Change, 60 Years of Facilitating Climate Change Research and Informing Policy" (International Council for Science [ICSU], 2015).

World Meteorological Organization (WMO), "Declaration of the World Climate Conference" (1979).

14. Richard Goody, "Global Changes: Impacts on Habitability. A Scientific Basis for Assessment" (1982).

W Henry Lambright, "Administrative Entrepreneurship and Space Technology: The Ups and Downs of 'Mission to Planet Earth'," *Public Administration Review* (1994).

15. Spencer R Weart, *The Discovery of Global Warming* (Harvard University Press, 2008).

16. Burton I Edelson, "Mission to Planet Earth," *Science* 227, no. 4685 (1985).

17. W Henry Lambright, *NASA and the Environment the Case of Ozone Depletion* (Diane Publishing, 2005).

18. NASA Advisory Council. Earth System Sciences Committee, United States. National Aeronautics, and Space Administration, *Earth System Science Overview: A Program for Global Change* (National Aeronautics and Space Administration, 1986).

19. S. Ride, United States National Aeronautics and Space Administration, *Leadership and America's Future in Space: A Report to the Administrator* (NASA, 1987).

20. T G E Sciences et al., *Mission to Planet Earth: Space Science in the Twenty-First Century—Imperatives for the Decades 1995 to 2015* (National Academies Press, 1988).

21. W Henry Lambright, "NASA and the Environment: Science in a Political Context," *Societal Impact of Spaceflight.* Washington, DC: NASA SP-2007-4801 (2007).

22. Norman R Augustine, "Report of the Advisory Committee on the Future of the US Space Program," Washington, DC: Government Printing Office (1990).

23. NASA, "Nasa Organizational Chart—February 1992" (1992).

"NASA Organizational Chart—January 1993" (1993).

24. William K Stevens, "Huge Space Platforms Seen as Distorting Studies of Earth," *New York Times*, 19 June 1990.

25. Ibid.

26. David I Lewin, "Bringing NASA Down to Earth," *Mechanical Engineering-CIME* 114, no. 12 (1992).

27. Stevens.

James Hansen, William Rossow, and Inez Fung, "The Missing Data on Global Climate Change," *Issues in Science and Technology* 7, no. 1 (1990).

28. Greg Williams, "A Washington Parable: EOS in the Context of Mission to Planet Earth," *The Earth Observer* 21, no. 2 (2009).

29. George Bush, "Space-Based Global Change Observation" (Executive Office of the President, 1992).

30. Garrett Culhane, "Mission to Planet Earth," *Wired*, 1 June 1993.

31. National Research Council, *Panel to Review EOSDIS Plans: Final Report* (National Academy Press, 1994).

32. James L Green, "The New Space and Earth Science Information Systems at NASA's Archive," *Government Information Quarterly* 7, no. 2 (1990).

33. Government Accountability Office (GAO), "Earth Observation System: MASA's EOSDIS Development Approach Is Risky" (1992).

34. Ibid.

35. National Research Council, *Panel to Review EOSDIS Plans: Final Report* (National Academy Press, 1994).

36. NASA, " NASA Comments on GAO Report "Earth Observing System: Concerns over NASA's Basic Research Funding Strategy" May 1996," in *Earth Observing System: Concerns over NASA's Basic Research Funding Strategy*, ed. Government Accountability Office (GAO) (1996).

37. Richard H. Truly, "Nasa Earth Science Program Data Program—Delegation of Authority" (NASA, 1991).

38. Lisa Robock Shaffer, "NASA Remote Sensing Data Policy," *International Archives Of Photogrammetry And Remote Sensing* 29 (1993).

39. Allen Bromley, "Data Management for Global Change Research Policy Statements" (Washington, DC: US Global Change Research Program (USGCRP) (1991).

40. Truly.

41. General Accounting Office (GAO), "Earth Observing System: Concerns over Nasa's Basic Research Funding Strategy" (1996).

42. Martha Maiden, interview by Mariel Borowitz, 10 August 2012, 2013.

43. Truly.

44. United States Congress, "Land Remote-Sensing Commercialization Act of 1984," ed. United States Congress (1984).

45. General Accounting Office (GAO), "ESO Data Policy: Questions Remain About U.S. Commercial Access" (1992).

46. *Land Remote Sensing Policy Act of 1992*, PL 102-555 (28 October 1992).

47. Ghassem Asrar and David Jon Dokken, "EOS Reference Handbook," *NASA Earth Science Support Office, Washington DC* (1993).

48. Eligar Sadeh, "A Failure of International Space Cooperation: The International Earth Observing System," *Space Policy* 18, no. 2 (2002).

49. Asrar and Dokken.

50. Shaffer.

51. Ibid.

52. Sadeh.

53. Ghassem Asrar and Reynold Greenstone, *"MTPE EOS Reference Handbook"* (1995).

54. Maiden.

55. NASA, "Mission to Planet Earth Enterprise Name Changed to Earth Science," news release, 21 January 1998,http://geo.arc.nasa.gov/sge/landsat/mtpe.html.

56. M King and R Greenstone, "EOS Reference Handbook: A Guide to Earth Science Enterprise and the Earth Observing System" (Greenbelt, MD: EOS Project Sci. Office, NASA/Goddard Space Flight Center, 1999).

57. Ibid.

58. Ibid.

59. Stanford B Hooker and Wayne E Esaias, "An Overview of the SeaWiFS Project," *EOS, Transactions American Geophysical Union* 74, no. 21 (1993).

NASA, "SeaWiFS Project—Detailed Description," http://oceancolor.gsfc.nasa.gov/SeaWiFS/BACKGROUND/SEAWIFS_970_BROCHURE.html.

60. Hooker and Esaias.

NASA, "SeaWiFS Project—Detailed Description."

61. National Research Council, "Resolving Conflicts Arising from the Privatization of Environmental Data," ed. Freeman Gilbert and William L Chameides (2001).

62. *Commercial Space Act of 1998*, P.L. 105-303.

63. Ronald J Birk et al., "Government Programs for Research and Operational Uses of Commercial Remote Sensing Data," *Remote Sensing of Environment* 88, no. 1 (2003).

Samuel N Goward et al., "Acquisition of Earth Science Remote Sensing Observations from Commercial Sources: Lessons Learned from the Space Imaging Ikonos Example," *Remote Sensing of Environment* 88, no. 1 (2003).

Samuel N Goward et al., "Empirical Comparison of Landsat 7 and Ikonos Multispectral Measurements for Selected Earth Observation System (EOS) Validation Sites," *Remote Sensing of Environment* 88, no. 1 (2003).

Vicki Zanoni et al., "The Joint Agency Commercial Imagery Evaluation Team: Overview and Ikonos Joint Characterization Approach," *Remote Sensing of Environment* 88, no. 1 (2003).

64. Samuel N Goward et al., "Empirical Comparison of Landsat 7 and Ikonos Multispectral Measurements for Selected Earth Observation System (EOS) Validation Sites," *Remote Sensing of Environment* 88, no. 1 (2003).

65. Goward et al.

66. NASA, "EOSDIS Annual Metrics Report FY 2004" (NASA, 2004).

"EOSDIS FY96-FY99 Statistics Collection and Reporting System (SCRS) Annual Report" (NASA, 1999).

67. Climate Change Science Program and Subcommittee on Global Change Research, "Strategic Plan for the Climate Change Science Program" (Executive Office of the President, 2003).

68. Bob Jacobs, "Administrator Unveils Future NASA Vision and a Renewed Journey of Learning," news release, 12 April 2002,http://www.nasa.gov/home/hqnews/2002/02-066.txt.

69. Sean O'Keefe, "Remarks by NASA Administrator O'Keefe at the Earth Observation Summit," in *Earth Observation Summit*, ed. NASA (Washington, DC, 2003).

70. NASA, "The Vision for Space Exploration" (2004).

71. "2006 MASA Strategic Plan" (NASA, 2006).

72. Andrew C. Revkin, "NASA's Goals Delete Mention of Home Planet," *New York Times*, 22 July 2006.

73. Robert Bauer and Michael Pasciuto, "Organizing to Implement Technology in the NASA Science Organization" (paper presented at the 2005 IEEE Aerospace Conference, 2005).

74. National Research Council. Committee on Earth Science et al., *Earth Science and Applications from Space: National Imperatives for the Next Decade and Beyond* (National Academies Press, 2007).

75. RA Anthes and A Charo, "Earth Science and Applications from Space: Urgent Needs and Opportunities to Serve the Nation" (Committee on Earth Science and Applications from Space, National Research Council, National Academy of Sciences, The National Academies Press, 2005).

76. Science et al.

77. Ibid.

78. CL Parkinson, A Ward, and MD King, "Earth Science Reference Handbook: A Guide to NASA's Earth Science Program and Earth Observing Satellite Missions," *National Aeronautics and Space Administration* (2006).

79. NASA, "EOSDIS Annual Metrics Report FY 2006" (NASA, 2006).

80. Office of Science and Technology Policy (OSTP), "Achieving and Sustaining Earth Observations: A Preliminary Plan Based on a Strategic Assessment by the U.S. Group on Earth Observations" (Executive Office of the President, 2010).

81. NASA, "2011 NASA Strategic Plan" (NASA, 2011).

82. L Y Cureton and E Robinson, *NASA Open Government Plan* (Diane Publishing).

83. Martha Maiden, "Accessing NASA Earth Science Data/Open Data Policy" (NASA, 2013).

NASA, "Open Data and the Importance of Data Citations: The NASA EOSDIS Perspective," news release, 15 October 2015, https://earthdata.nasa.gov/open-data -and-the-importance-of-data-citations-the-nasa-eosdis-perspective.

84. "EOSDIS Annual Metrics Reports FY 2015" (NASA, 2015).

10 US Defense, Intelligence, and Commercial Satellites

1. US National Security Council, "U.S. Scientific Satellite Program" (1955).

2. R L Perry, J D Outzen, and Center for the Study of National Reconnaissance, *A History of Satellite Reconnaissance: The Robert L. Perry Histories* (Center for the Study of National Reconnaissance, 2012).

3. William J. Broad, "Spy Data Now Open for Studies of Climate," *New York Times*, 23 June 1992.

4. Ibid.

5. Scott Pace, Kevin M O'Connell, and Beth E Lachman, "Using Intelligence Data for Environmental Needs: Balancing National Interests" (RAND, 1997).

6. Geoff Brumfiel, "Military Surveillance Data: Shared Intelligence," *Nature* 477 (2011).

7. William Clinton, "Release of Imagery Acquired by Space-Based National Intelligence Reconaissance Systems" (Executive Office of the President, 1995).

8. Karen Irby, "President Orders Declassification of Historic Satellite Imagery Citing Value of Photography to Environmental Science," news release, 24 February 1995.

USGS Long Term Archive, "Declassified Satellite Imagery—1," https://lta.cr.usgs.gov/ declass_1.

"Declassified Satellite Imagery—2," https://lta.cr.usgs.gov/declass_2.

"Declassified Satellite Imagery—3," https://lta.cr.usgs.gov/declass_3.

9. H B Stelling, "Data Declassification" (US Air Force, 1972).

10. Office of Technology Assessment (OTA), "Remotely Sensed Data: Technology, Management and Markets" (Washington, DC, 1994).

11. US Air Force, "Air Force Weather Strengthens International Relations," news release, 10 July 2001, http://www.557weatherwing.af.mil/news/story.asp?id= 123443000.

12. Comprehensive Large Array-data Stweardship System (CLASS), "DMSP," NOAA, https://www.class.ngdc.noaa.gov/data_available/dmsp/index.htm.

National Centers for Environmental Information (NCEI), "DMSP Data Availability," NOAA, http://ngdc.noaa.gov/eog/availability.html.

National Center for Environmental Information (ngdc.dmsp@noaa.gov), Email, 23 July 2012.

13. Jimmy Carter, "PD/NSC-37 National Space Policy" (Executive Office of the President, 1978).

14. Office of Technology Assessment, "Remote Sensing and the Private Sector: Issues for Discussion" (Washington, DC: US Congress, 1984).

15. Office of Technology Assessment (OTA), "Remotely Sensed Data from Space: Distribution, Pricing, and Applications" (Washington, DC: US Congress, 1992).

16. *Land Remote Sensing Policy Act of 1992*, PL 102-555 (28 October 1992).

17. United States Congress, "Land Remote-Sensing Commercialization Act of 1984," ed. United States Congress (1984).

Land Remote Sensing Policy Act of 1992.

18. Joanne Irene Gabrynowicz, "The Promise and Problems of the Land Remote Sensing Policy Act of 1992," *Space Policy* 9, no. 4 (1993).

19. William Clinton, "Fact Sheet: PDD-23 Foreign Access to Remote Sensing Space Capabilities," news release, 10 March 1994.

20. George Bush, "U.S. Commercial Remote Sensing Policy Fact Sheet" (2003).

21. DigitalGlobe, "National Geospatial-Intelligence Agency Awards $24 Million Clearview Contract to Digitalglobe," news release, 2005, http://investor.digitalglobe .com/phoenix.zhtml?c=70788&p=irol-newsArticle&ID=1178561.

Ron Stearns, "The Billion-Dollar Promise: Clearview Contract and Greater Imagery Availability Move U.S. Satellite Commercial Imaging Market Forward," *Frost & Sullivan Aerospace & Defense Market Insight*, 21 February 2003.

Dan Caterinicchia, "NIMA Seeks 'Clearview' of the World," *Federal Computer Week*, 16 January 2003.

22. ORBIMAGE, "Orbimage Receives Clearview Contract Award from NGA," news release, 29 March 2004, 2004, http://www.prnewswire.com/news-releases/orbimage -receives-clearview-contract-award-from-nga-72252317.html.

23. Peter Paquette, "National Geospatial-Intelligence Agency Commercial Remote Sensing" (2006).

24. Rick Akers, "Nextview Will Provide the Vision and Solutions for New U.S. Policy on Commercial Imagery," *NIMA Pathfinder* (2003).

25. Warren Ferster, "NGA-Commercial Partnership Could Hatch More Satellites," *Space News*, 7 November 2005.

26. Missy Frederick, "GeoEye Sets Sights on Expanding the Government Market," ibid., 16 January 2006.

27. J J McCoy, "DigitalGlobe Orders Worldview 2 Satellite," *Satellite Today*, 3 January 2007.

28. Peter B de Selding, "EnhancedView Contract Awards Carefully Structured, NGA Says," *Space News*, 10 September 2010.

"DigitalGlobe Awards $307M in Contracts for WorldView-3 Satellite," ibid., 31 August 201.

29. Marcia Smith, "EnhancedView News Not So Rosy for GeoEye," *Space Policy Online*, 23 June 2012.

30. Peter B de Selding, "GeoEye to Merge with DigitalGlobe," *Space News*, 23 July 2012.

31. DigitalGlobe, "DigitalGlobe 2015 Annual Report" (2016).

32. Planet Labs, "Planet Labs," https://www.planet.com/company/approach/.

33. Jeff Foust, "Planet Labs Buying BlackBridge and Its RapidEye Constellation," *Space News*, 15 July 2015.

34. Terra Bella, "Terra Bella," https://terrabella.google.com/?s=about-us&c=about -history.

35. NGA, "Commercial GEOINT Strategy" (National Geospatial Intelligence Agency, 2015).

Mike Gruss, "NGA to Weigh Smallsat Options under New Commercial Strategy," *Space News*, 26 October 2015.

36. DigitalGlobe, "Firstlook Data Sheet" (2016).

"DigitalGlobe Opens Access to Satellite Data to Support Disaster Response Efforts in Nepal," news release, 26 April 2015, http://blog.digitalglobe.com/2015/04/26/digitalglobe-opens-access-to-satellite-data-to-support-disaster-response-efforts-in-nepal/.

"Open Imagery and Data to Support Ecuador Earthquake Response" (2016).

37. "DigitalGlobe Foundation" (2016).

38. Will Marshall, "Planet Platform Beta & Open California: Our Data, Your Creativity," news release, 15 October 2015, https://www.planet.com/pulse/planet-platform-beta-open-california-our-data-your-creativity/.

39. Planet Labs, "RapidEye Science Archive (RESA)," https://www.planet.com/markets/rapideye-science-archive/.

11 European Organization for the Exploitation of Meteorological Satellites

1. EUMETSAT, *Eumetsat 25 Years: Foundations for the Future* (Darmstadt, Germany: EUMETSAT, 2011).

2. Gerald S. Schatz, *The Global Weather Experiment: An Informal History* (National Academy of Sciences, 1978).

3. John Krige, "Chapter 7: The European Meteorological Satellite Programme," in *A History of the European Space Agency 1958–1987, Volume II: The Story of ESA 1973 to 1987*, ed. A Russo, John Krige, and L Sebesta (European Space Agency, 2000).

4. Ibid.

5. United States Committee for the Global Atmospheric Research Program et al., *The Global Weather Experiment, Perspectives on Its Implementation and Exploitation: Report of the FGGE Advisory Panel to the U.S. Committee for the Global Atmospheric Research Program, Assembly of Mathematical and Physical Sciences, National Research Council* (National Academy of Sciences, 1978).

6. Krige.

ESA, "Meeting of Meteosat Scientific User: Darmstadt from 19/06 to 20/06/1979" (Darmstadt, Germany, 1979).

"Meeting for Meteosat Scientific Users: London from 26/03 to 27/03/1980" (ESA, 1980).

7. Krige.

8. Ibid.

9. US House of Representatives, ""Transfer of Civil Meteorological Satellites," House Concurrent Resolution 168, November 14, 1983," in *Exploring the Unknown Volume III: Using Space*, ed. Roger Launius, John Logsdon, David Onkst, and Stephen Garber (1983).

10. EUMETSAT, "Convention for the Establishment of a European Organization for the Exploitation of Meteorological Satellites (EUMETSAT)" (Geneva, Switzerland, 1986).

11. Ibid.

12. *EUMETSAT 25 Years: Foundations for the Future.*

13. Kenneth D Hodgkins et al., "International Cooperation in Assuring Continuity of Environmental Satellite Data," *Space Policy* 1, no. 4 (1985).

John A Leese, Peter F Noar, and Claude Pastre, "Operational Continuity of the Global Meteorological Satellite Network," ibid. 5, no. 1 (1989).

14. World Meteorological Organization (WMO), "Exchanging Meteorological Data Guidelines on Relationships in Commercial Meteorological Activities: WMO Policy and Practice" (Geneva, Switzerland: World Meteorological Organization, 1996).

15. EUMETSAT, *EUMETSAT 25 Years: Foundations for the Future.*

16. Ibid.

17. "EUMETSAT Principles on Distribution and Charging Adopted at the 7th Meeting of the EUMETSAT Council" (EUMETSAT, 1988).

18. Ibid.

19. Ibid.

20. "First EUMETSAT Workshop on Legal Protection of Meteorological Satellite Data" (EUMETSAT, 1989).

21. Volker Thiem, "EUMETSAT Practice for Protection of Its Satellite Data," *Earth Observation Space Programmes, SAFISY Activities* (1992).

EUMETSAT, "Resolution on Technical Protection of EUMETSAT Data Adoped at the 13th Meeting of the EUMETSAT Council" (EUMETSAT, 1990).

22. European Commission, "Directive 96/9/EC of the European Parliament and of the Council of 11 March 1996 on the Legal Protection of Databases," *Official Journal of the European Union* 77 (1996).

23. EUMETSAT, "Resolution on Distribution Policy" (EUMETSAT, 1990).

24. *EUMETSAT 25 Years: Foundations for the Future.*

25. "Resolution on Distribution Policy."

26. Thiem.

Office of Technology Assessment (OTA), "The Future of Remote Sensing from Space: Civilian Satellite Systems and Applications" (OTA, 1993).

27. EUMETSAT, *EUMETSAT 25 Years: Foundations for the Future*.

28. Thiem.

29. EUMETSAT, "Amendments to the Convention" (EUMETSAT, 1991).

30. European Commission, "Council Directive 90/313/EEC of 7 June 1990 on the Freedom of Access to Information on the Environment," *Official Journal of the European Communities* 158 (1990).

31. D Brent Smith, David F Williams, and Akihiro Fujita, "Satellite Missions, Global Environment, and the Concept of a Global Satellite Observation Information Network: The Role of the Committee on Earth Observation Satellites (CEOS)," *Acta Astronautica* 34 (1994).

CEOS, "CEOS Consolidated Report 1992" (London, United Kingdom: Committee on Earth Observation Satellites, 1992).

32. M G Phillips, "Meteosat Operational Programme Current Missions and Services," *Earth Observation Space Programmes, SAFISY Activities* (1992).

33. CEOS, "1998 CEOS Consolidated Report" (Committee on Earth Observation Satellites, 1998).

34. EUMETSAT, *EUMETSAT 25 Years: Foundations for the Future*.

35. (WMO).

36. Ibid.

37. WMO, "Resolution 40 (Cg-XII) WMO Policy and Practice for the Exchange of Meteorological and Related Data and Products Including Guidelines on Relationships in Commercial Meteorological Activities," ed. World Meteorological Organization (1995).

38. EUMETSAT, "Resolution on the Conditions of Real Time Access to HRI Data" (EUMETSAT, 1994).

"Resolution on the Completion of the Conditions of Real Time Access to HRI Data Outside the EUMETSAT Member States" (EUMETSAT, 1994).

39. "Resolution on the Conditions of Real Time Access to HRI Data."

"Resolution on the Competion of the Conditions of Real Time Access to HRI Data Outside the EUMETSAT Member States."

40. Commission of the European Communities, "Notice Pursuant to Article 19 (3) of Council Regulation No. 17 Concerning Case No. IV/34.563–Ecomet," *Official Journal of the European Communities* 38, no. C 223 (1995).

41. Ibid.

42. EUMETSAT, "Resolution on EUMETSAT Principles on Data Policy Adopted at the 28th Meeting of the EUMETSAT Council" (1998).

43. Tillmann Mohr, Letter, 4 August 1998.

44. EUMETSAT, "Resolution on Amending Resolution EUM/C/98/Res. IV on EUMETSAT Principles on Data Policy" (1998).

45. "Resolution on Access to MSG Data and Products" (EUMETSAT, 1999).

46. Commission of the European Communities, "Public Sector Information: A Key Resource for Europe: Green Paper on Public Sector Information in the Information Society" (Brussels, Belgium, 1999).

47. Pirkko Saarikivi, Daniel Söderman, and Harry Newman, "Free Information Exchange and the Future of European Meteorology: A Private Sector Perspective" (2000).

48. Pira International and European Commission. Information Society DG, *Commercial Exploitation of Europe's Public Sector Information: Executive Summary* (Office for Official Publications of the European Communities, 2000).

49. Peter Weiss and Y Pluijmers, *Borders in Cyberspace: Conflicting Public Sector Information Policies and Their Economic Impacts* (Edward Elger Publishing, 2004).

John W Zillman and John W Freebairn, "Economic Framework for the Provision of Meteorological Services," *Bulletin of the World Meteorological Organization* 50, no. 3 (2001).

National Research Council, "Resolving Conflicts Arising from the Privatization of Environmental Data," ed. Freeman Gilbert and William L Chameides (2001).

50. European Union, "Directive 2003/4/EC of the European Parliament and of the Council of 28 January 2003 on Public Access to Environmental Information and Repealing Council Directive 90/313/EEC," *Official Journal of the European Union* 41, no. 26 (2003).

51. European Commission, "Directive 2003/98/EC of the European Parliament and Council on the Re-Use of Public Sector Information (PSI Directive)" ibid. 345, no. 90.

52. EUMETSAT, "The Optional EUMETSAT Jason-2 Altimetry Programme" (2001).

53. "Resolution on the Preparation of a Jason Follow-on Optional Programme" (EUMETSAT, 2008).

54. *EUMETSAT 25 Years: Foundations for the Future.*

55. United States, "Agreement between the United States National Oceanic and Atmospheric Administration and the European Organization for the Exploitation of Meteorological Satellites on an Initial Joint Polar-Orbiting Operational Satellite System" (Washington, DC, 1998).

56. "Agreement between NOAA and EUMETSAT on Transition Activities Regarding Polar-Orbiting Operational Environmental Satellite Systems" (2003).

Bob Hopkins, "U.S., European Satellite Agencies Sign Cooperation, Data-Sharing Accord: Agreement Builds on Partnership to Provide Enhanced Earth Observing Capabilities," news release, 24 June 2003.

57. EUMETSAT, "Data Policy for Metop Data and Products" (EUMETSAT, 2006).

58. European Commission, "Directive 2007/2/EC of the European Parliament and of the Council of 14 March 2007 Establishing an Infrastructure for Spatial Information in the European Community (Inspire)," *Official Journal of the European Union* 108 (2007).

59. Ibid.

60. EUMETNET, "Oslo Declaration" (2009).

61. EUMETSAT, "Eumetsat Data Policy" (2016).

62. "Satellite Archive Rebranded EUMETSAT Data Centre," *Image*, May 2010.

63. "EUMETSAT 2011 Annual Report" (2011).

64. European Commission, "Communication from the Commission to the European Parliament and the Council Global Monitoring for Environment and Security (GMES): Establishing a GMES Capacity by 2008—(Action Plan [2004–2008])," *Official Journal of the European Union* 92 (2004).

65. EUMETSAT, *EUMETSAT 25 Years: Foundations for the Future.*

66. "EUMETSAT Contribution to GMES" (2010).

67. "Member States," http://www.eumetsat.int/website/home/AboutUs/WhoWeAre/MemberStates/index.html.

68. "Current Satellites." http://www.eumetsat.int/website/home/Satellites/CurrentSatellites/index.html.

69. "Eumetsat Data Policy."

70. Ibid.

71. "Eumetsat Annual Report 2015" (2015).

12 European Space Agency

1. ESA, "ESA Convention and Council Rules of Procedure" (European Space Agency, 2010).

2. John Krige, "Chapter 7: The European Meteorological Satellite Programme," in *A History of the European Space Agency 1958–1987, Volume II: The Story of ESA 1973 to 1987*, ed. A Russo, John Krige, and L Sebesta (European Space Agency, 2000).

EUMETSAT, "EUMETSAT Principles on Distribution and Charging Adopted at the 7th Meeting of the EUMETSAT Council" (EUMETSAT, 1988).

3. ESA.

4. "Resolution on a European Remote-Sensing Satellite Programme" (ESA, 1977).

5. L Marelli, "The Earthnet Programme" (paper presented at the ESA Special Publication, 1979).

6. ESA, "ESA Convention and Council Rules of Procedure."

7. K. Madders, *A New Force at a New Frontier: Europe's Development in the Space Field in the Light of Its Main Actors, Policies, Law and Activities from Its Beginnings up to the Present* (Cambridge University Press, 2006).

8. Guy Duchossois, "The ERS-1 Mission Objectives," *ESA Bulletin*, no. 65 (1991).

9. M Fea, "The ERS Ground Segment," *ESA Bulletin-European Space Agency*, no. 65 (1991).

10. UNOOSA, "International Agreements and Other Available Legal Documents Relevant to Space-Related Activities" (Vienna, Austria: United Nationals Office of Outer Space Affairs, 1999).

11. Madders.

12. Ann M Florini, "The Opening Skies: Third-Party Imaging Satellites and US Security," *International Security* 13, no. 2 (1988).

13. Fea.

14. Fea.

15. Ray Harris and Roman Krawec, "Some Current International and National Earth Observation Data Policies," *Space Policy* 9, no. 4 (1993).

16. Fea.

17. Harris and Krawec.

18. S Bruzzi, "The Processing and Exploitation of ERS-1 Payload Data," *ESA Bulletin* 65 (1991).

19. Evert Attema, Yves-Louis Desnos, and Guy Duchossois, "Synthetic Aperture Radar in Europe: ERS, Envisat, and Beyond," *Johns Hopkins APL Technical Digest* 21, no. 1 (2000).

20. V Beruti S D'Elia, and M Albani, "Two Years of ERS-1 Data Exploitation," *ESA Bulletin*, no. 77 (1994).

21. G Kohlhammer, "The Envisat Exploitation Policy," *ESA Bulletin*, no. 106 (2001).

22. Harris and Krawec.

23. Kohlhammer.

24. CEOS, "1998 CEOS Consolidated Report" (Committee on Earth Observation Satellites, 1998).

25. Marelli.

P Goldsmith and C J Readings, "The Plans of the European Space Agency for Earth Observation," *Advances in Space Research* 14, no. 1 (1994).

26. M Albani, V Beruti, and S D'Elia, "Evolution of the ERS-2 Data-Processing Ground Segment," *ESA Bulletin-European Space Agency*, no. 83 (1995).

Kohlhammer.

27. E P W Attema, G Duchossois, and G Kohlhammer, "ERS-1/2 Sar Land Applications: Overview and Main Results" (paper presented at the Geoscience and Remote Sensing Symposium Proceedings, 1998. IGARSS'98. 1998 IEEE International, 1998).

28. G Duchossois and R Zobl, "ERS-2—A Continuation of the ERS-1 Success," *ESA Bulletin-European Space Agency*, no. 83 (1995).

29. Y L Desnos et al., "Research Activities in Response to the Envisat Announcement of Opportunity," *ESA Bulletin* 106 (2001).

30. National Research Council, "Resolving Conflicts Arising from the Privatization of Environmental Data," ed. Freeman Gilbert and William L Chameides (2001).

31. Commission of the European Communities, "Public Sector Information: A Key Resource for Europe: Green Paper on Public Sector Information in the Information Society" (Brussels, Belgium1999).

32. Pira International and European Commission Information Society DG, *Commercial Exploitation of Europe's Public Sector Information: Executive Summary* (Office for Official Publications of the European Communities, 2000).

33. Jacques Louet and S Bruzzi, "Envisat Mission and System" (paper presented at the Geoscience and Remote Sensing Symposium, 1999. IGARSS'99 Proceedings. IEEE 1999 International, 1999).

34. F Martin Crespo et al., "The Payload Data Segment," *ESA Bulletin* (2001).

35. ESA, "Envisat Data Policy" (European Space Agency Earth Observation Programme Board, 1998).

36. Ibid.

37. European Commission, "Directive 96/9/ec of the European Parliament and of the Council of 11 March 1996 on the Legal Protection of Databases," *Official Journal of the European Union* 77 (1996).

38. ESA, "Envisat Data Policy."

39. Council.

40. Desnos et al.

41. ESA, "Envisat Data Policy."

42. Kohlhammer.

43. ESA, "Envisat Data Policy."

44. Ibid.

45. "Envisat Data Policy: Implementation Guidelines" (Paris, France: European Space Agency Earth Observation Programme Board, 1998).

46. Kohlhammer.

47. Kohlhammer.

48. ESA, "Envisat Data Policy."

Kohlhammer.

49. ESA, "Envisat Data Policy."

50. Kohlhammer.

51. Andrea Celentano, "ERS/Envisat ASAR Data Products and Services" (Eurimage, 2002).

52. ESA, "Envisat Data Policy."

53. Crespo et al.

54. ESA, "Envisat Data Policy."

55. Ibid.

Kohlhammer.

56. Kohlhammer.

57. ESA, "Envisat Data Policy."

58. Ibid.

59. Atsuyo Ito, "Issues in the Implementation of the International Charter on Space and Major Disasters," *Space Policy* 21, no. 2 (2005).

60. David Southwood and Bruce T Battrick, *Introducing the "Living Planet" Programme: The ESA Strategy for Earth Observation* (ESA Publications Division, 1999).

61. C Readings, "The Science and Research Elements of ESA's Living Planet Programme" (ESA Publications Division, Noordwijk, The Netherlands, 1998).

62. Southwood and Battrick.

63. Ibid.

64. B Battrick, "The Changing Earth," *New Scientific Challenges for ESA's Living Planet* (ESTEC, 2006).

65. ESA, "Earth Explorer Data Policy" (European Space Agency Earth Observation Programme Board, 2003).

66. Josef Aschbacher et al., "GMES Space Component: Status and Challenges," *ESA Bulletin*, no. 142 (2010).

67. Ibid.

68. European Union, "Directive 2003/4/EC of the European Parliament and of the Council of 28 January 2003 on Public Access to Environmental Information and Repealing Council Directive 90/313/EEC," *Official Journal of the European Union* 41, no. 26 (2003).

69. European Commission, "Directive 2003/98/EC of the European Parliament and Council on the Re-Use of Public Sector Information (PSI Directive)," ibid. 345, no. 90.

70. Volker Liebig and Josef Aschbacher, "Global Monitoring for Environment and Security—Europe's Next Space Initiative Takes Shape," *ESA Bulletin* 123 (2005).

71. European Commission, "Directive 2007/2/EC of the European Parliament and of the Council of 14 March 2007 Establishing an Infrastructure for Spatial Information in the European Community (Inspire)," *Official Journal of the European Union* 108 (2007).

72. J A Yoder et al., "Assessing Requirements for Sustained Ocean Color Research and Operations. Committee on Assessing Requirements for Sustained Ocean Color Research and Operations" (National Research Council, 2011).

73. Cara Wilson, "The Rocky Road from Research to Operations for Satellite Ocean-Colour Data in Fishery Management," *ICES Journal of Marine Science: Journal du Conseil* 68, no. 4 (2011).

74. IOCCG, "Tenth IOCCG Committee Meeting Minutes" (Isla de Margarita, Venezuela: International Ocean-Color Coordinating Group, 2005).

75. Yves-Louis Desnos and Michel Verbauwhede, "ESA Earth Observation Satellites—Opportunities for Use of Data for Environmental Research in FP7" (ESA, 2007).

ESA, "Access to Esa Earth Observation Data" (ESA, 2007).

76. Henri Laur and Veronique Amans, "Access to Envisat Data" (ESA, 2008).

77. ESA, "ESA 2010 Budget" (2010).

78. "Joint Principles for a GMES Sentinel Data Policy" (European Space Agency Earth Observation Programme Board, 2009).

79. European Commission, "Legal Notice on the Use of Copernicus Sentinel Data and Service Information" (European Commission, 2013).

80. ESA, "Joint Principles for a GMES Sentinel Data Policy."

81. "ESA Data Policy for ERS, Envisat and Earth Explorer Missions (Simplified Version)" (ESA, 2012).

82. "Envisat and ERS Missions Data Access Guide" (ESA, 2011).

83. European Commission, "Regulation (EU) No. 911/2010 of the European Parliament and of the Council of 22 September 2010 on the European Earth Monitoring Programme (GMES) and Its Initial Operations (2011 to 2013)," *Official Journal of the European Union* 276 (2010).

84. Peter B de Selding, "ESA Data Access Policy Draws Mixed Reviews," *Space News*, 12 July 2010 2010.

"EC Still Undecided on 'Free and Open' Access Data Policy," *Space News*, 29 November 2010.

85. Geoff Sawyer and Marc de Vries, "About GMES and Data: Geese and Golden Eggs: A Study on the Economic Benefits of a Free and Open Data Policy for GMES Sentinels Data Final Report" (European Association of Remote Sensing Companies [EARSC], ESA, Citadel Consulting, European Union, 2012).

86. Ibid.

87. Ibid.

88. Ibid.

89. Ibid.

90. EARSC, "EARSC Position Paper on Industry Access to Copernicus Sentinel Data" (EARSC, 2013).

91. European Commission, "Commission Delegated Regulation (EU) No. 1159/2013 of 12 July 2013 Supplementing Regulation (EU) No. 911/2010 of the European Parliament and of the Council on the European Earth Monitoring Programme (GMES)

by Establishing Registration and Licensing Conditions for GMES Users and Defining Criteria for Restricting Access to GMES Dedicated Data and GMES Service Information," *Official Journal of the European Union* 309 (2013).

92. Ibid.

93. Ibid.

94. Ibid.

95. Peter B de Selding, "Euro Soyuz Launches Sentinel-1B Earth Observation Satellite, Einstein-Challenging Physics Experiment," *Space News*, 26 April 2016.

96. Google, "Google Earth Engine," https://earthengine.google.com/.

97. Peter B de Selding, "Europe Grants U.S. Special Access to Copernicus Earth-Observation Data," *Space News*, 23 October 2015.

13   Japan Meteorological Agency

1. JMA, "Japan Meteorological Agency: The National Meteorological Service of Japan" (JMA, 2012).

2. S T Tamura, "Japanese Meteorological Service in Korea and China," *Science* (1906).

Tetsu Tamura, "Meteorology in Japan: Recent Advances in Meteorology and Meteorological Service in Japan," *The Popular Science Monthly*, February 1906.

3. JMA.

4. Kazuto Suzuki, "Transforming Japan's Space Policy-Making," *Space Policy* 23, no. 2 (2007).

5. JMA.

6. "U.S.-Japan Space Agreement" (1969).

7. Boeing, "GMS—Japan's Geostationary Meteorological Satellite in Orbit," http://www.boeingimages.com/archive/GMS—Japan's-Geostationary-Meteorological-Satellite-in-Orbit-2F3XC5JCJCJ.html.

8. Y Kuroda, A Kubozono, and M Miyazawa, "NASDA Space Program in Japan" (1978).

9. B A Walton, "Radiocommunications for Meteorological Satellite Systems" (1975).

10. Masaichi Hirai et al., "Development of Experimental and Applications Satellites," *Acta Astronautica* 7, nos. 8–9 (1980).

11. Gerald S. Schatz, *The Global Weather Experiment: An Informal History* (National Academy of Sciences, 1978).

12. Head of Planning Division for the Japan Meteorological Agency Yoshiro Sekiguchi, Letter, 6 February 1978.

13. Kiyoshi Tsuchiya, Kohei Arai, and Tamotsu Igarashi, "Present Status and Future Plans of the Japanese Earth Observation Satellite Program," *Advances in Space Research* 9, no. 1 (1989).

14. JMA, "Meteorological Stellite Center Technical Note No. 7: Announcement—Improved or New Products" (1983).

15. "Japan Meteorological Agency: The National Meteorological Service of Japan."

16. JMBSC, "Japan Meteorological Business Support Center (JMBSC) Brochure" (JMBSC, 2015).

17. Mitsuhiko Hatori, "Review of Japanese Meteorological Services and Lessons for Developing Countries," in *Disaster Risk Management Public Seminar No. 7: Modernizing Weather, Climate and Hydrological Services and Early Warning Systems: Experience of World Bank and Japan*, ed. Japan Meteorological Business Support Center (JMBSC) (Tokyo, Japan, 2016).

18. Pirkko Saarikivi, Daniel Söderman, and Harry Newman, "Free Information Exchange and the Future of European Meteorology: A Private Sector Perspective" (2000).

19. Tsuchiya, Arai, and Igarashi.

20. Kazuto Suzuki, "A Brand New Space Policy or Just Papering over a Political Glitch? Japan's New Space Law in the Making," *Space Policy* 24, no. 4 (2008).

21. Ibid.

22. JMA, "MTSAT Data Access" (World Meteorological Organization [WMO], 2011).

23. Suzuki, "A Brand New Space Policy or Just Papering over a Political Glitch? Japan's New Space Law in the Making."

24. Strategic Headquarters for Space Policy, "Basic Plan for Space Policy: Wisdom of Japan Moves Space" (2009).

25. Setsuko Aoki, "Current Status and Recent Developments in Japan's National Space Law and Its Relevance to Pacific Rim Space Law and Activities," *Journal of Space Law* 35 (2009).

26. David Cyranoski, "Japan's Tsunami Warning System Retreats: Lessons from Tohoku Wave Lead to Drop in Early Warning Precision," *Nature* (2011).

27. Kotaro Bessho et al., "An Introduction to Himawari-8/9—Japan's New-Generation Geostationary Meteorological Satellites," 気象集誌. 第 2 輯 94, no. 2 (2016).

28. JMA, "MTSAT Data Access."

29. WMO, "WMO Information System (WIS)," https://www.wmo.int/pages/prog/www/WIS/centres_en.html.

30. GISC Tokyo, "WIS Portal: Registration," http://www.wis-jma.go.jp/cms/about-wis/registration/.

31. Bessho et al.

32. JMBSC, "Dissemination of Meteorological Data, Products and Information," http://www.jmbsc.or.jp/en/meteo-data.html.

33. JMA, "Himawari 8 Data" (2015).

14 Japan Aerospace Exploration Agency

1. Shigebumi Saito, "Japan's Space Policy: Background and Outlook," *Space Policy* 5, no. 3 (1989).

2. "U.S.-Japan Space Agreement" (1969).

3. *Law Concerning the National Space Development Agency of Japan*, Law No. 50 (23 June 1969).

4. Office of Technology Assessment (OTA), "The Future of Remote Sensing from Space: Civilian Satellite Systems and Applications" (OTA, 1993).

5. Yukio Hakura, "Activities and Future Plan of Earth Observation by Satellites," *Acta Astronautica* 7, nos. 8–9 (1980).

6. Kiyoshi Tsuchiya, "Selection of Sensors and Spectral Bands of Marine Observation Satellite (MOS)—1," *Advances in Space Research* 3, no. 2 (1983).

7. Kiyoshi Tsuchiya, Kohei Arai, and Tamotsu Igarashi, "Present Status and Future Plans of the Japanese Earth Observation Satellite Program," ibid. 9, no. 1 (1989).

8. Ibid.

9. Patrick A Salin, "Proprietary Aspects of Commercial Remote-Sensing Imagery," *Northwestern Journal of International Law and Business* 13 (1992).

10. (OTA).

11. CEOS, "1998 CEOS Consolidated Report" (Committee on Earth Observation Satellites, 1998).

12. Tsuchiya, Arai, and Igarashi.

13. (OTA).

14. JAXA Official, interview by Mariel Borowitz, 20 August 2012.

15. Takashi Hamazaki, "Overview of the Advanced Land Observing Satellite (ALOS): Its Mission Requirements, Sensors, and a Satellite System" (paper presented at the Proceedings of the ISPRS Joint Workshop—Sensors and Mapping from Space, Hannover, Germany, 1999).

16. Masami Onoda, "Japanese Earth Observation Program and Data Policy" (paper presented at the The First International Conference on the State of Remote Sensing Law, Oxford, MS, 18–19 April 2002).

17. Daisuke Saisho, interview by Mariel Borowitz, 9 September 2016.

18. Tsuchiya, Arai, and Igarashi.

19. Christian Kummerow et al., "The Tropical Rainfall Measuring Mission (TRMM) Sensor Package," *Journal of Atmospheric and Oceanic Technology* 15, no. 3 (1998).

20. Yasushi Yamaguchi et al., "Overview of Advanced Spaceborne Thermal Emission and Reflection Radiometer (ASTER)," *IEEE Transactions on Geoscience and Remote Sensing* 36, no. 4 (1998).

21. United States State Department, "Agreement between the United States of America and Japan: Effected by Exchange of Notes" (Washington, DC, 1996).

22. Eligar Sadeh, "A Failure of International Space Cooperation: The International Earth Observing System," *Space Policy* 18, no. 2 (2002).

23. Kazuto Suzuki, "Administrative Reforms and the Policy Logics of Japanese Space Policy," ibid. 21, no. 1 (2005).

24. *Law Concerning Japan Aerospace Exploration Agency*, Law No. 161 (13 December 2002).

25. JAXA, Powerpoint Presentation Shared in Telecon, 19 August 2012.

26. Junichiro Koizumi, "Welcome Remarks by Prime Minister Junichiro Koizumi," in *Earth Observation Summit II* (Tokyo, Japan, 2004).

27. Japan Council for Science and Technology Policy, "Earth Observation Promotion Strategy (Provisional Translation from the Official Document in Japanese)" (2004).

28. Japan Space Activities Commission, "Japan's Earth Observation Satellite Development Plan and Data Utilization Strategy" (2005).

29. JAXA, "ALOS Data Users Handbook" (Japan Aerospace Exploration Agency, 2008).

30. Japan Space Activities Commission, "Japan's Earth Observation Satellite Development Plan and Data Utilization Strategy" (2005).

31. Japan Aerospace Exploration Agency, National Institute for Environmental Studies, and Ministry of the Environment of Japan, "Greenhouse Gases Observing Satellite (GOSAT) Data Policy" (2009).

32. JAXA and NIES, "GOSAT/Ibuki Data Users Handbook, 1st Edition" (2011).

33. Kazuto Suzuki, "Transforming Japan's Space Policy-Making," *Space Policy* 23, no. 2 (2007).

34. Nobuaki Hashimoto, "Establishment of the Basic Space Law–Japan's Space Security Policy," *The National Institute for Defense Studies News*, no. 123 (2008).

35. Suzuki, "Transforming Japan's Space Policy-Making."

36. Strategic Headquarters for Space Policy, "Basic Plan for Space Policy: Wisdom of Japan Moves Space" (2009).

37. Paul Kallender-Umezu, "Enacting Japan's Basic Law for Space Activities: Revolution or Evolution?," *Space Policy* 29, no. 1 (2013).

38. Japan IT Strategic Headquarters, "Open Government Data Strategy" (Japan 2012).

39. Soichiro Takagi, "Member Blog: Japan's Growing Open Data Movement," news release, 15 January 2014, 2014, https://theodi.org/blog/japan-open-data.

40. JAXA, "Jaxa's Earth Observation Satellite Highly Praised Worldwide: Q&A with Osamu Ochiai, Associate Senior Administrator of Satellite Applications and Promotion Center, JAXA Space Applications Mission Directorate," news release, 2 February 2011, http://global.jaxa.jp/article/special/geo/ochiai_e.html.

41. Paul Kallender-Umezu, "Japan Passes Overhaul of Space Management Structure," *Space News*, 2 July 2012.

42. Japan Strategic Headquarters for Space Policy, "Basic Plan on Space Policy" (Japan, 2013).

43. JAXA, "G-Portal (Beta)," https://www.gportal.jaxa.jp/gp/top.html.

44. Osamu Ochiai, "JAXA's Earth Observation Data and Information System," in *Copernicus Big Data Workshop*, ed. Japan Aerospace Exploration Agency (JAXA) (Brussels, Belgium, 2014).

45. JAXA and NIES.

46. NASA, "NASA, Japan Make ASTER Earth Data Available at No Cost," news release, 1 April 2016, http://www.nasa.gov/feature/jpl/nasa-japan-make-aster-earth-data-available-at-no-cost/.

47. Saisho.

15   Brazil, Russia, China, India, South Africa

1. US Department of State, "Memorandum of Understanding between the Brazilian Instituto De Pesquisas Espaciais and the United States National Aeronautics and

Space Administration," in *United States Treaties and Other International Agreements*, ed. United States Department of State (Department of State, 1973).

2. Carlos De Oliveira Lino, Maury Gonçalves Rodrigues Lima, and Genésio Luiz Hubscher, "CBERS—an International Space Cooperation Program," *Acta Astronautica* 47, no. 2 (2000).

3. INPE, "Brazil and China Set Up Policy for Space Data Distribution," news release, 15 April 2010, http://www.cbers.inpe.br/ingles/news.php?Cod_Noticia=152.

GEO, "CBERS," in *GEO-VII Plenary* (2010).

4. Gilberto Câmara (INPE) and Guo Jian Ning (CRESDA), "CBERS Data Policy Version 1.8," ed. CBERS JPC (2004).

5. GEO, "INPE Distributes 1 Million Free Satellite Images," news release, 2009, https://www.earthobservations.org/art_005_001.shtml.

Sérgio de Paula Pereira and Ivan Barbosa, "Info on CBERS-4," in *Agricab Project Final Meeting* (Antwerp, Belgium: INPE, 2015).

6. Hilcéa Ferreira, "Benefits of Data Sharing: The CBERS Programme," in *Symposium on the Data Sharing Action Plan for GEOSS and the Benefits of Data Sharing* (Beijing, China: National Institute for Space Research [INPE] Brazil, 2010).

7. Luis Geraldo Ferreira, 29 June 2016.

INPE, "INPE Data Portal," http://www.dgi.inpe.br/CDSR/.

8. Bart Hendrickx, "A History of Soviet/Russian Meteorological Satellites," *Journal of the British Interplanetary Society* 57, no. 1 (2004).

9. United States Department of State Historical Office, United States Congress Senate Committee on Aeronautical, and Space Sciences, *Documents on International Aspects of the Exploration and Use of Outer Space, 1954–1962: Staff Report* (US Government Printing Office, 1963).

10. Hendrickx.

11. Ibid.

12. Ibid.

13. Ibid.

14. V V Asmus et al., "Meteorological Satellites Based on Meteor-M Polar Orbiting Platform," *Russian Meteorology and Hydrology* 39, no. 12 (2014).

15. Ibid.

SRC Planeta Rosgidromet and IKI RAN, "'Sputnik' Server Satellite Archive Catalogues," http://sputnik.infospace.ru/catalog_eng.html.

16. Ministry of Natural Resources and Environment of the Russian Federation and European Organization for the Exploitation of Meteorological Satellites, "Agreement on Cooperation and Exchange of Data and Products from Meteorological Satellites in Support of Weather Analysis and Forecast" (2015).

17. A Karpov, "Hydrometeorological, Oceanographic and Earth-Resources Satellite Systems Operated by the USSR," *Advances in Space Research* 11, no. 3 (1991).

18. D. Gorobets, "The Use of Russian Space Means of Remote Sensing of the Earth for the Benefit of Developing Countries " in *UN Committee on the Peaceful Uses of Outer Space (COPUOS) Scientific and Technical Subcommittee Forty-ninth session* (Vienna, Austria: UN Office of Outer Space Affairs [UNOOSA], 2012).

19. Roscosmos, "Roscosmos Geoportal: Service Space Images," http://gptl.ru/index .php/welcome.

20. Ann M Florini, "The Opening Skies: Third-Party Imaging Satellites and US Security," *International Security* 13, no. 2 (1988).

21. Vipin Gupta, "New Satellite Images for Sale," ibid. 20, no. 1 (1995).

22. George J Tahu, John C Baker, and Kevin M O'Connell, "Expanding Global Access to Civilian and Commercial Remote Sensing Data: Implications and Policy Issues," *Space Policy* 14, no. 3 (1998).

23. LI Zhilin, "High-Resolution Satellite Images: Past, Present and Future," *Journal of Geospatial Engineering* 2, no. 2 (2000).

24. SOVZOND Company, "Remote Sensing," http://en.sovzond.ru/services/remote -sensing/#planirovanie-novoy-syemki.

25. SpaceNews Editor, "Spotlight—Dauria Aerospace," *Space News*, 24 February 2014.

26. PanGeo Alliance, "Pangeo Alliance: A Unique Earth Observation Constellation," http://www.pangeo-alliance.com/.

27. Matthew Bodner, "Russia's First Commercial Space Firm Abandons Offices in the West," *The Moscow Times*, 10 February 2015.

28. Y S Rajan, P P Nageswara Rao, and G Behera, "Indian Remote Sensing Programme," *Acta Astronautica* 18 (1988).

ISRO, "Bhaskara-I," Indian Space Research Organization (ISRO), http://www.isro.gov .in/Spacecraft/bhaskara-i.

"Bhaskara-II," Indian Space Research Organization, http://www.isro.gov.in/ Spacecraft/bhaskara-ii.

29. Rajan, Rao, and Behera.

30. K S Jayaraman, "US and India Sign Ocean Data Deal," *Nature* 375 (1995).

31. John A Leese, Peter F Noar, and Claude Pastre, "Operational Continuity of the Global Meteorological Satellite Network," *Space Policy* 5, no. 1 (1989).

32. Jayaraman.

33. "India and United States to Share Satellite Data," *Nature* 391, no. 6663 (1998).

Pallava Bagala, "US-India Pact to Plug Data Gap," *Science* 278, no. 5337 (1997).

34. Meteorological & Oceanographic Satellite Data Archival Centre (MOSDAC), "Data Access Policy," ISRO Space Applications Center, http://www.mosdac.gov.in/content/data-access-policy.

35. K Kasturirangan et al., "Indian Remote Sensing Satellite IRS-1C—the Beginning of a New Era," *Current Science* 70, no. 7 (1996).

36. ISRO, "OceanSat-2," Indian Space Research Organization, http://www.isro.gov.in/Spacecraft/oceansat-2.

37. Joanne Irene Gabrynowicz, "The Land Remote Sensing Laws and Policies of National Governments: A Global Survey" (The National Center for Remote Sensing, Air, and Space Law at the University of Mississippi School of Law, 2007).

38. ISRO, "IRS-P6/ResourceSat-1," Indian Space Research Organization, http://www.isro.gov.in/Spacecraft/irs-p6-resourcesat-1.

39. "Cartosat-1," Indian Space Research Organization, http://www.isro.gov.in/Spacecraft/cartosat-%E2%80%93-1.

40. NRSC Data Center, "IRS Satellite Data Products Price List" (Balanagar, Hyderabad: Indian Space Resarch Organization [ISRO], 2016).

41. India, "Remote Sensing Data Policy (RSDP)," ed. Department of Space (2011).

42. "National Data Sharing and Accessibility Policy 2012," in *NDSAP-2012*, ed. Department of Science & Technology (2012).

43. "National Geospatial Policy," in *NGP 2016*, ed. India (2016).

44. Wenjian Zhang, "Meteorological Satellite Program of China" (paper presented at the Asia-Pacific Symposium on Remote Sensing of the Atmosphere, Environment, and Space, 1998).

45. Ibid.

46. Wenjian Zhang et al., "China's Current and Future Meteorological Satellite Systems," in *Earth Science Satellite Remote Sensing* (Springer, 2006).

47. Ibid.

48. Chaohua Dong et al., "An Overview of a New Chinese Weather Satellite FY-3a," *Bulletin of the American Meteorological Society* 90, no. 10 (2009).

49. SpaceNews Editor, "Report: Chinese Weather Sats Could Fill U.S. Gap," *Space News*, 26 August 2013 2013.

50. Dong et al.

51. China National Satellite Meteorological Center (NSMC), "Fengyun Satellite Data Center," http://satellite.cma.gov.cn/PortalSite/Default.aspx.

52. Ya-Qiu Jin, Naimeng Lu, and Minseng Lin, "Advancement of Chinese Meteorological Feng-Yun (Fy) and Oceanic Hai-Yang (Hy) Satellite Remote Sensing," *Proceedings of the IEEE* 98, no. 5 (2010).

53. China National Satellite Ocean Application Service (NSOAS), "Distribution Policy: Hy-1b Satellite, Hy-2 Satellite Data Product International User Releasing Policy and Guide," http://www.nsoas.gov.cn/NSOAS_En/Products/3.html.

54. Ibid.

55. Wen Xu, Jianya Gong, and Mi Wang, "Development, Application, and Prospects for Chinese Land Observation Satellites," *Geo-spatial Information Science* 17, no. 2 (2014).

56. Ibid.

57. Peter B de Selding, "China Launches High-Resolution Commercial Imaging Satellite," *Space News*, 7 October 2015.

58. "Domestic Satellites Providing 80 Pct of China's Satellite Data," *China Daily*, 10 March 2016.

59. Xu, Gong, and Wang.

60. Huadong Guo et al., "Earth Observation Satellite Data Receiving, Processing System and Data Sharing," *International Journal of Digital Earth* 5, no. 3 (2012).

61. Gu Xingfa and Tong Xudong, "Overview of China Earth Observation Satellite Programs [Space Agencies]," *IEEE Geoscience and Remote Sensing Magazine* 3, no. 3 (2015).

62. SSTL, "Beijing-1: The Mission," http://www.sstl.co.uk/Missions/Beijing-1-Launched-2005/Beijing-1/Beijing-1-The-Mission.

63. Selding.

Twenty First Century Aerospace Technology (21AT), "Imagery Products," http://www.21at.com.cn/en/Imagery/ImageryProducts/.

Ltd. Chang Guang Satellite Technology Co., "Products: Satellite Images," http://www.charmingglobe.com/cpzx/wxyx/.

64. Guo Huadong and Wang Changlin, "Building Up National Earth Observing System in China," *International Journal of Applied Earth Observation and Geoinformation* 6, no. 3 (2005).

65. SANSA, "South African Satellites," South African National Space Agency (SANSA), http://atlas.sansa.org.za/atlas-sa_satellites.html.

Hickok, "Sumbandilasat EO Satellite Shows South Africa's Space Mettle," *Earthzine*, 11 January 2011, 2010.

66. AARSE, "Overview," African Association of Remote Sensing of the Environment, http://www.africanremotesensing.org/page-1611993.

"AARSE Johannesburg Declaration" (Johannesburg, South Africa: African Association of Remote Sensing of the Environment, 2014).

Ashlea du Plessis and Jide Kufoniyi Islam Abou El-Magd, "AARSE Survey on Utilisation of African National Earth Observation Satellites Products" (African Association of Remote Sensing of the Environment [AARSE], 2015).

S Mostert, "The African Resource Management (ARM) Satellite Constellation," *African Skies* 12 (2008).

16   Sharing Satellite Data

1. General Accounting Office (GAO), "EOS Data Policy: Questions Remain About U.S. Commercial Access" (1992).

2. Steven Ashley, "Bringing Launch Costs Down to Earth," *Mechanical Engineering* 120, no. 10 (1998).

3. Landsat Advisory Group, "The Value Proposition for Ten Landsat Applications" (National Geospatial Advisory Committee, 2012).

Dr. Victoria (Tori) Adams, "Improving the Way the Government Does Business: The Value of Landsat Moderate Resolution Satellite Imagery in Improving Decision-Making" (Booz Allen Hamilton, 2012).

4. Joanne Irene Gabrynowicz, "The Promise and Problems of the Land Remote Sensing Policy Act of 1992," *Space Policy* 9, no. 4 (1993).

Peter B de Selding, "Indian Rocket Lofts Spot 6 Earth-Observing Satellite," *Space News*, 18 September 2012 2012.

5. Peter B de Selding, "Indian Rocket Lofts Spot 6 Earth-Observing Satellite," *Space News*, 18 September 2012.

6. Rolf Werninghaus, Stefan Buckreuß, and Wolfgang Pitz, "TerraSAR-X Mission Status" (paper presented at the 2007 IEEE International Geoscience and Remote Sensing Symposium, 2007).

Peter B de Selding, "Public Funding on the Table for Germany's Next Radar Satellite," *Space News*, 8 July 2011 2011.

7. Kevin M O'Connell et al., "US Commercial Remote Sensing Satellite Industry: An Analysis of Risks" (DTIC Document, 2001).

8. Peter B de Selding, "France to Make Older Spot Images Available to Researchers for Free," *Space News* 2014.

17 Future of Data Sharing

1. National Research Council, "Resolving Conflicts Arising from the Privatization of Environmental Data," ed. Freeman Gilbert and William L Chameides (2001).

Appendix A   Global Satellite Data Sharing

1. Alex da Silva Curiel et al., "First Results from the Disaster Monitoring Constellation (DMC)," *Acta Astronautica* 56, no. 1 (2005).

2. DMCii, "DMC Constellation," Disaster Monitoring Constellation International Imaging, http://www.dmcii.com/?page_id=9275.

3. Agence Spatiale Algerienne, "AlSat-2A Second Observation Satellite Earth of the National Space Program-Horizon 2020 (NSP)," http://www.asal.dz/Alsat%202A.php.

4. Marcos Machado et al., "SAC-A Satellite," in *10th Annual AIAA/USU Conference on Small Satellites* (1996).

5. FR Colomb et al., "SAC-C Mission, an Example of International Cooperation," *Advances in Space Research* 34, no. 10 (2004).

6. NASA, "Earthdata Search," https://search.earthdata.nasa.gov/; CONAE, "Product Catalog Satellite SAC-C," http://catalogos.conae.gov.ar/SAC_C/default3.asp.

7. ICARE, "ICARE Data and Services Center: ICARE on-Line Data Archive," University of Lille Cloud-Aerosol-Water-Radiation Interactions (ICARE), http://www.icare.univ-lille1.fr/archive.

8. NASA, "Missions: Aquarius," NASA Jet Propulsion Laboratory, http://podaac.jpl.nasa.gov/aquarius.

9. CONAE, "Data & Product Catalogs Sac-D," http://www.conae.gov.ar/index.php/espanol/catalogos-de-datos-productos.

10. Ann M Florini, "The Opening Skies: Third-Party Imaging Satellites and US Security," *International Security* 13, no. 2 (1988).

11. Kevin Rokosh, "Radarsat," Canadian Space Agency and IEEE, https://www.ieee.ca/millennium/radarsat/radarsat.pdf.

12. Ibid.

13. CSA, "Science and Operational Applications Research for Radarsat-2: Interferometry (INSAR)," CSA, http://www.asc-csa.gc.ca/eng/ao/2015-soar-insar.asp.

14. Canada Government Consulting Services, "Evaluation of the Radarsat-2 Major Crown Project" (2009).

15. Canada, "Remote Sensing Space Systems Act," in *Statutes of Canada 2005, chapter 45*, ed. Government of Canada (2005).

16. Canada Government Consulting Services, "Evaluation of the Radarsat-2 Major Crown Project" (2009).

17. Prepared by the Audit and Evaluation Directorate, "Evaluation of the Earth Observation Data and Imagery Utilization Program" (Canada: Canada Space Agency, 2011).

18. Canada, "Canada's Action Plan on Open Government" (Government of Canada, 2012).

19. "Canada's Action Plan on Open Government 2014–16" (2014).

20. MDA, " Surveillance and Intelligence Satellites," http://mdacorporation.com/geospatial/international/satellites.

21. Stéphane Chalifoux, "Radarsat Constellation Mission Update," in *POLinSAR 2015 &1st BIOMASS Science Workshop* (Frascati, Italy: Canadian Space Agency, 2015).

22. Ibid.

23. Christopher D Boone and Peter F Bernath, "SciSat-1 Mission Overview and Status" (paper presented at the Optical Science and Technology, SPIE's 48th Annual Meeting, 2003).

24. University of Waterloo, "ACE: Atmospheric Chemistry Experiment Data," http://www.ace.uwaterloo.ca/data.html.

ESA, "ESA Earth Online Data Access," https://earth.esa.int/web/guest/data-access.

25. Andrew W Yau and H Gordon James, "Scientific Objectives of the Canadian Cassiope Enhanced Polar Outflow Probe (E-Pop) Small Satellite Mission," in *The Sun, the Solar Wind, and the Heliosphere* (Springer, 2011).

26. Department of Physics and Astronomy Andrew Yau, University of Calgary, Email, 4 July 2016.

27. University of Calgary, "E-Pop Enhanced Polar Outflow Probe Data," http://mertensiana.phys.ucalgary.ca/data.html.

"Welcome to the E-Pop Data Set," http://epop-data.phys.ucalgary.ca/.

28. Yong Xue et al., "Small Satellite Remote Sensing and Applications–History, Current and Future," *International Journal of Remote Sensing* 29, no. 15 (2008).

ESA, "FASat-Bravo (Fuerza Aerea Satellite—Bravo)," https://directory.eoportal.org/web/eoportal/satellite-missions/f/fasat-bravo.

Craig I Underwood et al., "Initial in-Orbit Results from a Low-Cost Atmospheric Ozone Monitor Operating on Board the FASat-Bravo Microsatellite," *Philosophical Transactions of the Royal Society of London A: Mathematical, Physical and Engineering Sciences* 361, no. 1802 (2003).

Craig Underwood, Email, 29 June 2016.

29. C Mattar et al., "A First in-Flight Absolute Calibration of the Chilean Earth Observation Satellite," *ISPRS Journal of Photogrammetry and Remote Sensing* 92 (2014).

Major Juan Carlos Reyes, "FASat Charlie: A Contribution to the Development of Chile," in *Committee on the Peaceful Uses of Outer Space 57th Session* (Vienna, Austria: Fuerza Aérea de Chile [Chilean Air Force], 2014).

Fuerza Aérea de Chile Grupo de Operaciones Espaciales, "Sistema Satelital Para Observación De La Tierra SSOT)/FASat Charlie," http://www.ssot.cl/index.php.

30. Danish Meteorological Institute, "The Ørsted Satellite Data Exchange Page," https://web.archive.org/web/20070610050916/http://web.dmi.dk/fsweb/cgi-bin/webin_oer_data.sh?objtype=satellite&satellite=%26Oslash%3brsted.

31. Mosalam Shaltout, "The Egyptian Space Programme," *Space Research Today* 170 (2007).

Mohammed Shokr, "Potential Directions for Applications of Satellite Earth Observations Data in Egypt," *The Egyptian Journal of Remote Sensing and Space Science* 14, no. 1 (2011).

NARSS, "Space Sciences and Strategic Studies: Egypt Space Program," National Authority for Remote Sensing & Space Sciences, http://www.narss.sci.eg/divisions/view/5/%20Space%20Sciences%20and%20Strategic%20Studies/16/Egypt%20Space%20Program.

32. F. Barlier, "Early Satellite Laser Ranging for Geodesy at CNRS, CNES, and ONERA in France: First Geodetic Junctions Europe-Africa 1965–1975," in *19th International Workshop on Laser Ranging: Celebrating 50 Years of SLR* (Annapolis, MD, 2014).

M R Pearlman E C Pavlis, C E Noll, W Gurtner, and J Mueller, "International Laser Ranging Service (ILRS)" (2008).

33. NASA, "Data and Derived Products Overview," Crustal Dynamics Data Information System (CDDIS), http://cddis.nasa.gov/Data_and_Derived_Products/index.html.

34. CNES, "Demeter," Centre National d'Etudes Spatiales (CNES), https://demeter.cnes.fr/en/DEMETER/index.htm.

35. Ph Lier and M Bach, "PARASOL a Microsatellite in the A-Train for Earth Atmospheric Observations," *Acta Astronautica* 62, no. 2 (2008).

36. Werner Schmutz et al., "The Premos/Picard Instrument Calibration," *Metrologia* 46, no. 4 (2009).

37. Jacques Verron et al., "The Saral/Altika Altimetry Satellite Mission," *Marine Geodesy* 38, no. suppl. 1 (2015).

38. CNES CDPP, "Centre de Données de la Physique des Plasmas (CDPP) Data Archive," Centre National d'Etudes Spatiales (CNES), https://cdpp-archive.cnes.fr/.

CDPP Team, Email, 30 June 2016 2016.

39. ICARE.

40. CNES, "10/21/2013: The Picard Data Are Now Available to the Scientific Community," CNES, https://picard.cnes.fr/en/PICARD/GP_actualites.htm.

B.USOC, "Picso-Picard Search Tool," Belgian User Support and Operations Center (B.USOC) Picard Scientific Mission Centre, http://picso.busoc.be/.

41. CNES, "Aviso Satellite Altimetry Data," CNES, http://www.aviso.altimetry.fr/en/home.html.

42. Pierre Tran, "Space Intel Gives France Policy Independence," *Defense News*, 26 February 2015.

43. Jeffrey T Richelson and Mark Strauss, "The Whole World Is Watching," *Bulletin of the Atomic Scientists* 62, no. 1 (2006).

CNES, "Helios," CNES, https://helios.cnes.fr/en/helios-0.

44. Peter D Zimmerman, "From the Spot Files: Evidence of Spying," *Bulletin of the Atomic Scientists* 45, no. 7 (1989).

45. USGS, "SPOT Historical," United States Geological Survey, https://lta.cr.usgs.gov/SPOT_Historical.

46. "SPOT Controlled Image Base 10 Meter (CIB-10)," United States Geological Survey, https://lta.cr.usgs.gov/SPOT_CIB-10.

47. "Spot—North American Data Buy," United States Geological Survey, https://lta.cr.usgs.gov/SPOT_NADB.

SPOT Image Corporation, "NIMA Licenses SPOT Imagery for Public Domain," news release, 31 May 2000, 2000, http://www.prnewswire.com/news-releases/nima-licenses-spot-imagery-for-public-domain-73367192.html.

48. Peter B de Selding, "ESA, SPOT Image Sign Multiyear Contract," *Space News*, 23 January 2006.

49. "After 20 Years, SPOT Image Reinvents Itself," *Space News*, 14 November 2006.

Staff Writer, "Astrium Purchases Majority Share in SPOT Image," *Via Satellite*, July 15, 2008.

50. Peter B de Selding, "Indian Rocket Lofts SPOT 6 Earth-Observing Satellite," *Space News*, 18 September 2012 2012.

51. "France to Make Older Spot Images Available to Researchers for Free," *Space News*, 23 January 2014.

52. ESA, "Guidelines for the Submission of Project Proposals" (Frascati, Italy: European Space Agency, 2015).

53. Airbus Defence & Space, "Products & Services: Satellite Imagery," http://www.intelligence-airbusds.com/en/65-satellite-imagery.

54. Volker Liebig, "Small Satellites for Earth Observation—the German Small Satellite Programme," *Acta Astronautica* 46, no. 2 (2000).

55. DLR, "Earth Observation Center (EOC) Data Guide," http://www.dlr.de/eoc/en/desktopdefault.aspx/tabid-5356/9015_read-37549/.

56. GFZ Potsdam, "Champ Ground Segment and Data/Product Flow," http://op.gfz-potsdam.de/champ/operation/operation_CHAMP.html.

GFZ Potzdam, "Information System and Data Center for Geoscientific Data," http://isdc.gfz-potsdam.de/index.php.

57. Liebig.

58. M Krischke, W Niemeyer, and S Scherer, "RapidEye Satellite Based Geo-Information System," ibid.

George Tyc et al., "The RapidEye Mission Design," ibid. 56, no. 1 (2005).

59. Peter B de Selding, "Earth Imagery Firm RapidEye Seeking Bankruptcy Protection," *Space News*, 3 June 2011.

60. "Canadian Earth Observation Firm Buys Bankrupt RapidEye," *Space News*, 2 September 2011.

61. Jeff Foust, "Planet Labs Buying BlackBridge and Its RapidEye Constellation," ibid., 15 July 2015.

62. Planet Labs, "RapidEye Science Archive (RESA)," https://www.planet.com/markets/rapideye-science-archive/.

63. Rolf Werninghaus, Stefan Buckreuß, and Wolfgang Pitz, "TerraSAR-X Mission Status" (paper presented at the 2007 IEEE International Geoscience and Remote Sensing Symposium, 2007).

Peter B de Selding, "Public Funding on the Table for Germany's Next Radar Satellite," *Space News*, 8 July 2011.

64. DLR, "TerraSAR-X—Radar Technology from Germany: Financing Via Public-Private Partnership," news release, 18 September 2006, 2006, http://www.dlr.de/en/desktopdefault.aspx/tabid-24/82_read-4668/.

65. Bernhard Schmidt-Tedd, "PPP between DLR and Infoterra: The SatDSiG—German Satellite Data Security Act," in *Workshop on Trans-Atlantic Issues in Earth Observations* (Washington, DC: DLR, 2008).

66. Manfred Zink et al., "The TanDEMem-X Mission Concept" (paper presented at the Synthetic Aperture Radar [EUSAR], 2008 7th European Conference on, 2008).

67. BMWi, "German National Data Security Policy for Space-Based Earth Remote Sensing Systems," in *United Nations Office of Outer Space Affairs (UNOOSA) Legal Sub-committee 2010 IISL/ECSL Symposium on National Space Legislation—crafting legal engines for the growth of space activities*, ed. Bundesministerium fur Wirtschaft und Technologie (Vienna, Austria, 2010).

68. DLR, "TerraSAR-X Science Service System," http://terrasar-x.dlr.de/.

"TanDEM-X Science Service System," https://tandemx-science.dlr.de/.

69. "COFUR Prices for TerraSAR-X—Products (Scientific Use)" (TerraSAR-X Science Service System).

70. "COFUR Prices for Scientific TanDEM-X Products ".

71. "User License for the Utilisation of TerraSAR-X/TanDEM-X Data and Products for Scientific Use between DLR and the Principal Investigator" (2011).

72. "TerraSAR-X Mission Remarks" (TerraSAR-X Science Service System).

73. Rebecca Johnson, "Europe's Space Policies and Their Relevance to ESDP," *Policy Department External Policies, EP-ExPol-B-2005-14* (2006).

74. Peter B de Selding, "Germany's 2nd Military Radar Satellite Launched from Russia," *Space News*, 11 July 2007.

Sascha Lange, "SAR-Lupe Satellites Launched," *Strategie & Technik—International Edition II* (2007).

75. Agus Nuryanto Abdul Rahman, Ayom Widipaminto, Eriko N. Nasser, Mahdi Kartasasmita, Mohammad Mukhayadi, Mochamad Ichsan, Robertus Heru Triharjanto, Soewarto Hardhienata, Toto Marnanto Kadri, and Wahyudi Hasbi, *LAPAN-TUBSAT: From Concept to Early Operation* (Jakarta, Indonesia, 2007).

76. R H Triharjanto et al., "LAPAN-TUBSAT: Micro-Satellite Platform for Surveil-lance and Remote Sensing" (paper presented at the Proceedinga of the 4S Small Sat-ellite System and Services Symposium, La Rochelle, France, 2004).

Abdul Rahman.

77. Robertus Heru Triharjanto Soewarto Hardhienata, and Mohamad Mukhayadi "Lapan-A2 : Indonesian near-Equatorial Surveilance Satellite," in *18th Asia-Pacific Regional Space Agency Forum (APRSAF)* (Singapore: Indonesia National Institute of Aeronautics & Space [LAPAN] Center for Satellite Technology 2011).

78. Space News Staff, "Iran Lofts Imaging Satellite into Orbit Atop Safir Rocket," *Space News*, 20 June 2011.

IUST, "Navid Satellite Launched Successfully into Space," news release, 3 February 2012, http://www.iust.ac.ir/find-1.11658.23398.en.html.

79. E L Zorn, "Israel's Quest for Satellite Intelligence," *Space*, no. 1 (1991).

80. Ruth Eglash, "Introducing Ofek 10, the Just-Launched Israeli Satellite Likely Designed to Keep Tabs on Iran," *The Washington Post*, 10 April 2014.

81. Barbara Opall-Rome, "Israel MoD Considers Commercial Sale of Ofeq Imagery," *Space News*, 17 July 2006.

82. "Israel's Eros B Imaging Satellite Reaches Orbit," *Space News*, 1 May 2006.

83. Rani Hellerman, "Growing Demand for Commercial Very High Resolution Imaging Satellites," in *International Commercial Remote Sensing Symposium* (Washing-ton, DC: ImageSat International, 2010).

ImageSat International, "Satellite Imagery Services," http://www.imagesatintl.com/ satellite-imagery-services/.

84. Moshe Guelman et al., "Gurwin-Techsat: Still Alive and Operational after Nine Years in Orbit," *Acta Astronautica* 65, no. 1 (2009).

85. E C Pavlis.

86. F Covello et al., "Cosmo-SkyMed an Existing Opportunity for Observing the Earth," *Journal of Geodynamics* 49, no. 3 (2010).

Italian Space Agency (ASI), "Cosmo-SkyMed Mission and Products Description" (2016).

87. Peter B de Selding, "France Reluctant yet Hopeful on Cooperative Military Space Programs," *Space News*, 5 April 2011.

88. e-GEOS, "Cosmo-SkyMed," http://www.e-geos.it/cosmo-skymed.html.

89. "Price List" (2016).

90. Peter B de Selding, "Profile: Marcello Maranesi, Chief Executive, E-Geos," *Space News*, 5 March 2012.

91. ASI, "License to Use Cosmo-SkyMed Products" (Italian Space Agency, 2014).

92. MJPM Lemmens, "Geomatics Developments in Asia," *GIM International, 28 (December), 2014* (2014).

93. Airbus Defence and Space, "Kazakhstan's First Earth Observation Satellite to Be Launched at the End of April," news release, 24 April 2014, http://www.space -airbusds.com/en/press_centre/kazakhstan-s-first-earth-observation-satellite-to-be -launched-at-the-end-of.html.

Peter B de Selding, "Europe's Vega Rocket Lofts Kazakhstan's First Reconnaissance Satellite," *Space News* 2014.

Kazakhstan Gharysh Sapary, "Kazeosat Web Site," http://cof.gharysh.kz/ COFWelcome/COFWelcomePageEN.html.

Yerken Ospanov, Email, 30 June 2016.

94. H J Chun et al., "Razaksat–a High Performance Satellite Waiting for Its Mission in Space" (2006).

Dr. Tuan Hj Norhizam Hamzah and Dr. Rosymah, "Introduction to RazakSAT (Malaysia's Remote Sensing Satellite)," news release, 2009, https://www.aprsaf.org/ interviews_features/features_2009/feature_93.php.

Malaysia Space Agency (ANGKASA), "RazakSAT," http://www.angkasa.gov.my/?q=en/ node/188.

95. Ummahani Abd Rashid, Email, 21 August 2016.

Ibid.

96. Stephan Roemer and Udo Renner, "IAC-03-IAA. 11.2. 07 Flight Experience with the Micro Satellite Maroc-TUBSAT" (2003).

TUB, "Maroc-Tubsat," Technical University of Berlin, https://www.raumfahrttechnik .tu-berlin.de/menue/forschung/abgeschlossene_projekte/tubsat/v_menue4/tubsat/ maroc_tubsat/.

CRTS, "Geospatial Data and Information," Centre Royale de Télédétection Spatiale, http://www.crts.gov.ma/produits-services/donnees-et-informations-geospatiales.

97. NASRDA, "Missions," National Space Research and Development Agency, http://www.nasrda.gov.ng/?q=mission#sat1.

SSTL, "Nigeriasat-1: The Mission," Surrey Satellite Technology Limited, https://www .sstl.co.uk/Missions/NigeriaSat-1-Launched-2003/NigeriaSat-1/NigeriaSat-1-The -Mission.

"Nigeriasat-2: The Mission," Surrey Satellite Technology Limited, http://www.sstl.co
.uk/Missions/NigeriaSat-2–Launched-August-2011/NigeriaSat-2/NigeriaSat-2–The
-Mission.

"Nigeriasat-X: The Mission," Surrey Satellite Technology Limited, https://www.sstl
.co.uk/Missions/NigeriaSat-X–Launched-August-2011/NigeriaSat-X/NigeriaSat
-X–The-Mission.

98. UNOOSA, "Nigerian Space Program and African Regional Perspectives," in *Scientific and Technical Subcommittee 2012* (Vienna, Austria: United Nations Office for Outer Space Affairs, 2012).

99. GeoApps Plus, "About Us," http://www.geoappsplus.com/aboutus.html.

100. DMCii.

101. Space and Upper Atmosphere Research Commission (SUPARCO), "Badr-B," http://suparco.gov.pk/webroot/pages/badrb.asp?badrblinksid=1.

Nicholas R Waltham et al., "Development of a Compact Low-Mass Wide-Angle CCD Camera for Earth Observation" (paper presented at the Satellite Remote Sensing II, 1995).

102. SSTL, "PoSAT-1: The Mission," https://www.sstl.co.uk/Missions/PoSAT-1
-Launched-1993/PoSAT-1/PoSAT-1-The-Mission.

Fernando Carvalho Rodrigues, "PoSAT-1," http://www.fernandocarvalhorodrigues
.com/posat/imarch0.html.

103. KACST, "Strategic Priorities for Space and Aeronautics Technology Program" (Riyadh, Saudi Arabia: Kingdom of Saudi Arabia King Abdulaziz City for Science and Technology and the Ministry of Economy and Planning, 2009).

104. Haithem Altwaijry, "Saudi Space Activities National Satellite Technology Program King Abdulaziz City for Science and Technology (KACST)," in *Space Weather Workshop 2010* (Boulder, Colorado: KACST, 2010).

105. KACST.

106. T Bretschneider et al., "X-Sat Mission Progress," *Small Satellites for Earth Observation.(Special issue from the International Academy of Astronautics)* (2005).

Yin-Liong Mok, Cher-Hiang Goh, and R Chantiraa Segaran, "Redundancy Modeling for the X-Sat Microsatellite System" (paper presented at the Reliability and Maintainability Symposium (RAMS), 2013 Proceedings-Annual, 2013).

107. CRISP, "About Us," Singapore Centre for Remote Imaging, Sensing and Processing (CRISP), https://crisp.nus.edu.sg/crisp_top.html.

108. AgilSpace, "Geoservices Teleos-1: Singapore's First Commercial Neqo Earth Observation Satellite," AgilSpace, http://www.agilspace.com/geoservice.php?pc=teleos1.

109. Asian Scientist Newsroom, "Singapore Universities Launch Satellites," *Asian Scientist*, 22 December 2015 2015.

Tom Segert et al., "Kent Ridge 1–A Hyper Spectral Micro Satellite to Aid Disaster Relieve" (2014).

Yung-Fu Tsai et al., "Hardware-in-the-Loop Validation of GPS/GNSS Based Mission Planning for Leo Satellites," *Journal of IPNT* 1, no. 1 (2015).

110. Hyun-Ok Kim et al., "Space-Based Earth Observation Activities in South Korea [Space Agencies]," *IEEE Geoscience and Remote Sensing Magazine* 3, no. 1 (2015).

111. ESA, "Kompsat-1 EOC European Cities Dataset," European Space Agency Earth Online, https://earth.esa.int/web/guest/-/kompsat-1-eoc-european-cities-dataset -3792.

112. Yu Hwan Ahn, "Status of Kompsat-1 Osmi," in *Tenth IOCCG Committee Meeting* (Isla de Margarita, Venezuela: Satellite Ocean Research Laboratory Korea Ocean Research and Development Institute, 2005).

113. Kim et al.

114. SIIS, "Products," Satrac International Imaging Services, http://www.si-imaging .com/ds2_1_1.html.

115. ESA, "3rd Party Missions: Kompsat-2," European Space Agency Earth Online, https://earth.esa.int/web/guest/missions/3rd-party-missions/current-missions/ kompsat-2.

116. Kim et al.

117. KOSC, "Ocean Satellite Data Service," Korea Ocean Satellite Center, http:// kosc.kiost.ac/eng/p10/kosc_p11.html#.

118. NMSC, "Data Services," National Meteorological Satellite Center, http:// nmsc.kma.go.kr/html/homepage/en/dataservice/system_info.do#none.

119. Deimos Imaging, "Satellite Imagery," http://www.deimos-imaging.com/ imagery.

120. Elecnor Deimos, "Launch of the First Global Alliance of Earth Observation Satellites Operators," news release, 9 November 2014, 2014, http://www.deimos -space.com/en/noticias/showNew.php?idNew=160.

121. Swedish National Space Board, "Space Activities in Sweden: Odin," http:// www.snsb.se/en/Home/Space-Activities-in-Sweden/Satellites/Odin/.

122. OdinSMR, "Level1 Data Dashboard," Odin Sub Millimeterwave Radiometer, http://odin.rss.chalmers.se/level1.

123. ESA, "Odin/Smr Site," European Space Agency, http://amazonite.rss.chalmers .se:8280/OdinSMR.

Odin Osiris, "Odin-Osiris Level 2 Data Policy and Access," University of Saskatche-wan, http://osirus.usask.ca/?q=node/247.

124. Hsien-Wen Li et al., "An Overview of Rocsat-1 OCI Science Team and Science Data Distribution Center" (paper presented at the COSPAR Colloquia Series, 1999).

HC Yeh et al., "Scientific Mission of the Ipei Payload Onboard Rocsat-1," *Terrestrial, Atmospheric and Oceanic Sciences*, no. 1 (1999).

Chi-Kuang Chao, Email, 1 July 2016.

125. Airbus Defence & Space, "FormoSat-2: High-Resolution Change Detection," http://www.intelligence-airbusds.com/en/160-formosat-2.

126. International Society for Photogrammetry and Remote Sensing, "Research Announcement: Free FormoSat-2 Satellite Imagery," http://www.isprs.org/news/ announcements/140214-FS2CFP-ISPRS.pdf.

127. TACC, "FormoSat-3/Cosmic Data Distribution Policy," Taiwan Analysis Center for COSMIC, http://tacc.cwb.gov.tw/service/policy.htm.

CDAAC, "CDAAC Data Products," COSMIC Data Analysis and Archive Center, http://cdaac-www.cosmic.ucar.edu/cdaac/products.html.

128. M N Sweeting and Sitthichai Pookyaudom, "TMSAT: Thailand's First Microsat-ellite for Communications and Earth Observation," *Acta Astronautica* 40, no. 2 (1997).

Sujate Jantarang, "Thai-Paht—a Small Satellite for Education" (paper presented at the Cooperation in Space, 1999).

Assoc. Prof. Dr. Sujate Jantarang, Email, 3 July 2016.

129. T Choomnoommanee, M Kaewmanee, and R Fraisse, "Thailand Earth Observa-tion System: Mission and Products," in *ISPRS Commission I Symposium "From Sensors to Imagery"*, eds. Nicolas Paparoditis and Alain Baudoin (Paris, France, 2006).

Suchit Leesa-Nguansuk, "Theos Creates New Roles for Thailand," *Bangkok Post*, 1 September 2010.

130. GISTA, "Satellite Data Products Price List," Geo-Informatics and Space Tech-nology Development Agency (GISTA), http://www.gistda.or.th/main/en/node/1085.

131. Leesa-Nguansuk.

132. Andy Bradford et al., "BILSAT-1: A Low-Cost, Agile, Earth Observation Micro-satellite for Turkey," *Acta Astronautica* 53, no. 4 (2003).

133. DMCii.

134. Mustafa Teke et al., "Geoportal: Tübïta Uzay Satellite Data Processing and Sharing System" (paper presented at the Recent Advances in Space Technologies (RAST), 2015 7th International Conference on, 2015).

135. TUBITAK UZAY, "Geoportal," TUBITAK UZAY, http://uzay.tubitak.gov.tr/en/uydu-uzay/geoportal.

Gezgin, "Gezgin Geoportal," https://gezgin.gov.tr/geoportal/app/main?execution=e2s1.

136. TUBITAK UZAY, "Gokturk-2," Turkey Space Technologies Research Institute, http://uzay.tubitak.gov.tr/en/uydu-uzay/gokturk-2.

TAI, "Gokturk-2: High Resolution Remote Sensing Satellite Design Heritage," Turkish Aerospace Industries, Inc., https://www.tai.com.tr/en/project/gokturk-2.

137. Yuzhnoye State Design Office, "Activity of Ukraine in Satellite Development Domain," *EO Mag: The European Association of Remote Sensing Companies Newsletter,* Issue 14 (2008).

138. Geoportal Sich-2, "Geoportal Sich-2," Service of Space Research Insittute NASU-NSAU, http://sich2.ikd.kiev.ua/eng/index.php.

139. S AL-Mansoori, A Bushahab, and O AL-Shehhi, "Application of DubaiSat-1 Imagery" (paper presented at the International Symposium on Remote Sensing [ISRS], 2012).

Ali Al Suwaidi, "DubaiSat-2 Mission Overview" (paper presented at the SPIE Remote Sensing, 2012).

140. Mohammed Bin Rashid Space Centre, "Satellite Imagery Services: Providing High-Resolution Satellite Images for Government and Commercial Use," http://mbrsc.ae/en/page/satellite-imagery-services.

141. PanGeo Alliance, "Pangeo Alliance: A Unique Earth Observation Constellation," http://www.pangeo-alliance.com/.

142. M N Sweeting, "UoSAT Microsatellite Missions," *Electronics & Communication Engineering Journal* 4, no. 3 (1992).

143. United Kingdom House of Commons Science and Technology Committee, "The UK Space Agency VolumeII Additional Written Evidence: Annex 1 to Annex A—Mosaic Programme" (London, UK: UK House of Commons, 2010).

144. DMCii.

145. United Kingdom House of Commons Science and Technology Committee, "The UK Space Agency VolumeII Additional Written Evidence: Annex 1 to Annex A—Mosaic Programme" (London, UK: UK House of Commons, 2010).

Steve Cawley, "Topsat: Low Cost High-Resolution Imagery from Space," *Acta Astronautica* 56, no. 1 (2005).

146. Airbus Defence & Space GEOInformation Services, "Products: TopSat," http://geosurveysolutions.com/topsat.

147. R Acevedo et al., "Space Activities in the Bolivarian Republic of Venezuela," *Space Policy* 27, no. 3 (2011).

148. Doan Minh Chung, "Space Technology Development in Vietnam 2012-2013," in *APRSAF-20* (Hanoi, Vietnam: Space Technology Institute (STI) Vietnam Academy of Science and Technology [VAST], 2013).

149. Koh Swee Lean Collin, "Vietnam's Master Plan for the South China Sea," *The Diplomat*, 4 February 2016.

Appendix B   Satellite Data Sharing Database

1. WMO OSCAR, "Space-Based Capabilities (Oscar/Space)," World Meteorological Organization (WMO) Observing Systems Capability Analysis and Review Tool (OSCAR), https://www.wmo-sat.info/oscar/spacecapabilities.

2. NASA, "Nasa Space Science Data Coordinated (NSSDC) Archive," http://nssdc.gsfc.nasa.gov/nmc/SpacecraftQuery.jsp.

US Federal Aviation Administration (FAA) Commercial Space Transportation (AST) and the Commercial Space Transportation Advisory Committee (COMSTAC), "2015 Commercial Space Transportation Forecasts" (2015).

Index